Advanced Methods in Protein Microsequence Analysis

Edited by
B. Wittmann-Liebold, J. Salnikow
V. A. Erdmann

With 165 Figures

Springer-Verlag
Berlin Heidelberg New York
London Paris Tokyo

CHEMISTRY

Professor Dr. BRIGITTE WITTMANN-LIEBOLD
Max-Planck-Institut für Molekulare Biologie
Abteilung Wittmann
Ihnestraße 63–73
D-1000 Berlin 33

Professor Dr. JOHANN SALNIKOW
Institut für Biochemie und Molekulare Biologie
Technische Universität Berlin
Franklinstraße 29
D-1000 Berlin 10

Professor Dr. VOLKER A. ERDMANN
Institut für Biochemie
Freie Universität Berlin
Otto-Hahn-Bau, Thielallee 63
D-1000 Berlin 33

Cover illustration:
Separation of DABTH-amino acid derivatives two-dimensional polyamide
thin-layer sheets (see also Fig. 2 on p. 85)

ISBN 3-540-16997-0 Springer-Verlag Berlin Heidelberg New York Tokyo
ISBN 0-387-16997-0 Springer-Verlag New York Berlin Heidelberg Tokyo

Library of Congress Cataloging-in-Publication Data. Advanced methods in protein microsequence analysis. Includes index. 1. Amino acid sequence. I. Wittmann-Liebold, Brigitte,
1931– . II. Salnikow, Johann, 1938– . III. Erdmann, Volker A., 1941– . [DNLM: 1. Amino
Acid Sequence. 2. Proteins – analysis. QU 60 A2438] QP551.A325 1986 547.7′5 86-22122

Typesetting: K. u. V. Fotosatz, Beerfelden
Offsetprinting and bookbinding: Konrad Triltsch, Graphischer Betrieb, Würzburg
2131/3130-543210

Contents

Chapter 3
Gas-Phase and Radio-Sequence Analysis

Chapter 4
Phenylthiohydantoin Identification, On-Line Detection,
Sequences Control, and Data Processing

Chapter 5
Analysis of Cysteine Residues in Proteins

Chapter 6
Methods of Analyzing Protein Conformation

Chapter 7
Strategies and Specific Examples of Sequencing Proteins and
Peptides

Contributors

You will find the addresses at the beginning of the respective contribution

AMONS, R. 352
ASHMAN, K. 219
BAUW, G. 179
BEEUMEN, J. VAN 256
BENNETT, J.C. 208
BERGMAN, T. 45
BEYREUTHER, K. 276
BHOWN, A.S. 208
BLÖCKER, H. 387
BORCHART, U. 91
BOULANGER, Y. 291
BRANDT, W.F. 161
BROCKMÖLLER, H.-J. 34
BUSE, G. 340
CARLQUIST, M. 45
CHANG, JUI-YOA 56, 265
CRABB, J.W. 64
DAMME, J. VAN 179
DIGWEED, M. 364
ECKART, K. 403
ERDMANN, V.A. 364
FRANK, RAINER 149
FRANK, RONALD 387
FRIEDRICH, J. 226
GAUSEPOHL, H. 149
GIEGE, R. 291
GOTTO, A.M., Jr. 320
HENSCHEN, A. 244
HENSEL, S. 340
HIRANO, H. 77
HOLT, C. VON 161
JÖRNVALL, H. 45
KALKKINEN, N. 194

KAMP, R.M. 8, 21, 34
KERN, D. 291
KIMURA, M. 77
KNECHT, R. 56
KRAUHS, E. 316
KUHN, C.C. 64
LITTLE, M. 316
MACHLEIDT, W. 91
MAIER, G. 316
MEINECKE, L. 126, 340
MEJDOUB, H. 291
MONTAGU, M. VAN 179
PIELER, T. 364
PONSTINGL, H. 316
POWNALL, H.J. 320
PRINZ, H. 276
PUYPE, M. 179
REIMANN, F. 118
REINBOLT, J. 291
REUMKENS, J. 340
RITONJA, A. 91
SALNIKOW, J. 108
SCHULZE-GAHMEN, U. 276
SPIESS, J. 302
STEFFENS, G.C.M. 340
STEFFENS, G.J. 340
TROSIN, M. 149
TSCHESCHE, H. 126
VANDEKERCKHOVE, J. 179
WITTMANN-LIEBOLD, B.
    77, 118
YANG, CHAO-YUH 320
YANG, TSEMING 320

Introduction

Much of the recent spectacular progress in the biological sciences can be attributed ot the ability to isolate, analyze, and structurally characterize proteins and peptides which are present in cells and cellular organelles in only very small amounts. Recent advances in protein chemistry and in particular the application of new micromethods have led to fruitful advances in the understanding of basic cellular processes. Areas where protein-chemical studies have resulted in interesting discoveries include the peptide hormones and their release factors, growth factors and oncogenes, bioenergetics, proton pumps and ion pumps and channels, topogenesis and protein secretion, molecular virology and immunology, membrane protein analysis, and receptor research. In fact, the key methods are now on hand to unravel many of the major outstanding problems of molecular biology and in particular questions of fundamental interest which relate to developmental biology and specificity in cell-cell interaction.

In this volume we have assembled descriptions of procedures which have recently been shown to be efficaceous for the isolation, purification, and chemical characterization of proteins and peptides that are only available in minute amounts. Emphasis is placed on well-established micromethods which have been tested and found useful in many laboratories by experienced investigators. The chapters are written by specialists, and describe a range of sensitive techniques which can be used by researchers working in laboratories with only modest resources and equipment. The book is also a compilation of experimental protocols which are suitable for use in the laboratory for student courses at the advanced undergraduate and graduate level, as well as for use by researchers who are new to the field of protein microsequence analysis. Furthermore, modifications of some newer manual microsequence methods are described that demonstrate that low picomole amino acid sequence analysis may not always require sophisticated equipment. It is our objective to describe these procedures with sufficient clarity that even researchers without prior experience in protein chemistry and especially without experience in protein microsequence analysis can use these methods. By employing the techniques as described the reader is able to avoid errors that cause substance losses at the isolation stages or diminish sequence information.

However, it is not our intention to cover the whole field of contemporary protein chemistry. Conventional methodology which can be adapted readily to the microscale level is not discussed in this book except selected applications at the microlevel, e.g., the performance of the main chemical and enzymatic cleavages of proteins. A comprehensive manual of protein chemistry, *Practical Protein Chemistry,* edited by A. Darbre, John Wiley & Sons, Chichester, 1986, has recently been published, and review articles on protein-analytical and physico-chemical determinations are covered in the series *Modern Methods in Protein Chemistry,* edited by H. Tschesche, Walter deGruyter, Berlin, 1983 and 1985, these volumes providing information where additional basic knowledge in this field is needed. The many gel electrophoresis techniques which have found wide application in protein analysis are detailed in the recent volumes *Gel Electrophoresis of Proteins,* edited by B. D. Hames and D. Rickwood, IRL Press, Oxford, England, 1981; in *Proteins* edited by J. W. Walker, Humana Press, Clifton, New Jersey, 1984; and in *Two-Dimensional Gel Electrophoresis of Proteins, Methods and Applications,* edited by J. E. Celis and R. Bravo, Academic Press Inc., New York, 1984. Conventional manual sequencing techniques such as the dansyl-Edman degradation for peptides are described in detail elsewhere; therefore only more recent and sensitive manual microsequencing techniques which can be applied for peptides as well as for proteins in the picomole range are included here.

Most methods discussed in this book were demonstrated during the FEBS Advanced Course on Microsequence Analysis of Proteins, held at Berlin (West) in September 1985, and organized by our research groups. This course was followed by an International Symposium on *Novel Techniques in Protein Sequence Analysis* and additional information on micromethods in this book is based on the presentations made at this workshop. At this meeting it became obvious that the demand for highly sensitive and practical methods for basic protein research for a wide range of investigations has increased dramatically. We obtained so many applications for attendance at this course that we could not accept most of these researchers, although they all had urgent reasons for learning these techniques for their present research work. Therefore, we decided to assemble and carefully describe the methods reported or demonstrated at the course so that they might be available to a wider scientific community and especially to young students who will need these techniques for future scientific work. Nowadays, protein analytical methods are not the exclusive preserve of a few specialists who are well experienced with protein analysis; since the recent developments in gene technology and immunology the use of sophisticated protein analysis techniques has become widespread.

Protein Sequence Analysis as a Complementary Technique to Nucleotide Sequencing

The recent advances in the purification of proteins and peptides by HPLC and the development of several very sensitive microsequencing techniques have opened new vistas and possibilities in molecular biology and medicine. It has become possible to isolate a gene based on partial protein sequence data by synthesizing oligonucleotide probes for hybridization with gene libraries, isolation of the specific gene, and subsequently to derive the sequence of the entire protein by recombinant DNA techniques. Other approaches use synthetic peptides manufactured based on partial protein sequences to produce specific antibody, isolate the protein in larger quantity, and conduct functional studies. However, as only a negligible number of the total proteins which can be potentially coded for by eukaryotic genomes are known at this time, there remain major possibilities for research at the protein level and even for the development of a new generation of even more powerful and sensitive protein micro-analytical techniques.

With the development of rapid nucleotide sequencing techniques, however, problems with large-scale isolation of proteins and certain difficulties in the sequence determination are simplified since, by this means, partial amino acid sequences can be extended and peptide fragments easily aligned. Thus, the protein-chemical methods are complemented by nucleotide sequencing once the gene becomes available. Therefore, two chapters of the book are devoted to a description of useful RNA- and DNA-sequencing methods.

However, as has often been emphasized, the direct sequencing of a protein often provides a characterization of the polypeptide structure which is not possible with nucleic acid sequencing alone. In addition, the direct comparison of the nucleotide and protein sequence is especially valuable in the study of organisms where introns may make derivation of the entire protein sequence difficult, based on the nucleotide sequence alone. By N-terminal sequence analysis and determination of the C-terminal amino acids, the putative sequence based on DNA analysis can be confirmed. Furthermore, protein fragmentation can be used to identify functional domains and selective modifications of amino acid residues to locate functionally important regions and active sites of the protein.

Also, by direct amino acid sequencing in contrast to gene sequencing it can be established whether modifications, such as methylations, acetylations, or phosphorylations, occur in a native protein. Proteins frequently contain covalently attached carbohydrate moieties or lipids; the protein itself may be blocked at the N- or C-terminus of the chain, as usually found for short peptide hormones. It is estimated that in eukaryotic cells at least one third of all proteins may be blocked. This number, however, is still uncertain; often inadequate isolation procedures, especially when purifying scarce substance amounts, cause blockage at the N-terminus. Therefore, with well-tested methods on hand, errors at the isolation level of the peptides and proteins can be avoided and the appropriate strategy for sequencing can more easily be selected.

Elucidation of the Protein's Secondary or Tertiary Structure

Information about the secondary or three-dimensional structure of proteins can only be gained if the amino acid sequence is available and the protein isolated in pure form. Establishing the type of secondary structural elements involved in a certain protein, e.g., helical, beta-sheet and beta-turn structures, can be predicted from its amino acid sequence with some certainty (at most with about 60% reliability); however, confirmation of the predicted structure by physicochemical experiments still needs quite large amounts of protein.

Also, the definitive determination of cysteine or cystine groups in a protein and the elucidation of the sulfur bridges is still laborious and rather complicated. Here, the book provides options for the determination of the cyst(e)ine content of the protein and the location of disulfide linkages using small amounts of polypeptide. Without the direct isolation of the protein under appropriate conditions the S-S bridges cannot be established.

Knowledge of the amino acid sequence is also, of course, a prerequisite for the determination of the protein's three-dimensional structure by X-ray analysis. However, crystallization still needs considerable quantities of protein, often 50 to 100 mg at least. With a substantial reduction in the amounts required for sequence analysis using newer methods more material is now available for these purposes.

After elucidation of the native structure and in combination with gene technology, it is possible to manufacture synthetic proteins, e.g., enzymes with slightly altered properties for special applications or study of functional domains. Further, with the knowledge of the three-dimensional structure, it becomes feasible to synthesize peptides that reproduce the arrangement of the amino acid residues at the protein's surface (topographic antigenic determinants), and to use these peptides for the production of antibodies specific for the native structure of that protein. This is of theoretical and practical importance in immunology and may have medical applications.

Topographical Protein-Chemical Studies on Complexes

Protein-chemical approaches permit investigations on the topography of proteins in organelles or the arrangement of subunits in multi-enzyme complexes. They facilitate the study of DNA- and RNA-protein interactions as well by reaction of the native complexes with bifunctional reagents. After purification of the protein-protein or nucleic acid-protein crosslink, the interacting components can be analyzed and the amino acids and nucleotides involved can be determined.

Strategies for Primary Structural Analysis of Scarce Polypeptide Amounts

Amino acid sequence analysis of increasing numbers of proteins combined with other structural and functional studies illuminates the great diversity and unique properties of proteins. Unlike that of nucleic acids, protein behavior is hard to predict and varies considerably due to the large differences in size, net charge, amino acid composition, solubility, and native secondary and tertiary structure. Hence, generally applicable isolation and characterization schemes cannot be given. This makes protein chemistry rather difficult, and much experience is necessary to manipulate small amounts of protein optimally. On the other hand, this makes dealing with peptides and proteins more challenging compared to other compounds. In order to provide the reader with diverse possible strategies for sequence analysis of structurally disparate polypeptides, one section of the book gives examples of selected strategies for isolation, chemical and enzymatic fragmentations, amino acid analysis, and microsequencing of such diverse molecules as, for example, membrane proteins, or small-sized peptide hormones. These chapters enable the reader to select the most suitable methods for his own project by reference to methods which were found useful in a similar case.

In addition to the chemical methods for micro-analysis of polypeptides, an outline of the application of mass spectrometry for peptide investigation is included in this book. Recently, mass spectrometry has become very important for the structure analysis of modified peptides and of peptides of unusual structure or composition.

Acknowledgments. We acknowledge the help of Mrs. I. Brauer and Mrs. L. Teppert for co-editing and of Dr. Michael J. Walsh for carefully reading several chapters of the book.

BRIGITTE WITTMANN-LIEBOLD
JOHANN SALNIKOW
VOLKER A. ERDMANN

Chapter 1
**Separation and Amino Acid Analysis of Proteins
and Peptides for Microsequencing Studies**

1.1 Separation of Peptides

Roza Maria Kamp[1]

Contents

1 Introduction

Peptide mapping is a very useful technique for the characterization of proteins. Various rather simple methods can be employed for the separation of peptides, e.g., fingerprinting on thin-layer sheets or a combination of gel filtration or ion exchange chromatography with one- or two-dimensional thin-layer chromatography.

The advantages of these methods are their good resolution and easy production of peptides suitable for direct microsequencing analysis; disadvantages are the low recovery (40–70%), depending on the type of peptide and the solvent used for the elution.

---

1 Max-Planck-Institut für Molekulare Genetik, Abteilung Wittmann, Ihnestraße 63–73, D-1000 Berlin 33

Advanced Methods in Protein Microsequence Analysis
Ed. by B. Wittmann-Liebold et al.
© Springer-Verlag Berlin Heidelberg 1986

The development of high performance liquid chromatography has revolutionized separation technology of biomolecules. This enabled new purification and fingerprinting techniques for peptide mixtures. The advantages of this method are the very quick separation, high resolution and excellent separation of hydrophobic peptides. The amounts necessary for sequencing peptides separated by HPLC are approximately five times less than that previously necessary by thin-layer fingerprinting; they are 30 times less than that used for a combined open column and thin-layer technique.

The use of these techniques depends on the properties of the peptide mixtures and the equipment of the laboratory. General features and details of sensitive separations of peptides are given for purification by thin-layer fingerprinting or HPLC.

2 Thin-Layer Fingerprints

Peptide mixtures can be separated with good resolution by two-dimensional fingerprinting [1, 2]. The first-dimension peptide separation depends on the peptide's net charges and molecular masses [3]. In the second dimension the peptides are separated by ascending chromatography depending on their individual distribution coefficient. As an example, Fig. 1 shows the fingerprint of tryptic peptides of cytochrome c.

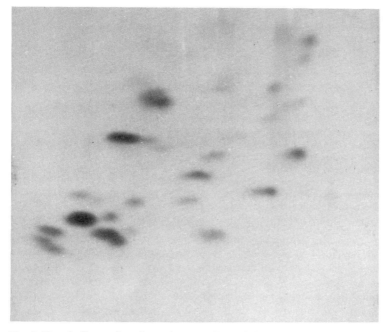

Fig. 1. Tryptic fingerprint of cytochrome c (5 nmol detected with ninhydrin)

2.1 First-Dimension Electrophoresis

The peptide mixtures are separated on coated cellulose sheets, Cel 300, 20×20 cm, (Macherey-Nagel, Düren, Germany) or, in the case of hydrophobic peptides, Cel 400 sheets are used. For the first run the peptide mixture is applied in the middle of the sheet 3 cm from the bottom and electrophoresed for 1 h at 400 V. Depending on the peptide spot distribution after detection with ninhydrin (see below), the next sample is placed near the anode (with positively charged peptides), or near to the cathode (with negatively charged peptides). The procedures for performing thin-layer fingerprints for peptides are given in Tables 1 and 2.

Table 1. Electrophoresis of cytochrome c tryptic peptides on thin-layer sheets

Electrophoresis is always performed prior to chromatography in order to free the sample from traces of salts. These salts disturb the fingerprinting and should be totally avoided.

1. Dissolve $2-5$ nmol peptide mixture in $2-5$ µl water and centrifuge shortly (2 min) at low speed.
2. Wet the thin layer sheet by dipping into a tank filled with electrophoresis buffer and dry quickly with filter paper.
3. Open water cooling to the electrophoresis chamber (about 14 °C).
4. Place the wet cellulose sheet into the electrophoresis chamber (CAMAG), fill the electrode tanks with electrophoresis buffer pH 4.4 (pyridine/acetic acid/acetone/bidistilled water, 50:100:375: 1975, v/v).
5. Proof wet grade of thin-layer sheet by preconditioning at 400 V for 5 min. Optimal stream strength is $12-18$ mA for a Cel 300 and $10-15$ mA for a Cel 400 sheet.
6. Add $2-5$ µl of 5 nmol tryptic peptides at the corner of the sheet (usually at the corner in the 4×3 cm position). Add DNP-OH (dinitrophenylsulfonic acid) and amido black as marker at the top of the sheet above the sample.
7. Start the electrophoresis at 400 V for 2 h with cooling.
8. Dry cellulose sheets after electrophoresis for 1 h at room temperature (never use high temperatures or warm fan).

Table 2. Ascending chromatography and fluorescamine and ninhydrin reactions

1. Place the cellulose sheet in a chromatography tank for $6-7$ h.
   Chromatography buffer: PBEW pH 4.4,
   pyridine: butanol: acetic acid: water = 50:75:15:60, v/v.
2. Dry the sheet for 1 h at room temperature.
3. Wet cellulose sheet with 5% pyridine solution in acetone; after short drying ($1-2$ min) wet thin-layer with 0.05% fluorescamine in acetone.
4. Dry the cellulose sheets for 15 min at room temperature.
5. Mark the peptides under UV light at 366 nm with a soft pencil, copy peptide pattern on transparent paper and scrape out (see below).
6. Spray the remainder of the cellulose sheet with 0.3% or 0.15% ninhydrin to detect peptide spots weakly staining with fluorescamine and dry for some hours for complete development of spots (keep in aluminum or plastic foil overnight in the dark).

0.3% Ninhydrin solution
3 g ninhydrin in collidine/acetic acid/ethanol, 30:100:870, v/v)

0.15% Ninhydrin solution
The 0.3% solution is diluted with ethanol.

2.2 Second Dimension and Detection of Peptides

After electrophoresis the cellulose sheets are placed in chromatography tanks (filled with 100 ml solvent mixture; ascending chromatography is performed). The peptides are separated according to their hydrophobicity. The sheets are dried at room temperature and developed by staining with fluorescamine and ninhydrin (Table 2).

Ninhydrin reacts with amino groups of peptides, proteins, and free amino acids and results in blue colored spots, see Fig. 2. This reaction is not reversible and the N-terminal amino acids and lysine side chains are partially destroyed. If 0.15% ninhydrin is applied first and the sheet only gentle sprayed, about 70% of the N-terminal groups may be recovered and sequencing of the eluted peptide is possible. With the 0.3% reagent the endgroups are almost completely reacted and the recovery of the peptide is low. However, some peptides give a specific

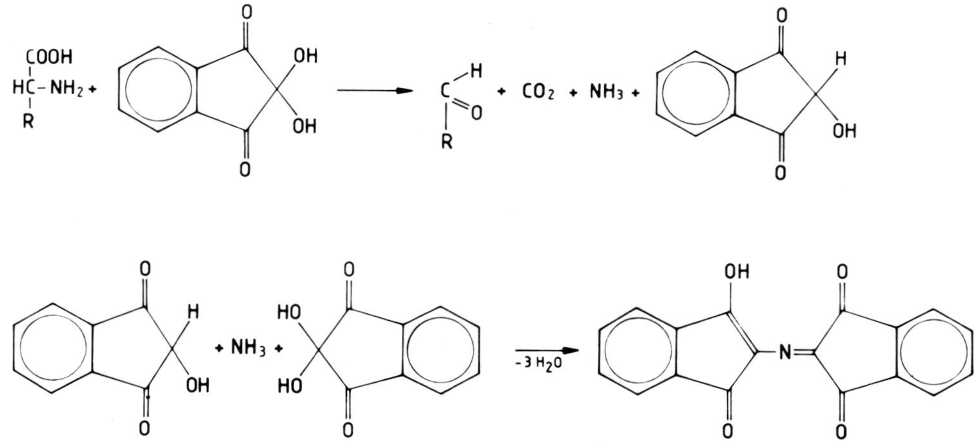

Fig. 2. Reaction with ninhydrin

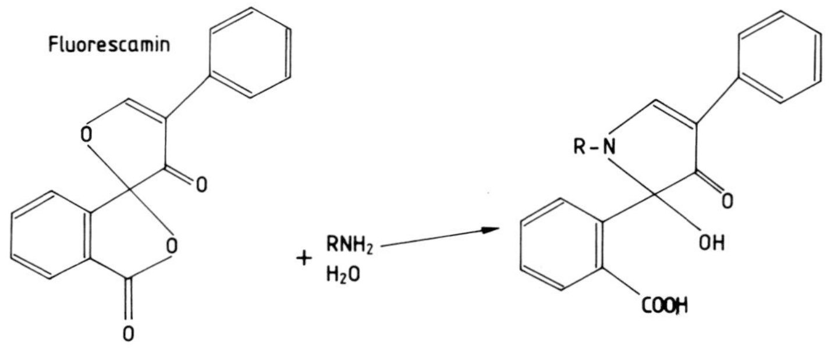

Fig. 3. Reaction with fluorescamine

Table 3. Arginine test

1. Prepare 0.02% phenanthrene quinone in ethanol (A) and 10% NaOH in 60% ethanol (B).
2. Mix solution A and B (1 : 1, v/v) directly before application.
3. Spray thin-layer sheet with prepared mixture and dry for 20 min at room temperature. Arginine-containing peptides are visible in UV (254 nm) as yellowish spots.

Table 4. Tyrosine test

1. Spray thin-layer sheet with 0.1% α-nitroso-β-naphthol in ethanol and dry at room temperature.
2. Spray thin-layer with 10% HNO_3 and develop for 3 min at 100 °C.

Tyrosine-containing peptides form red spots on a yellow background.

Table 5. Tryptophan test

1. Prepare (1% p-dimethylaminobenzaldehyde in 2 N HCl in acetone.
2. Spray thin-layer with freshly prepared solution and dry.

Tryptophan peptides result in red spots.

color with ninhydrin; e.g., N-terminal glycine, threonine, and serine peptides produce yellow spots, N-terminal tyrosine and histidine stain brown.

Fluorescamine reacts with proteins and peptides at pH 9.0 and forms a fluorescent complex visible at 366 nm, see Fig. 3. This reaction is reversible and allows elution of the peptides for quantitative amino acid analysis or sequencing.

Spray tests for the presence of arginine, tyrosine, and tryptophan are given in Tables 3 – 5; after applying these tests, the sheets may be dipped shortly in acetone and sprayed with ninhydrin.

2.3 Elution of Peptides from Thin-Layer Sheet

The peptide spots from three thin-layer sheets are eluted and used for

- amino acid analysis after OPA derivatization: 1/20 of amounts recovered,
- endgroups analysis: 1/20 of amounts recovered,
- manual or automatic sequencing: remainder.

The peptide elution procedure is detailed in Table 6.

3 High Performance Liquid Chromatography

HPLC is very useful for isolation of complex mixtures of smaller-sized peptides. In most cases, reversed-phase HPLC is applied and is well suited to the purification of smaller-sized peptides, even if hydrophobic fragments are contained in the mixture. The advantage of volatile and UV-transparent HPLC buffers allows

Table 6. Elution of peptides

1. Scrape out peptide spots with a sharpened spatula (which forms rolls) and transfer to Eppendorf plastic tubes (1.5 ml).
2. Make elution of peptides with:
   - 2×200 µl of 50% acetic acid for basic peptides,
   - 2×200 µl of 20% pyridine or 0.07% ammonia for acidic peptides,
   - 2×100 µl of 5.7 N HCl for peptides used only for amino acid analysis,
   - 2×100 µl of 70% formic acid for peptides used for oxidation with performic acid (cysteine determination).
3. Stir scraped peptides with a Vortex and shake the suspension for 30 min in an Eppendorf mixer 5432 (Eppendorf Gerätebau, Netheler + Hinz GmbH, Hamburg, Germany).
4. Centrifuge the suspended cellulose 5 min in a Beckman Microfuge B (Beckman Instruments, USA) and collect the supernatant into hydrolysis or sequencing tubes.
5. Repeat the elution of peptides for a second time and pool with the first extract.
6. Dry the peptides in vacuum (Speed Vac Concentrator) and store at $-20\,°C$ for further use.

high sensitivity runs, and subsequent direct amino acid analysis and microsequencing [4].

We employed three different buffer systems, at pH 2.0, 4.4, and 7.8, taking into account the different solubilities of the proteins and their peptides. Mixtures which tend to precipitate are injected freshly prepared, avoiding drying or concentrating of the digest. To avoid losses of insoluble peptides, these were recovered by centrifugation, and instead of further purification steps, directly subjected to another enzymatic digest or chemical cleavage. The partial peptides obtained were purified by thin-layer fingerprinting or HPLC techniques.

The following separation systems were employed:

System I.   Buffer A: 0.05% trifluoroacetic acid in water (v/v) at pH 2.0
             Buffer B: acetonitrile with 0.05% TFA
System II.   Buffer A: 1.5 ml 25% ammonia + 0.25 ml 98% formic acid/2 l water, pH 7.8
             Buffer B: methanol + 20% buffer A
System III. Buffer A: ammonium formate pH 4.4 made from 0.4 ml 25% ammonia and 0.25 ml 98% formic acid
             Buffer B: methanol with 20% buffer A

As an example for the separation of peptides by HPLC on reversed-phase C_{18} support the separation of the tryptic peptides of cytochrome c is presented in Fig. 4. The separation procedure is given in Table 7.

The tryptic peptide mixture of cytochrome c serves as a good test for the resolving capacity of new HPLC-columns for peptides. Depending on the hydrophobicity of the peptide mixture C_4-, C_8- or C_{18}-alkylated supports may be chosen; the pore size of the support can be selected according to the length and hydrophobicity of the peptides, e.g., short hydrophilic peptides are separated on 75 Å support, more hydrophobic and large peptides are obtainable in better yields on 300 Å material. Different batches of supports and/or material from different suppliers vary considerably. Best suited for reproducible separations are spherically shaped, uniformly sized particles.

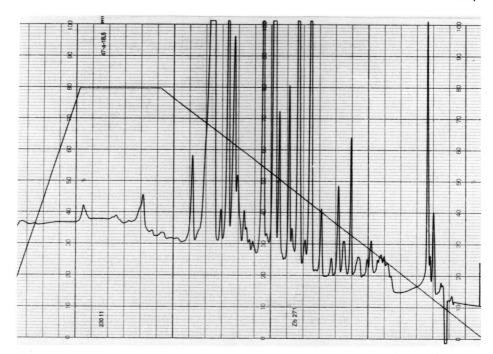

Fig. 4. Separation of tryptic peptides of cytochrome c on HPLC Shandon ODS column. Amounts of 100 μg peptide mixture were injected in 10 μl water. The eluents were: buffer A, 0.05% TFA in water; buffer B, 0.05% TFA in acetonitrile. The gradient applied was held at 0% B for 10 min, 0% B to 80% B for 40 min, 80% B to 0% B for 10 min. Measurements were made at 220 nm, 0.2 AUFS, flow rate 1.0 ml min^{-1}

Table 7. Separation of tryptic peptides of cytochrome c by reversed-phase HPLC

1. Equilibrate the column filled with Shandon C18-ODS, 60 Å pore size, 5 μm particle size) with start-ing eluent (0.05% TFA) for half an hour.
2. Inject 100 μg tryptic peptide mixture in 50 μl 2% acetic acid after short centrifugation to remove nonsoluble particles. The sensitivity of detection is 0.16 AUFS at 220 nm.
3. Start the gradient as follows:
      0% B for 10 min
      0% B to 80% B in 60 min
    80% B to 0% B in 10 min.
4. Dry the peptide-containing fractions in a Speed Vac Concentrator, dissolve in 100 μl water or 50% pyridine/water and transfer to hydrolysis or sequencing tubes.

4 Peptide Fragmentation by Enzymatic and Chemical Cleavages in the Micro-Scale

Different enzymatic and chemical cleavages may be employed for protein and peptide fragmentation [2, 5, 6]. In this paper practicable conditions for the frag-mentation of small amounts of polypeptides are given. Which enzyme or chemical cleavage should be selected is dictated by the properties of the investigated protein, e.g., by its solubility, amino acid composition and whether manual or automatic sequencing is applied. It is recommended to perform several fragmentations on a micro-scale and to compare the separation profiles which can be obtained by (1) thin-layer fingerprinting, (2) by HPLC methods, and (3) by slab gel chromatography (for bigger peptides). Such comparison allows an easy decision of the appropriate cleavages and separation techniques and to select an optimal sequencing strategy.

4.1 Enzymatic Cleavages

4.1.1 Digestion with Trypsin

Cleavage with trypsin occurs with the C-terminal peptide bond of arginines and lysines. The bond Lys-Pro is not cleaved, that of Arg-Pro can be cleaved and bonds involved in repeated basic residues are cleaved partially. The procedure of the tryptic digestion is given in Table 8.

Table 8. Tryptic digestion

---

1. Dissolve 100 µg protein in 100 µl of bidistilled water in glass tubes tempered at 500 °C, and add 100 µl of 0.2 M N-methyl-morpholine acetate buffer pH 8.1 (add dropwise and proof solubility).
2. Add 2 µg TPCK-chloro[N-tosyl-L-phenyl(alanyl)methane]-treated trypsin (Worthington) dissolved in 2 µl of bidistilled water, cover with parafilm and keep stirring at 37 °C with micro-rod for 4 h (enzyme to substrate ratio 1:50).
3. Freeze and lyophilize the sample (with small holes in the parafilm cover), or inject directly into the HPLC column (detection at 220 nm, range 0.2).

4.1.2 Digestion with Chymotrypsin

Use TLCK-(n-tosyl-L-lysyl-chloromethane) chymotrypsin.

Cleavage with chymotrypsin occurs at: Leu ↓
                             ↓Tyr ↓
                             ↓Phe ↓
                             ↓Trp ↓
                             ↓Met ↓
                             Ala ↓

Conditions for the chymotrypsin digestion are the same as for trypsin; however, cleavage time is 2 – 4 h.

4.1.3 Digestion with Thermolysin

Buffers for thermolysin digestion are the same as for trypsin at pH 8.1; however, temperature is 50 °C, enzyme/substrate ratio 1:100 and cleavage time 2 h. For more details see [7].

Cleavage with thermolysin occurs at: ↓Ile
                                        ↓Val
                                        ↓Thr
                                        ↓Asx↓
                                        ↓Tyr↓
                                        ↓Phe↓
                                        ↓Trp↓
                                        ↓Leu
                                        ↓Ala↓

4.1.4 Digestion with Staphylococcus aureus Protease [8]

Staphylococcus aureus protease V-8 (Miles, Slough UK) cleaves proteins and peptides at carboxyl termini of glutamic acid at pH 4.0 and additionally at carboxyl termini of aspartic acid at pH 7.8. Cleavage at pH 4.0 is advantageous to yield specific digestion fragments and to avoid unspecific reaction of other contaminated enzymes active at basic milieu.

4.1.4.1 Digestion with Staphylococcus aureus Protease at pH 4.0

The procedure for the digestion with SP-enzyme is presented in Table 9.

Table 9. Digestion with *Staphylococcus aureus* protease at pH 4.0

1. Lyophilize 100 μg protein sample and dissolve in 200 μl of 0.1 M ammonium acetate buffer pH 4.0 (diluted acetic acid in bidistilled water adjusted to pH 4.0 with dilute ammonia).
2. Add 2 μg of SP protease V-8 enzyme dissolved in 2 μl bidistilled water, cover with parafilm and keep stirring at 37 °C for 24 – 48 h.

Cleavage with SP-enzyme at pH 4.0 occurs at: Glu↓
                                               Glu↓Gln↓
not at Glu-Gly, Glu-Pro, Glu-Lys, Glu-Arg and near the C-terminus.

4.1.4.2 Digestion with Staphylococcus aureus Protease at pH 8.1

Conditions and buffer at pH 8.1 are the same as for tryptic digestion but cleavage time is 24 – 48 h.

Cleavage with SP enzyme at pH 8.1 occurs at: Glu↓
                                               Gln↓
                                               Asp↓
                                               Asn↓

and other unspecific cleavage sites.

4.1.5 Digestion with Pepsin

The conditions for the digestion with pepsin are given in Table 10.

Table 10. Digestion with pepsin

1. Dissolve 100 µg dry protein in 100 µl of 0.2% acetic acid or dilute formic acid at pH 2.0.
2. Add 1 µg of pepsin dissolved in 1 µl bidistilled water and keep stirring at 37 °C for 2 h.
3. Freeze the sample and lyophilize.

Cleavage with pepsin occurs at: ↓Phe↓
                                     ↓Trp
                                     ↓Tyr
                                     ↓Met
                                     ↓Ala
                                     ↓Leu↓
                                     ↓Val
                                     ↓Thr

4.1.6 Digestion with Clostripain [9]

Conditions for clostripain (Boehringer or Sigma) digestion are as follows:
Buffer: water or dilute buffer in the range of pH 7.8
Enzyme/substrate ratio: 1 : 100
Temperature: 37 °C
Cleavage with clostripain occurs at: Arg↓
    Depending on the enzyme preparation used it may also cleave after lysine [10].

4.1.7 Digestion with Armillaria mellea Protease [11]

The enzyme was obtained from V. B. Petersen, Copenhagen, Inst. f. Biokemisk Genetik and is available through Boehringer, Mannheim.
Conditions:
Buffer: pH 8.1, same as for tryptic digestion
Enzyme/substrate ratio: 1 : 1000
Temperature: 37 °C
Time: 4 – 6 h

Cleavage with *Armillaria mellea* protease occurs at: ↓Lys
                                                          ↓LysLys

Examples of protein cleavages employing this enzyme are given in [10, 12].

4.2 Chemical Cleavages

4.2.1 Cyanogen Bromide Cleavage [13]

Cleavage with cyanogen bromide occurs at carboxyl termini of methionine, see Fig. 5. Cyanogen bromide reacts with the sulfur of the side chains of methionine and forms mixtures of homoserine lactone, homoserine, and methylthiocyanate. The cleavage conditions are given in Table 11.

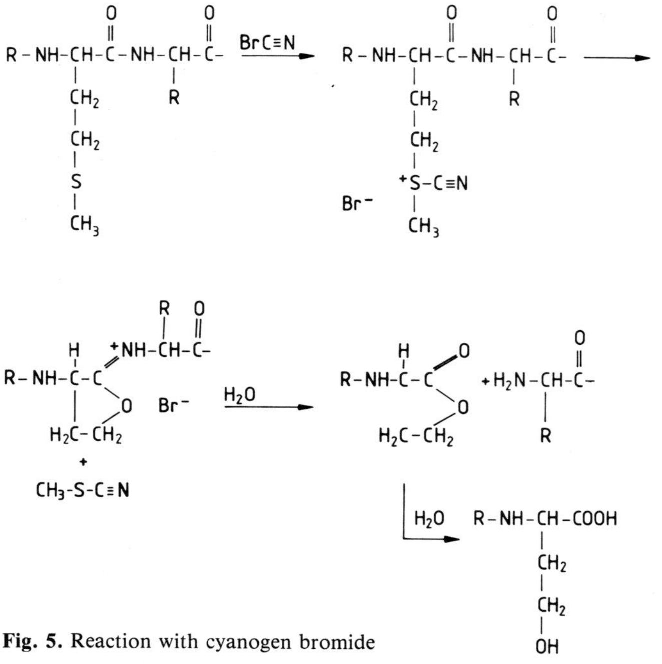

Fig. 5. Reaction with cyanogen bromide

Table 11. Cleavage with cyanogen bromide

1. Dissolve 200 µg lyophilized protein in 200 µl 70% formic acid and 2 µl mercaptoethanol.
2. Dissolve 1 mg cyanogen bromide in 200 µl 70% formic acid and add to protein solution.
3. Keep the sample in the dark under nitrogen at 25 °C for 24 h.
4. Subsequently add 3 ml water and lyophilize.
5. Check cleavage by slab gel electrophoresis.
6. Desalt sample prior to use.

4.2.2 Cleavage with BNPS-SKATOLE [14]

BNSP-SKATOLE [3-brom-3 methyl-2-(nitrophenylthio)-indolenine] cleaves peptide bonds of tryptophan, tyrosine and histidine, see Fig. 6. The conditions are given in Table 12.

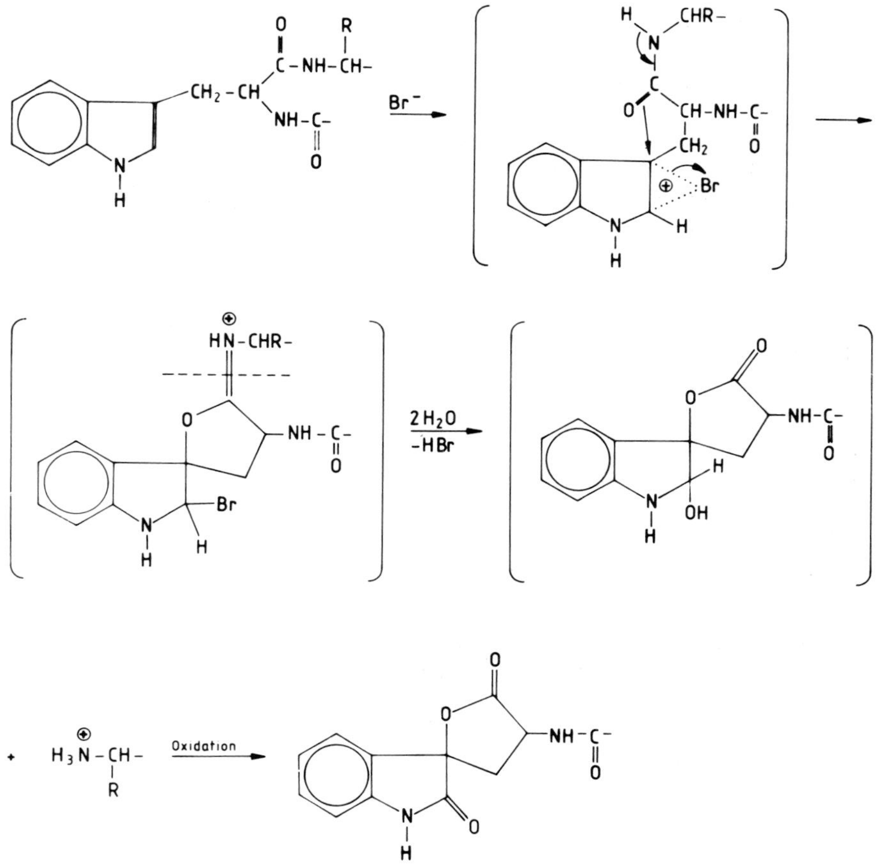

Fig. 6. Reaction with BNPS-SKATOLE

Table 12. Cleavage with BNPS-SKATOLE

1. Dissolve 100 µg protein in 100 µl 50% acetic acid.
2. Add 100 µg tyrosine in 10 µl 50% acetic acid for protection of tyrosine in the protein.
3. Add 400 µg BNPS-SKATOLE (recrystallized shortly before use from acetone) dissolved in 40 µl acetic acid (1 : 1, v/v) mixture.
4. Keep the sample in the dark under stirring for 48 h.
5. After cleavage, lyophilize the sample, check the cleavage in polyacrylamide gel electrophoresis and desalt.

4.2.3 Partial Acid Hydrolysis

Dilute acid (2% acetic acid or 0.03 N HCl) cleaves proteins at the peptide bond before and after aspartic acid, and to a lesser extent after asparagine [12, 15 – 17]. The cleavage conditions are given in Table 13.

Table 13. Mild acid cleavage of polypeptides

1. Dissolve 100 µg protein in 100 µl 2% acetic acid in a tempered glass (0.6 × 8 cm).
2. Flush the solution with nitrogen and close the ampoule under vacuum.
3. Keep the glass for 15 h at 110 °C.
4. Dry the sample after cleavage in a Speed Vac Concentrator and analyze (slab gel electrophoresis, thin-layer fingerprinting and/or HPLC reverse-phase chromatography).

References

1. Wittmann-Liebold B, Kamp RM (1980) Biochem Int 1:436–445
2. Wittmann-Liebold B, Lehmann A (1980) In: Birr Chr (ed) Methods in peptide and protein sequence analysis. Elsevier/North Holland Biomedical Press, Amsterdam New York Oxford, pp 49–72
3. Offord RE (1966) Nature 211:591–593
4. Kamp RM, Yao ZJ, Wittmann-Liebold B (1983) Biol Chem. Hoppe Seyler's 364:141–155
5. Kamp RM, Wittmann-Liebold B (1982) FEBS Lett 149:313–319
6. Keil B (1982) In: Elzinga M (ed) Methods in protein sequence analysis. Humana, Clifton, NJ, pp 291–304
7. Heinrikson RL (1977) In: Hirs CHW, Timasheff SN (eds) Methods in enzymology, vol 47. Academic Press, London New York, pp 175–188
8. Drapeau GR (1977) In: Hirs CHW, Timasheff SN (eds) Methods in enzymology, vol 47. Academic Press, London New York, pp 189–194
9. Mitchell WM (1977) In: Hirs CHW, Timasheff SN (eds) Methods in enzymology, vol 47. Academic Press, London New York, pp 165–169
10. Kimura M, Foulaki K, Subramanian A-R, Wittmann-Liebold B (1982) Eur J Biochem 123:37–53
11. Lewis WG, Bassford ZM, Walton PL (1978) Biochem Biophys Acta 522:551–560
12. Rombauts W, Feytons V, Wittmann-Liebold B (1982) FEBS Lett 149:320–327
13. Gross E, Witkop B (1961) J Am Chem Soc 83:1510
14. Fontana A (1972) In: Hirs CHW, Timasheff SN (eds) Methods in enzymology, vol 25. Academic Press, London New York, pp 419–423
15. Kamp RM, Wittmann-Liebold B (1980) FEBS Lett 121:117–122
16. Wittmann-Liebold B, Bosserhoff A (1981) FEBS Lett 129:10–16
17. Inglis AS (1983) In: Hirs CHW, Timasheff SN (eds) Methods in enzymology, vol 91. Academic Press, London New York, pp 324–334

1.2 High Performance Liquid Chromatography of Proteins

ROZA MARIA KAMP[1]

Contents

1 Introduction

In recent years high performance liquid chromatography has been applied widely to the separation of biomolecules. Since new HPLC supports have become available for the separation of high molecular mass compounds, the rapid purification of proteins has become possible (see for details [1 – 8]).

Separation time can be reduced from days or weeks to hours, in contrast to conventional methods. The HPLC techniques are useful for the isolation of small sample quantities, and the separations established on an analytical scale can be transferred to larger quantities by the use of preparative columns.

HPLC methods have been shown to offer unrivalled advantages in terms of speed, resolution, sensitivity, and recovery.

---

1 Max-Planck-Institut für Molekulare Genetik, Abteilung Wittmann, Ihnestraße 63 – 73, D-1000 Berlin 33 (Dahlem)

Advanced Methods in Protein Microsequence Analysis
Ed. by B. Wittmann-Liebold et al.
© Springer-Verlag Berlin Heidelberg 1986

Rapid chromatographic separation of proteins results through high linear flow rates of $0.1 - 5$ cm s^{-1} at pressures of $10 - 400$ atm compared with 0.02 cm s^{-1} at hydrostatic pressure by open column chromatography.

Several methods of separating protein mixtures can be employed: size exclusion chromatography on hydrophilic phases, ion exchange and reversed-phase chromatography on hydrocarbon bonded supports, or combined modes of chromatography.

The resolution and recovery of the proteins varies considerably, depending on the type of support and gradient used. Best recoveries of $90 - 100\%$ were found with size exclusion columns. The yields after reversed-phase chromatography are lower, depending on the hydrophobicity and molecular masses and the charges of the proteins. The hydrophilic, basic, and small proteins are eluted first, while the acidic or hydrophobic ones are more retarded.

Reversed-phase chromatography is most useful for the separation of ribosomal proteins. Volatile and UV-transparent buffers allow high detection sensitivities at 220 nm and enable further micro-scale protein-chemical investigations. Purification of proteins on ion-exchange columns is applicable to proteins which are not well resolved by reversed-phase chromatography. However, in general, this chromatography is disadvantageous, since the measurements have to be made at 280 nm, where the sensitivity is lower and the detection of proteins with no aromatic amino acids is not possible.

The combination of the three techniques enables, in most cases, the purification of complex protein mixtures. Further, HPLC can be used for desalting of protein samples, as discussed below.

2 Size Exclusion Chromatography

Separation of proteins on size exclusion columns depends on the molecular mass of the molecules. Smaller proteins penetrate into the pores of the particles of the column and will be retarded longer than larger proteins.

Usually, size exclusion columns are applied for prefractionations, analytical and preparative isolations of different sized proteins, and may be applied for desalting procedures or molecular mass determinations. As an example, Fig. 1 shows the separation of 50S ribosomal proteins from *E. coli* on size exclusion columns.

2.1 Columns

The commercially available columns are alkylated silica-based or organic-based supports. The most popular gel filtration supports are from Toyo Soda (Japan):

- TSK gel SW type is spherical porous silica with bonded hydrophilic polar groups stable in the pH range of $2.0 - 8.0$. Table 1 lists properties of different types of TSK columns.
- TSK gel PW type is hydroxylated ether stable in the pH range of $2 - 12$.

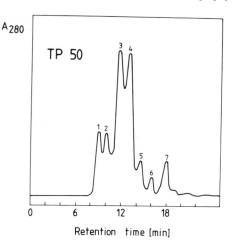

Fig. 1. Group separation of 50S ribosomal proteins from *E. coli* by HPLC on TSK 2000 SW column. An amount of 1 mg protein mixture was injected in 100 μl 2% acetic acid in water. The eluent was 0.1 M ammonium acetate, pH 4.1, flow rate 1.0 ml min^{-1} at room temperature, the eluate was monitored at 280 nm, 0.02 AUFS

Table 1. Size exclusion TSK columns

TSK-Gel	MW range (Daltons)	Pore size (Å)	Particle size (μm)
G 2000 SW	500 – 60,000	125	10
G 3000 SW	1000 – 300,000	250	10
G 4000 SW	5000 – 1,000,000	400	13

Table 2. Maintenance of size exclusion columns

1. Use TSK-gel columns only with a guard column to separate unpurified samples.
2. Replace the guard column from time to time, if the separation does not suffice (wide peaks) or the pressure increases. Normal values for analytical column vary between 30 – 40 bar at 1 ml min^{-1}.
3. Rinse the new column with the mobile phase overnight prior to sample injection.
4. Keep the flow rate between 0.1 – 1.0 ml min^{-1} to obtain high resolution.
5. Pump solvent in indicated direction only. This is the same direction as for packing of the column and allows highest resolution.
6. Avoid using halide salts in the mobile phase if possible; these are deleterious to stainless steel.
7. For longer storage, wash column with Millipore-Q water to remove all salts followed by rinses with methanol or 0.05% sodium azide at a flow rate lower than 1 ml min^{-1}.
8. Keep the ends of the columns capped with the original nuts or seal with parafilm.

The separation on silica-based columns depends not only on the molecular mass, but also on the net charge of the proteins, as demonstrated with ribosomal proteins. Negatively charged free silanol groups absorb basic molecules and repulse acidic proteins. In this case acidic proteins are eluted earlier than the larger and more basic ones. High salt concentrations are often applied to suppress ionic interaction between the support and the proteins. The organic-based gel filtration supports are stable at a higher pH, but resolution of proteins is generally higher on silica-based columns.

The procedure for the use of TSK-columns is given in Table 2.

2.2 Mobile Phases

Typical eluents for separation on gel filtration columns are sodium phosphate buffer in the range of pH 6–8, ammonium acetate, ammonium formate, Tris acetate and citrate. The low salt concentration buffers (0.05–0.1 M) allow direct protein identification by gel electrophoresis and manual sequencing. For separation of water-insoluble membrane proteins 0.1% SDS, 6 M urea or guanidine chloride eluents are possible, but unfortunately, the column life is drastically reduced by using SDS, guanidine chloride or urea.

The preparation of eluents for the chromatography with TSK-columns is shown in Table 3.

Table 3. Eluent preparation for size exclusion chromatography

This is demonstrated for the use of 0.1 M ammonium acetate, pH 4.1; other buffers would be prepared and used in a similar manner.

1. Use only deionized water from a Mill-Q purification system (equipped with ion-Ex and super-C carbon cartridges) from Millipore (USA) or two times quartz-distilled water with no plastic connections.
2. Prepare mobile phases in glass only. Plastic containers cause ghost peaks.
3. Mix ammonium acetate buffer (0.1 M) from ammonia (pro analysis) and acetic acid (pro analysis), to obtain high purity mobile phases. Ammonium acetate salt (p.a.) is not pure enough for HPLC separations.
4. Add to the buffer 1 mg sodium azide per 1 l solution to inhibit microbial growth in the mobile phases.
5. Degas the solvent completely (water pump for about 20 min).
6. Place a 2.0 μm steel filter into the buffet reservoir.
7. Use ammonium acetate buffer no longer than 1 week.

2.3 Equipment

One HPLC pump, column, precolumn, detector (230 and 280 nm), oven, recorder, injection valve, one buffer vessel with fritte and a sonicator.

3 Reversed-Phase Chromatography

Reversed-phase chromatography is the HPLC technique most often applied for the separation of biological molecules. Table 4 lists the properties of spherical reversed phase supports. Figure 2 shows separation of a 50S *E. coli* protein mixture on a Vydac column.

3.1 Supports

The proteins are separated on hydrocarbonaceous (C_4, C_8, or C_{18} alkylated) silica-based supports. Recent experience shows that small particle (5 μm) and wide pore sizes (300–400 Å) are optimal for resolution and high recovery. Table 4 lists the best available columns for separation of proteins as tested on ribosomal protein mixtures. All these columns are based on uniformly spherical macro-porous particles. The resolution on irregularly shaped columns is lower and the reproducibility less good, especially if filled with support of different batches. Table 5 contains recommendations for the best maintenance of the columns.

Table 4. High performance reversed-phase supports for proteins

Name	Supplier	Support material	Bonded phase	Particle size (μm)	Pore size (Å)
Vydac TP-RP C$_4$	The Separation Group, Hesperia CA, USA	Hydrocarbon butyl phase	Silica	5, 10	330
Vydac TP-RP C$_{18}$		Hydrocarbon octadecyl phase	Silica	5, 10	330
Nucleosil 300-5 C$_4$	Macherey-Nagel, Düren, FRG	Hydrocarbon butyl phase	Silica	5	300
Nucleosil 300-10 C$_4$		Hydrocarbon butyl phase	Silica	10	300
TSK ODS-120T	Toya Soda, Japan	Hydrocarbon octadecyl phase	Silica	5, 10	120

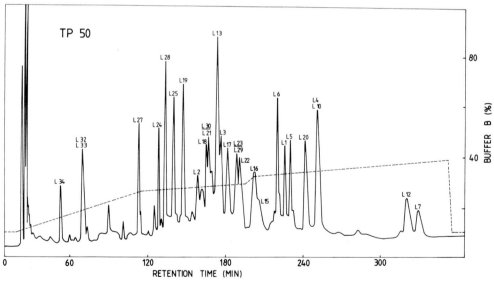

Fig. 2. Separation of 50S ribosomal proteins from *E. coli* on Vydac TP-RP column. Analytical column, 250 × 4.6 mm ID; 2 mg TP50 was injected in 200 μl 2% acetic acid. The eluents were: buffer A, 0.1% aqueous TFA; buffer B, 0,1% TFA in 2-propanol. The gradient applied was: 10% B to 27% B in 100 min, 27% B to 30% B in 80 min, 30% B to 33% B in 5 min, 33% B to 38% B in 170 min, 38% B to 10% B in 5 min and reconditioning for 30 min at initial conditions. The eluate was monitored at 220 nm, 0.64 AUFS, flow rate 0.5 ml min^{-1}, temperature 35 °C, recorder speed 2 mm min^{-1}

Table 5. Maintenance of columns

1. Use column only in connection with a precolumn (3 – 4 cm) packed with same material as in the separation column.
2. Prior to use wash column with 80% MeOH in water and then with the aqueous buffer for 2 h. Rinse overnight, by running a gradient of 0 – 80% B in 2 – 3 h with the buffers to be used the next day. Keep the flow rate at 0.5 ml min^{-1}.
3. Check the column purity by a gradient run without sample injection.
4. Test the column efficiency by running a protein reference solution, e.g., a ribosomal protein mixture or a mixture of standard proteins.
5. Inject sample solute in starting buffer or corresponding buffer.
6. In the case of "ghost" peaks appearing, regenerate the columns as follows:
   – Wash column with water to obtain straight baseline.
   – Inject 4 times 200 µl DMSO (dimethylsulfoxide) (p.a.).
   – Wash column with 100 ml water.
   – Replace water by methanol and wash the column to obtain a stable baseline (100 ml or more).
   – Replace methanol by chloroform and wash out all impurities.
   – Repeat column wash with methanol (100 ml or overnight) to obtain a stable baseline.
   – Equilibrate column with starting aqueous buffer.
   – Check the resolution by injection of test mixture.
7. Store column after washing with methanol.

3.2 Eluents

3.2.1 Aqueous Eluents

Different aqueous eluents can be applied as eluents for separation of proteins, e.g., trifluoroacetic acid (TFA), triethanolamine phosphate (TEAP), ammonium acetate, and formic acid. For protein investigation, volatile buffers of low salt concentrations are preferably used for reversed-phase HPLC; these allow direct microsequencing or identification of proteins by gel electrophoresis.

Application of 0.1% TFA as buffer A to separate ribosomal proteins results in sufficient resolution. The preparation of aqueous buffers is given in Table 6.

Table 6. Preparation of 0.1% TFA eluent

1. Use Milli-Q water.
2. Use only TFA sequence grade (or three times re-distilled technical TFA).
3. Degas the 0.1% TFA solution thoroughly (by water pump, sonication, or use solvent degaser from ERMA (Japan).
4. Add 1 mg sodium azide in 10 µl and 10 µl 2-mercaptoethanol per 1 l of 0.1% TFA solution to stabilize the buffer and proteins.
5. Prepare fresh 0.1% TFA solution every week.

3.2.2 Organic Modifiers

Many water-miscible organic solvents have found use for protein separation in reversed phase HPLC. However, at the wavelength at which the eluent will be recorded the solvent should show at least 70 – 80% transparency. The studies with

methanol, acetonitrile, and propanol show that the relative retention of proteins decreases in the order:
methanol, acetonitrile, 1-propanol or 2-propanol, and corresponds to the elutropic strength.

Experience with ribosomal proteins shows that gradients with propanol are superior to those with acetonitrile and methanol. The use of acetonitrile as organic modifier often causes proteins to unfold and denature, and might cause migration of some proteins in multiple peaks. This problem can be avoided by using 2-propanol as eluent. The ribosomal proteins separated in propanol gradients are fully active in reconstitution assays. Preparation of organic solvents is exemplified in Table 7.

Table 7. Preparation of organic modifier

1. Use 2-propanol (or other modifier) of spectroscopic or chromatographic grade.
2. Mix propanol with 0.1% TFA sequential grade.
3. Degas 0.1% TFA in propanol by sonication or automatically by ERMA (Japan) degaser.

3.3 Sample Preparation and Injection into the Column

Ribosomal proteins were extracted from ribosomes and stored in 2% acetic acid at $-20\,°C$ in concentration 1 mg/100 µl. Precipitation steps and lyophilization were avoided. In the case of dilute solutions, concentration (but not to dryness) was done in a Speed Vac Concentrator. Manual and automatic injection procedures are given in Tables 8 and 9.

Table 8. Manual injection

1. Wash injection syringe with methanol.
2. Fill the syringe slowly with sample of the protein mixture.
3. Turn the syringe at the top and remove air bubbles.
4. Inject the sample into the rheodyne valve in 20 – 200 µl aliquots.

Table 9. Automatic injection

1. Fill sample glass tube with protein solution.
2. Remove air bubbles from the bottom of the glass, mechanically or by short low speed centrifugation.
3. Place the sample glass in automatic sample injector (e.g., WISP of Millipore/Waters).
4. Program the injector.

3.4 Equipment

Two HPLC pumps in case of high pressure mixing system (or one low pressure system pump with mixing chamber), column, precolumn, column oven, variable wavelength detector (220 and/or 280 nm), two-channel recorder, injector and/or sampler, two buffer vessel with frittes, sonicator, degaser.

4 Ion-Exchange Chromatography

Protein separation on ion-exchange columns depends on the charge of the proteins and the ion exchange matrix. Figure 3 shows a typical separation a of 50S *E. coli* proteins on an ion-exchange column.

4.1 Columns

The ion-exchange columns are based on a silica or organic matrix gel, coated with sulfonic acid, carboxyl groups, or primary and quarternary amines.

Widely applied are ion exchange columns from Toyo Soda (Japan), of TSK type or the organic-based Mono Beads from Pharmacia. Table 10 lists the properties of different ion-exchange supports.

The choice of the column support depends on the properties of the protein mixture and pH range of the mobile phases. The loading capacity of typical analytical columns (25×4 cm) is about 10 mg. This is five times more than the load maximum for a corresponding reversed-phase column. The recovery on the ion exchange column depends on the interaction of the proteins and the support, their solubility in the mobile phases, and partial denaturation or accumulation of aggregates during separation. Instructions for the maintenance and cleaning of ion-exchange columns are given in Tables 11 and 12.

4.2 Eluents

The mobile phases used for separation on ion-exchange columns are similar to those of conventional chromatography. Ionic strength gradients and pH gradients have been successfully used for protein separations. The most frequently applied system is that based on ionic strength gradients. Proteins are displaced from the column in the order of their increasing charge. The procedure for preparation of buffers for ion-exchange chromatography is given in Table 13.

4.3 Equipment

Equipment as in case of reversed-phase HPLC (see Sect. 3.4).

5 Combined HPLC Chromatography

The described methods for isolation of proteins in most cases obtain sequence grade pure proteins. Sometimes it is necessary to repurify pooled HPLC fractions on another type of column or with a different elution system.

In general, one of the following combined chromatographic techniques is applied:

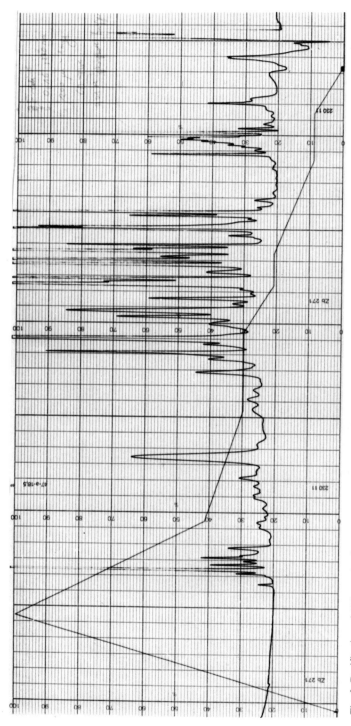

Fig. 3. Purification of 50S proteins from *E. coli* on TSK-IEX 535 column. Amounts of 20 mg TP50 were injected in 200 µl 2% acetic acid in water. The eluents were: buffer A, 0.01 M sodium phosphate in 5 M urea, pH 6.5; buffer B was made from buffer A and 1.0 M potassium chloride. The gradient applied was: hold at 0% B for 30 min, 0% B to 9% B in 30 min, hold at 9% B for 20 min, 9% B to 21% B in 60 min, hold at 21% B for 20 min, 21% B to 30% B in 30 min, hold at 30% B for 50 min, 30% B to 40% B in 60 min, 40% B to 100% B in 60 min, 100% B to 0% B in 30 min. Measurements were made at 230 nm, 0.2 AUFS, flow rate 1.0 ml min^{-1}

Table 10. High performance ion-exchange columns for proteins

Name	Supplier	Bonded phase	Support material	Particle size (μm)	Pore size (Å)
TSK 530 CM-SW	Toyo Soda (Japan)	Carboxylmethyl	Silica	5	130
TSK 535 CM-SW		Carboxylmethyl	Silica	10	240
TSK 540 DEAE-SW		Diethylaminoethyl	Silica	5	130
TSK 545 DEAE-SW		Diethylaminoethyl	Silica	10	240
TSK SP-PW		Sulphopropyl	Organic	10	130
Mono Q	Pharmacia Fine	Quarternary amine	Organic	10	Not available
Mono S	Chemicals (Sweden)	Sulfonic acid	Organic	10	Not available

Table 11. Maintenance of columns

1. Use Millipore-Q water for the preparation of the aqueous buffer.
2. Use wide pore size exclusion guard columns for trapping of impurities (the pore size of the guard column should be higher than that of the separation column).
3. Avoid any high pressure surges and pressure pulsation. Use only a pulseless pumping system. Avoid quick and frequent changes in the flow rate.
4. The column should be equilibrated overnight prior to use with starting buffer.
5. For storage, purge the column with redistilled water at low flow rate, close the column at its ends and store in methanol.

Table 12. Cleaning of column

Prolonged separation can cause adsorption of proteins on the support and thereby a loss of capacity and resolution. Adsorbed molecules may be stripped from the column by using gradients of pH, high ionic strength and organic modifier as follows:
- Purge the system overnight with Milli-Q water to dissolve salt particles from the column and capillaries.
- Rinse with a gradient from 0.1 to 0.5% TFA to 60% 2-propanol in 0.1% to 0.5% TFA.
- Wash separation system carefully with water.
- Change from water to the buffer solution to be used.

Table 13. Preparation of buffers for ion exchange HPLC

As example the preparation of 0.01 M sodium phosphate pH 6.5 is given.
1. Prepare a stock buffer solution of 0.05 M sodium phosphate and adjust pH to 6.0 with 5 N NaOH.
2. Check the pH the next day and readjust to pH 6.5.
3. Prepare a 6 M urea (p.a.) solution.
4. Prepare starting buffer by mixing 0.01 M sodium phosphate solution (1 vol 0.05 M stock buffer) and 6 M urea (4 vol).
5. Prepare second buffer from starting buffer and potassium chloride to a final concentration of 1 M KCl.
6. Filter both buffers first through a paper filter (Schleicher and Schüll) and secondly through a 0.45 μm Millipore filter.
7. Degas all buffer carefully with a water pump to ensure a continuous flow through the system.

1. Group separation of the complex protein mixture on gel filtration columns into distinct fractions followed by rechromatography on reversed-phase columns.
2. Separation on an ion-exchange column and rechromatography on a reversed-phase column using volatile buffers.

In combined chromatographic modes it is important to select the reversed-phase separation as the last purification step in order to obtain salt-free samples. The proteins then can be used directly for identification or sequencing. Purification of proteins using multiple columns can be carried out by coupling different columns with the aid of switching valves. In this technique the sample can be directly transferred from one column to another without collecting and preparing samples for the next separation step.

In case of a separate column system the fractions after size exclusion or ion-exchange HPLC are pooled, desalted on short reversed-phase columns (see Sect. 6) and reinjected on the analytical reversed-phase column.

6 Desalting by Reversed-Phase Chromatography

The commonly used desalting techniques are dialysis, chloroform-methanol, or trichloroacetic acid precipitation and open column chromatography.

In dialysis, which relies on osmotic and diffusion forces to drive molecules through a membrane, the rate of salt removal is slow. The proteins often stick to the dialysis membrane which causes loss of material.

The trichloroacetic acid precipitation is faster than dialysis, but can be employed only for protein in highly concentrated solutions (10 μg ml^{-1}). Recovery depends on protein concentration and individual handling and varies from 10 – 80%. Further disadvantages are solubility problems after desalting.

Open column desaltings made by gel filtration on Sephadex or Bio-gel permit separation of high molecular weight proteins from low molecular contaminants. An alternative method to these desalting techniques is application of reversed-phase chromatography to separate the protein from salt and other low weight impurities. The equipment for desalting by HPLC is demonstrated in Fig. 4.

The diluted fractions are injected in volumes up to 2 ml onto a small 0.4×4 cm reversed-phase column. The low molecular contaminants are eluted first by pumping with aqueous mobile phase; proteins are desorbed by replacing aqueous eluents with organic modifier (2-propanol).

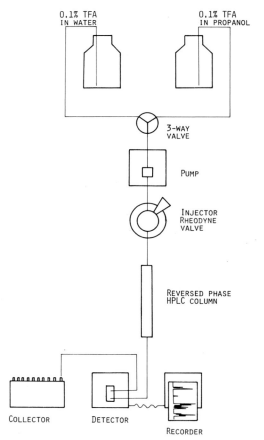

Fig. 4. Equipment for desalting of proteins

Since the separation time for desalting of protein fractions by HPLC is very short (5 – 10 min) in comparison with the classical methods, and since the purification can be performed under mild conditions (e.g., in volatile buffers and with a low percentage of alcohols as organic modifiers), the described method is well suited for the desalting of active proteins for functional investigations.

The recovery is excellent; it depends on the hydrophobicity and size of the protein and varies fom 90 – 100%.

Desalting on 300 Å pore-sized support can be applied for the purification of proteins in the molecular mass range of 50 – 60,000. Larger proteins cannot penetrate through the small pores and stick to the column.

References

1. Kamp RM, Yao ZY, Bosserhoff A, Wittmann-Liebold B (1983) Hoppe-Seyler's Z Physiol Chem 364:1777
2. Kamp RM, Wittmann-Liebold B (1984) FEBS Lett 167:59
3. Kamp RM, Bosserhoff A, Kamp D, Wittmann-Liebold B (1984) J Chromatogr 317:181 – 192
4. Gupta S, Pfannkoch E, Regnier FE (1982) Anal Biochem 196
5. Kato Y, Komiya K, Sasaki H, Hashimoto T (1980) J Chromatogr 190:297
6. Regnier FE, Googing KM (1980) Anal Biochem 130:1
7. Kerlavage AR, Weitzmann CJ, Hasan T, Cooperman BS (1983) J Chromatogr 266:225
8. Kehl M, Lottspeich F, Henschen A (1982) Hoppe-Seyler's Z Physiol Chem 363:1501

1.3 Two-Dimensional Polyacrylamide Gel Electrophoresis in Stamp-Sized Gels

HEINZ-JÜRGEN BROCKMÖLLER and ROZA MARIA KAMP[1]

Contents

1 Introduction

Two-dimensional polyacrylamide gel electrophoresis is a useful technique for screening complex protein mixtures. In the case of such mixtures, single bands obtained by any of the one-dimensional separation systems (such as SDS gel electrophoresis or isoelectric focusing) do not prove the purity of the sample. In contrast, single spots on an appropriate two-dimensional gel are more reliable, and they less frequently consist of two or more proteins.

Different two-dimensional polyacrylamide gel electrophoresis systems have been described adapted to the special nature of the protein mixture. This technique was first developed for the separation, identification, and correlation of ribosomal proteins [1]. Later the method was modified for the application of smaller sample amounts (e.g., to about 100 µg protein mixture) or to special ribosomal proteins and those of other organisms (for citations see [2]). The two-dimensional gel technique proved to be useful also for other protein classes and finally allowed the resolution of most of the many proteins from entire cells [3].

The usual two-dimensional gel procedures are well established in most protein laboratories (for more details see the chapters of the recent book on gel electro-

---

1 Max-Planck-Institut für Molekulare Genetik, Abteilung Wittmann, Ihnestraße 63 – 73, D-1000 Berlin 33

Advanced Methods in Protein Microsequence Analysis
Ed. by B. Wittmann-Liebold et al.
© Springer-Verlag Berlin Heidelberg 1986

phoresis of proteins [4], for other micro-gel techniques see [5]). We therefore restrict ourselves to the description of a new microtechnique which is valuable for the application of very limited protein amounts, e.g., of $100-200$ ng (≈ 10 pmol) of a purified ribosomal protein [2].

The advantages of this new micro-gel technique are obvious: The micro-gels of $3 \times 4 \times 0.05$ cm in size as opposed to the normal $10 \times 10 \times 0.3$-cm gels need much smaller substance amounts, and the developing and staining time is much reduced. Only aliquot fractions of the HPLC chromatogram are wasted for identification and purity check of the protein-containing fractions. Another advantage is that 20 samples can be run simultaneously and the results are obtained in a few hours.

Handling of micro-gels is difficult compared to macro-gels. Therefore, practical aspects as to how to avoid difficult handling steps and the appropriate equipment are described in detail.

The gel solutions given here are developed for the identification of basic ribosomal proteins. In the first dimension the proteins are separated mainly on the basis of charge. The higher the isoelectric point of a protein, the faster it moves toward the cathode in the wide-pore first-dimension gel (4% acrylamide). In the second dimension the proteins are separated on the basis of their molecular masses. The small-pore gel of this dimension, made from 20.5% acrylamide with 6 M urea, hinders migration of the larger proteins, which is the main separation mechanism.

With modified or different gel solutions and buffers the method described here can be adapted to other protein mixtures from other organisms, other organelles, or different protein complexes. At the end of this chapter a description about how to perform microdiagonal gels for investigation of chemically cross-linked organelles or multi-enzyme complexes is included.

2 Materials and Methods

2.1 Equipment for the First-Dimension Capillary Gels

1. The first dimension electrophoresis is carried out in 50 µl capillary pipettes (Karl Hecht, D-8741 Sondheim) which have an internal diameter of 0.9 mm. Similar capillaries with the same internal diameter from other manufacturers may also be applied. The capillaries are used without pre-cleaning or silylation.
2. The apparatus for the 1-D electrophoresis system is designed for 20 samples, as shown in Figs. 1 and 2. The anode A is set on top of the anode chamber (top chamber) and the cathode D is fixed at the bottom of the lower one. At the bottom of the anode chamber a 3-mm-thick silicon rubber seal C is placed between two Plexiglas plates B. Each plate has 20 holes to accommodate up to 20 samples simultaneously. Before its first use, the seal is pricked with a needle to ease the passage of the capillaries through it. The holes in the Plexiglas plates B serve to keep the capillaries in a vertical position. The silicon rubber seal C is easily changed to allow this chamber to be used for different sizes of capillaries.

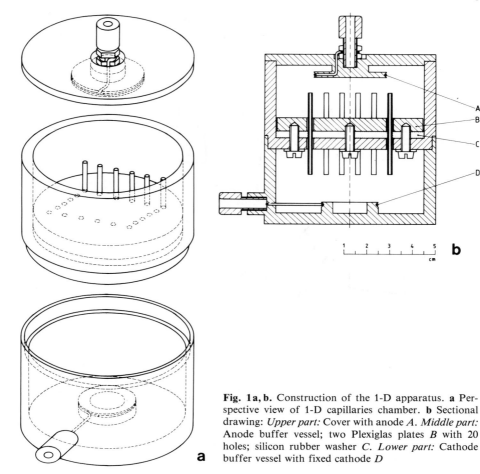

Fig. 1a, b. Construction of the 1-D apparatus. **a** Perspective view of 1-D capillaries chamber. **b** Sectional drawing: *Upper part:* Cover with anode *A*. *Middle part:* Anode buffer vessel; two Plexiglas plates *B* with 20 holes; silicon rubber washer *C*. *Lower part:* Cathode buffer vessel with fixed cathode *D*

3. For sample application onto the gels, a 5-µl microsyringe with a thin needle is used (outer needle diameter: 0.5 mm). Typically, 5 µg protein mixture (i.e., 100 – 200 ng per protein) are applied in sample buffer (see Sect. 2.3).
4. For extruding the first-dimensional gel, a piece of steel wire is used which fits closely into the capillaries.

2.2 2-D Chamber

2.2.1 Construction of 2-D Apparatus

The apparatus for the second dimension is shown in Fig. 3. This is a chamber for running 20 samples simultaneously under identical electrophoresis conditions.

The 2-D apparatus is constructed of 21 Plexiglas sandwiches E with 0.5-mm Plexiglas spacers stuck onto the Plexiglas plates on both sides. All these elements are tightly screwed together to ensure good sealing at the sides.

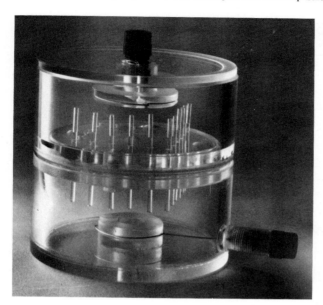

Fig. 2. Photo of 1-D apparatus with some gel capillaries

2.2.2 Holder for Polymerizing the Mini Slab-Gels

The bottom of the gel spaces is then closed by pressing the whole block of 21 sandwiches onto a silicon rubber seal. This is done by using a simple holder made for this purpose. Then, after the gels are polymerized and the first-dimension gels are embedded, the block with the gels is taken out of the holder, the upper buffer vessel is screwed onto it and the whole block is placed into the lower buffer vessel.

2.3 Gel Solutions and Buffers

2.3.1 First-Dimension Gel Solution

4.0 g acrylamide
0.1 g N,N′-methylene-bis-acrylamide
1.19 g bis-Tris
36.0 g urea
0.19 g EDTA
dissolve in 60 ml bidistilled water, adjust to pH 5.0 with acetic acid, fill up to 100 ml with bidistilled water, filter, store at 4 °C until use.

2.3.2 First-Dimension Upper-Electrode-Buffer (Stock Solution)

20.9 g bis-Tris (0.1 M)
adjust to pH 4.0 with acetic acid
fill up to 1000 ml
dilute 1 : 10 before use

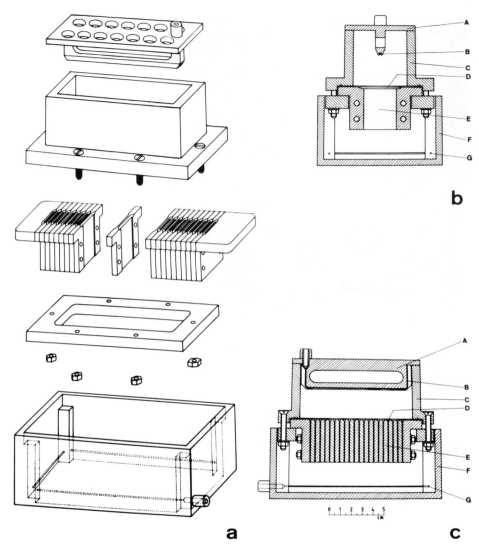

Fig. 3a – c. Construction of the 2-D apparatus. **a** Perspective view of 2-D apparatus. **b** Cross-section: Cover *A* with anode *B*; upper buffer vessel *C*; 21 Plexiglas plates *E* with V-shaped crevices *D* for the 1-D gel rod; lower buffer vessel *F* with cathode *G*. **c** Longitudinal section

2.3.3 First-Dimension Lower-Electrode Buffer (Stock Solution)

175.7 g potassium acetate
adjust to pH 5.0 with acetic acid
fill up to 1000 ml
dilute 1 : 10 before use

Fig. 4. Construction of the gel casting holder. Photo of the block of 21 Plexiglas plates within the gel casting holder

2.3.4 Sample Buffer

0.154 g dithioerythritol
36 g urea
50 mg pararosanilin as tracking dye (Pararosanilin from Sigma)
fill up to 100 ml with dilute first-dimension upper-electrode buffer

2.3.5 Second-Dimension Gel Solution for Micro-Gels (the same gel is also used for microdiagonal gels in both dimensions)

36.4 g urea
20.5 g acrylamide
0.52 g N,N'-methylene-bis-acrylamide
5.4 ml acetic acid
1 ml 5 M KOH
add bidistilled water to 90 ml, dissolve, fill up to 100 ml with bidistilled water, filter, and store at 4°C.

2.3.6 Second-Dimension Electrode Buffer (this buffer is used for micro 2D-gels and for diagonal gels)

140 g glycine
15 ml acetic acid
dissolve in 10 l bidistilled water; final pH is 4.0.

2.3.7 Second-Dimension Tracking Dyes

pararosanilin in 40% glycerin

2.3.8 Agarose-Gel for Embedding the First-Dimension Gels

5.4 ml acetic acid
1 ml 5 M KOH
fill up to 100 ml with bidistilled water, add 0.5 g agarose, boil until solution is completely clear and use at about 50 °C.

2.3.9 Staining Solution

·5 g Coomassie Blue R
2.5 l methanol
375 ml acetic acid
fill up to 5 l with bidistilled water, dissolve, filter.

2.3.10 Destaining Solution

1 l acetic acid
1 l methanol
fill up to 10 l with distilled water.

All reagents used were pro analysis grade, purchased from Merck; acrylamide (2 × crystallized) and bis-acrylamide (2 × crystallized) were purchased from Serva.

2.4 Description of the Procedures

2.4.1 Preparation of the First-Dimension Capillary Gels

1. Use 50 µl capillaries (0.9 mm inner diameter).
2. Prepare gel solution (1st dimension, see Sect. 2.3.1), degas before use, keep gel solution on ice.
3. Fill a small beaker with 10 ml first-dimension gel, add 35 µl TEMED (N,N,N′,N′-tetramethylethylenediamine) and 140 µl 5% ammonium peroxodisulfate and mix.
4. Fill a second beaker with the solution of (3) to a height of 1 cm, place 20 capillaries vertically into the gel solution. Thereby the capillaries are filled up to 2.7 cm by capillary attraction.
5. Polymerize the capillaries for approximately 20 min.
6. Add bidistilled water onto the gel of the beaker with the polymerized capillaries (for preventing air bubbles from penetrating into the lower part of the capillaries at the following step).
7. Remove the gel capillaries from the surrounding gel block by turning them individually.
8. Cut off the capillaries 5 mm above the upper gel surface (use a glass cutter); do not use any gel capillaries that show air bubbles.
9. Place the gel capillaries immediately into the first-dimension electrophoresis chamber.
10. Fill the buffer vessel with upper and lower electrode buffer (see Sects. 2.3.2 and 2.3.3).

11. Rinse the gel surfaces with upper electrode buffer using a small syringe (which removes small air bubbles).
12. Apply 1 µl of sample (typically 100 – 200 ng per protein) by using a Hamilton syringe. Note that the sample has to be as salt-free as possible. Use small dispensable micro test tubes (of 500 µl vol or less) for handling small sample amounts. Vortex thoroughly to dissolve the sample and centrifuge shortly prior to the application of the sample onto the capillaries.
13. Carry out the electrophoresis at ambient temperature, at 30 V for 5 min, then at 300 V for another 7 min, until the red tracking dye has reached the lower end of the gel (upper electrode +, lower electrode −).

2.4.2 Preparation of the Second-Dimension Micro-Gels

1. Screw the second-dimension gel apparatus tightly together. Screw the gel holder tightly onto the silicon rubber of the holder to close the bottom of the gel spaces.
2. Prepare 20 ml second-dimension gel solution (see Sect. 2.3.5) at 4 °C and degas.
3. Add 120 µl TEMED and 500 µl 5% ammonium peroxodisulfate.
4. Fill the gel spaces up to the beginning of the V-shaped crevice and cover with a solution of 4 M urea in water. For convenience cover all gels at the same time by using a sprayer.
5. Polymerize at ambient temperature for at least 1 h. Take care to avoid drying of the gel surfaces.
6. Suck off the remaining buffer on top of the polymerized gel with filter paper.
7. Push some normal candle wax into the lower end of the gel capillary by pressing it into a block of wax in order to allow gentle removal of the gel from the capillaries [see (9)].
8. Fill the V-shaped crevice of the respective gel space of the 2-D apparatus with 0.5% agarose gel.
9. Press out the gel directly onto the still warm and liquid agarose by means of a steel rod (with the wax serving as a tight sealing piston head).
10. Keep for 10 min at ambient temperature for polymerization.
11. Carry out electrophoresis at 100 V for 80 min until the red tracking dye has reached the bottom of the gels.
12. Remove the two-dimensional gels from the electrophoresis chamber and stain with Coomassie Brilliant Blue for 10 min.
13. Destain for 1 h using the destaining solution.

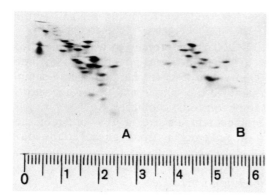

Fig. 5. Two-dimensional electrophero-grams of 50S (*A*) and 30S (*B*) ribosomal proteins from *Bacillus stearothermophilus* (gels are shown in original sizes)

3 Microdiagonal Electrophoresis

3.1 Application

Earlier diagonal procedures were developed as diagonal electrophoresis on paper for determining the location of disulfide bonds within a single polypeptide chain [6].

For the investigation of protein-protein neighborhoods in protein complexes the crosslinking with bifunctional reagents has been applied (for review see [7]). However, it is often difficult to differentiate between monovalently reacted groups within one protein and those residues which by the reaction with the re-agent really form a crosslink between two adjacent proteins. The detection of such crosslinked protein pairs can be made applying a two-dimensional diagonal polyacrylamide gel electrophoresis provided a crosslink is formed which is cleav-able thereafter ([8, 9], H.-J. Schönfeld, submitted). Again, it is advantageous to apply the micro-two-dimensional gel described above in order not to waste too much of the material isolated. Especially after HPLC isolation of a crosslinked protein pair, this method proved to be very useful [10]. We therefore describe details of this microdiagonal technique in the following.

The first step in crosslinking is the treatment of the respective organelles or multi-enzyme complexes in their native states with an appropriate reagent. It is very advantageous to use cleavable reagents such as those containing disulfide bonds as the cleavable linking group; in this case it is easy to cleave the complex by means of an SH reagent to recover the monomeric constituents of the complex [8]. Another class of reagent introduces a vicinal hydroxyl group between the crosslinked proteins, which can be cleaved by mild perjodate treatment [9].

3.2 Principles of the Method

In the two-dimensional gels described in Section 2, both directions of the electro-phoresis are performed in different gels under different conditions which sepa-rate the proteins by different criteria so as to achieve a maximum of separation.

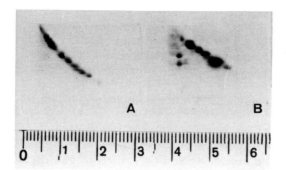

Fig. 6. Diagonal electropherograms: *A* Not crosslinked material (30S ribosomal proteins from *B. stearothermophilus*). *B* Lithiumchloride extract of diepoxybutane crosslinked 30S subunits of *E. stearothermophilus:* One crosslink can be seen as the *two darkly stained spots* below each other. Next to it another crosslink is present in smaller amounts

In contrast to this, the diagonal gels are made by the same gel solution, under the same buffer and voltage conditions in both directions. After the first dimension of electrophoresis the gel is incubated in an appropriate cleaving solution which diffuses through the gel and cleaves the complexes. Then the second dimension is carried out. Resulting from the identical migration distances in both dimensions, all monomeric proteins appear on a diagonal line; only pairs of crosslinked proteins migrate out of the diagonal as two spots below each other and below the diagonal line. SDS gels can be used for the diagonal technique; for stamp-sized gels we use acidic urea gels with 21% acrylamide that also separate the proteins on the basis of their sizes.

The excellent resolving power of the micro-gels is demonstrated in Fig. 6, which shows 30S ribosomes of *B. stearothermophilus* without and after crosslinking in a microdiagonal gel. The two proteins of the crosslink which was cleaved after the first dimension are clearly visible as two spots below the diagonal. Typically, after crosslinking only 1 – 10% of two neighboring proteins are found as crosslinked complexes.

3.3 Description of the Procedures

3.3.1 Preparation of the First-Dimension Capillary Gels

1. As high total protein amounts have to be applied onto these gels to make small amounts of cross-linked protein visible use capillaries with 1.5 mm internal diameter to prevent overloading of the gels.
2. Prepare 10 ml gel solution, degas, mix with 60 µl TEMED and 250 µl 5% ammonium peroxodisulfite.
3. Fill a second beaker with the solution prepared in (2) to a height of 2 cm and place ten capillaries vertically into the gel solution.
4. Keep for 10 min at ambient temperature for polymerization.
5. Carry out electrophoresis at 100 V for about 60 min.

3.3.2 Cleavage and Preparation of the Second-Dimension Gels

1. Remove the gel capillaries from the gel block as described in Section 2.4.1.
2. Keep the first-dimension gels in the cleaving solution for 1 h.
   Cleaving solution for reagents (e.g., diepoxybutane) that produce cleavable vicinal diol-groups:
   15 mM sodium(meta)periodate (Fluka) in 20 mM triethanolamine/HCl buffer, pH 7.5
   Cleaving solution for reagents that can produce disulfide bonds:
   5% β-mercaptoethanol in 10 mM Tris/HCl buffer, pH 8.0
3. Prepare second-dimension gels as described in Section 2.4.2 with one alteration:
   Use 1 mm-thick gels instead of 0.5 mm. In the 2-D chamber described (see Sect. 2.2) this can readily be done by screwing pairs of Plexiglas sandwiches together with their spacers facing each other. Then the block contains 10 1-mm-thick gel spaces.
4. Carry out electrophoresis for 80 min at 100 V (upper electrode +, lower electrode −).
5. Staining and destaining as given in Section 2.4.2 for sample amounts of 5 – 20 µg. With smaller protein amounts silver staining can be carried out, e.g., by the procedure of Wray et al. [11].

References

1. Kaltschmidt E, Wittmann HG (1970) Anal Biochem 36:401 – 412
2. Brockmöller J, Kamp RM (1985) Biol Chem Hoppe-Seyler 366:901 – 906
3. O'Farrell PH (1975) J Biol Chem 250:4007 – 4021
4. Hames BD (ed) (1986) Gel electrophoresis of proteins. IRL, Oxford Washington
5. Poehling H-M, Neuhoff V (1980) Electrophoresis 1:90 – 102
6. Brown JR, Hartley BS (1966) Biochem J 101:214 – 228
7. Wold F (1967) Methods Enzymol XI:617 – 640
8. Kenny W, Lambert JM, Traut RR (1979) Methods Enzymol LIX:534 – 550
9. Lutter LC, Ortanderl F, Fasold H (1974) FEBS Lett 48:288 – 292
10. Brockmöller J, Kamp RM (1986) Biol Chem Hoppe-Seyler 367 (in press)
11. Wray W, Bonlikas T, Wray VP, Hancock R (1981) Anal Biochem 116:197 – 203

1.4 Amino Acid Analysis by High Performance Liquid Chromatography of Phenylthiocarbamyl Derivatives

Tomas Bergman, Mats Carlquist, and Hans Jörnvall[1]

Contents

1 Introduction

Reliable and sensitive amino acid analyses are important steps in studies of protein structures. In this respect, high performance liquid chromatography (HPLC) has greatly increased speed and sensitivity. The use of ortho-phthalaldehyde [1] and subsequent fluorimetric detection is applicable also to HPLC. Thus, separation of underivatized amino acids by ion exchange HPLC and subsequent detection by post-column derivatization [2], as well as pre-column derivatization and subsequent separation of amino acid derivatives by reverse phase HPLC have been widely used. We have tested both methods and find, like others, that they are suitable. However, base line drift due to ammonia contamination is a serious problem when maximal sensitivity is attempted in the post-column mode of the ortho-phthalaldehyde method. Similarly, the lack of direct detection of proline requires one further step in this derivatization procedure [3] both in the pre- and

---

1 Departments of Chemistry I and Biochemistry II, Karolinska Institutet, S-104 01 Stockholm, Sweden

Advanced Methods in Protein Microsequence Analysis
Ed. by B. Wittmann-Liebold et al.
© Springer-Verlag Berlin Heidelberg 1986

post-column derivatization mode. Therefore, additional methods for analysis are of value. In this respect, pre-column derivatization with phenylisothiocyanate (PITC) to produce the phenylthiocarbamyl (PTC)-amino acids for subsequent separation by reverse phase HPLC has proved highly efficient and valuable [4, 5]. We have tested this method extensively and find it reliable, sensitive, and easy to use.

In the present work, protocols for PTC-amino acid analysis by reverse phase HPLC are given, together with actual analyses of several peptides, estimates of reproducibility, and comparisons with compositions from known structures or from conventional ninhydrin-based amino acid analyzers. Results show excellent correlation between PTC-amino acid analysis by HPLC and true compositions. This applies both to small peptides and large proteins. The PTC-amino acid HPLC analysis method can therefore be recommended for routine use in analysis of total compositions.

2 Materials and Methods

2.1 Chemicals and Glassware

Ortho-phosphoric acid, hydrochloric acid (Suprapur) and phenol were obtained from E. Merck (Darmstadt, FRG). Acetonitrile (HPLC-grade S) and phenyliso-thiocyanate (sequencer grade) were obtained from Rathburn Chemicals (Walker-burn, Scotland). Triethylamine (sequanal grade) and an amino acid standard mixture (Pierce H) were from Pierce Chemical Co. (Rockford, Illinois). Water used was deionized, glass-distilled, and filtered (0.2 µm). Glass tubes for hydro-lysis and subsequent derivatization, had an inner diameter of 5 mm and were 35 mm long. Those for mixing the derivatization solutions had dimensions 10×75 mm. All glass tubes used were submitted to pyrolysis (400 °C for $3-4$ h) before use to remove contaminating material.

2.2 HPLC Equipment

The high performance liquid chromatography equipment consisted of an M710B WISP auto injector, an M6000A and an M45 solvent delivery system controlled by an M720 system controller, an M441 absorbance detector (254 nm) and an M730 data module integrator/plotter, all obtained from Waters (Milford, Mas-sachusetts). The column was a C_{18} reverse phase HPLC column (Spherisorb S3 ODS2, 100×4.6 mm), packed with 3 µm spherical particles (Phase Separations, Queensferry, Clwyd, UK). It was held at constant temperature by a heating block and a temperature control unit (Waters).

2.3 Hydrolysis of Proteins and Peptides

Solutions of proteins and peptides in acetic acid, formic acid or other volatile sol-vents, were pipetted into the 5×35-mm glass tubes, and lyophilized. Hydrolysis was carried out in evacuated, sealed tubes with 40 µl 6 M HCl, containing 0.5% (w/v) phenol. After 20 – 24 h at 110 °C, the tubes were opened and dried under vacuum.

2.4 Derivatization of Amino Acids with PITC

Prior to derivatization, 40 µl 99.5% ethanol/water/triethylamine (2 : 2 : 1, by vol) [4] was added to each tube of the hydrolyzed samples, after which they were re-dried under vacuum. This step is essential to get rid of any residual acid. Samples were then derivatized by addition of 3 µl 50% ethanol to each tube [6], followed by vigorous shaking to obtain an aqueous film in the lower part of the tube, and subsequent addition of 7 µl 99.5% ethanol/triethylamine/PITC (7 : 2 : 1, by vol) to each tube. After mixing, with tubes covered by parafilm, the reaction was allowed to proceed for 15 – 30 min at room temperature. Excess reagent was then removed by high vacuum at room temperature overnight in a desiccator with solid NaOH. If not immediately subjected to chromatography, the sample tubes were then covered with parafilm and stored in a freezer.

The PITC-reagent and the redrying solution were made fresh daily. Stock PITC was stored at – 20 °C under nitrogen in screw-capped tubes. Triethylamine and 50% ethanol were stored at + 4 °C.

2.5 Chromatography

The dried PTC-amino acids were dissolved in 100 µl starting buffer (solvent A, below), and injected using the WISP (Waters) autosampler onto the C18-column in volumes of 1 – 90 µl. The column was thermostatically controlled at + 36 °C and the absorbance of the effluent was monitored at 254 nm. The mobile phase system consisted of two solvents in a gradient programmed as given in Table 1. Solvent A was 0.030 M sodium hydroxide titrated to pH 6.60 with 1 M ortho-phosphoric acid, Solvent B was 60% acetonitrile in water. The washing step at 100% B is essential to clean the column from several reagent-derived by-products. Samples dissolved or remaining parts of them can be stored at – 20 °C for analysis/re-analysis (cf. Sect. 3.2).

3 Results

3.1 Resolution and Sensitivity

Chromatography of derivatized amino acid standards generates HPLC-profiles as shown in Fig. 1 at the 5 – 50 pmol levels. All PTC-amino acids including those of carboxymethylcysteine, homoserine, and tryptophan, are well resolved. The peak between that of PTC-phenylalanine and PTC-lysine is reagent-related and is present in the analysis of blank derivatives. The baseline is excellent well below

Table 1. Analytical program for the separation of PTC-amino acids

Time (min)	A (%)	B (%)
0	100	0
1	95	5
26	64	36
28	0	100
32	0	100
35	100	0

Flow 1 ml min^{-1}, column temperature $+36\,°C$. The segments of the gradient were linear (curve No. 6 in the Waters system controller). Equilibration delay used was 10 min.

the 50-pmol level (Fig. 1, top) and acceptable even at the 5-pmol level (Fig. 1, bottom). The practical detection limit of the system is below 3 pmol for each amino acid derivative, and it is possible to detect down to 1 pmol, but at that level quantitation of hydrolyzed and derivatized protein or peptide samples is not fully reliable because of a high background.

It is possible to run a considerably faster separation program than that shown in Table 1, especially with large sample amounts and standards, when artifact peaks are largely absent. However, with true hydrolytic samples in the low pmol range and at highest sensitivity, the program in Table 1 is essential for separation from reagent-derived peaks. Therefore, this program has been used in this study throughout, also for the standard curves.

3.2 Stability of PTC Derivatives

Derivatized samples can be stored at $-20\,°C$, dissolved for at least 1 week and dried for at least 2 months without noticeable loss in chromatographic response. However, in solution and at ambient temperature a slow time-dependent degradation ($<2\%$ per h) will occur (Table 2) at different rates for different amino acids. The most sensitive derivatives appear to be PTC-half-cystine and to some extent PTC-valine, PTC-isoleucine and PTC-glutamic acid. On the other hand, PTC-serine, PTC-glycine, PTC-threonine and PTC-histidine seem to be quite stable in the dissolved state.

3.3 Response Linearity and Quantitation

The linearity of the derivatization reaction was investigated by injection of standards in amounts of $10-200$ pmol (constituting 10% of the amounts derivatized). The resulting chromatographic peak area of each PTC-amino acid is plotted as a function of the injected amount in Fig. 2. It is evident that there is a true response linearity in the range $10-200$ pmol for all amino acids. As noticed before, response yields are similar for most amino acids, but methionine,

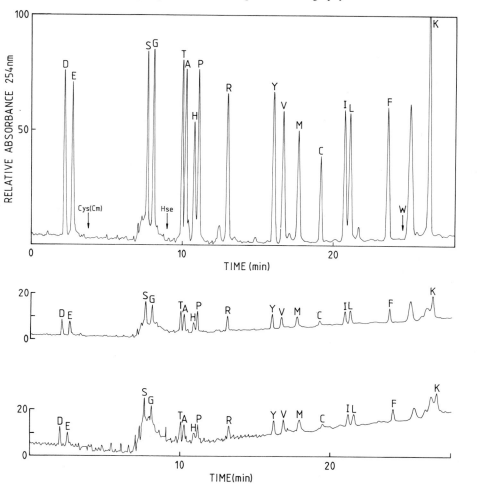

Fig. 1. HPLC-separation of a PTC-amino acid standard mixture. The chromatographic program used is given in Table 1. The amount injected was 50 pmol (*top*) with the absorbance detector (254 nm) set at 0.005 absorbancy units for full scale (AUFS), 5 pmol (*bottom*) at the same sensitivity, and 10 pmol (*middle*) at half that sensitivity (0.010 AUFS)

half-cystine, and to some extent histidine, give a weaker response, while lysine gives a stronger response, probably because it can form a disubstituted product [5].

3.4 Reproducibility

The reproducibility of the methodology was evaluated through standard derivatizations and analyses. Retention times and peak areas were compared for five nonconsecutive runs of five different samples. In each case, the injected amount was 50 pmol per PTC-amino acid. Results are shown in Table 3. The average

Table 2. Decrease in chromatographic peak area of dissolved PTC-amino acids at ambient temperature

Amino acid	Decrease (%)		
	After 3 h	After 9 h	After 18 h
Asp	1	2	5
Glu	2	5	10
Ser	−	−	−
Gly	−	−	−
Thr	−	−	−
Ala	−	1	2
His	−	−	−
Pro	−	−	1
Arg	−	1	2
Tyr	1	3	6
Val	2	8	15
Met	−	1	2
Cys	5	15	30
Ile	2	5	10
Leu	1	2	4
Phe	1	3	5
Lys	1	1	3

Derivatives dissolved in starting buffer (solvent A, cf. Sect. 2.5), pH 6.60. A dash indicates no significant decrease ($\leqslant 0.5\%$).

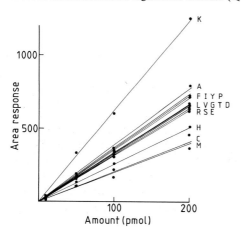

Fig. 2. Response curves. From $10-200$ pmol of standard amino acids were injected after derivatization of ten times larger amounts

retention time variation was 0.22% and the average peak area variation was 2.55%.

3.5 Samples from Peptide Hydrolysates

Analyses of hydrolysates are shown in Fig. 3. The spread of elution positions is slightly larger than in the case of standards, as shown in Table 3. This minor dif-

Table 3. Reproducibility of methodology

Amino acid	Percent relative standard deviation (n = 5)		
	Retention time		Peak area
Asp	0.26	(1.96)	3.29
Glu	0.20	(1.84)	2.73
Ser	0.17	(1.03)	3.60
Gly	0.14	(0.98)	1.21
Thr	0.24	(0.90)	2.70
Ala	0.28	(0.84)	3.45
His	0.37	(0.86)	1.58
Pro	0.22	(0.80)	2.83
Arg	0.29	(0.51)	1.72
Tyr	0.16	(0.39)	1.14
Val	0.14	(0.83)	3.13
Met	0.21	(0.81)	4.81
Cys	0.32	–	2.84
Ile	0.18	(0.39)	1.99
Leu	0.18	(0.43)	2.40
Phe	0.19	(0.29)	2.63
Lys	0.20	(0.25)	1.28
Average	0.22	(0.82)	2.55

Values were calculated from injection of 50 pmol standards in five non-consecutive runs. For the retention time, values within parentheses show corresponding calculations from five nonconsecutive runs of actual hydrolysates of 3.4 – 119 pmol.

ference is probably due to a larger spread in amounts and in the presence of other compounds in the hydrolysates versus the standards. At moderate to high sensitivities, hydrolytic samples give separations with equally good baselines as the standards, as shown in Fig. 3, top frames. At the highest sensitivity, a few additional peaks are noticed in the hydrolytic samples. Thus, a large peak is eluting in front of lysine, a usually smaller peak in front of arginine, a still smaller after leucine, and often one or two additional peaks in front of aspartic acid. These extra peaks are visibile in Fig. 3, bottom frames. They are probably derived from reaction by-products, and are also visible in other reports using the present methodology [4, 5]. However, the chromatographic conditions now chosen are such that these artifacts do not interfere with the separation or quantitation of the true amino acid derivatives.

3.6 Comparisons with Results from Other Methods

The present separations give full identification and quantitation of the PTC derivatives of all common amino acids. Elution patterns are essentially as shown before [4, 5] but analytical curves are now throughout shown at high sensitivity, still demonstrating acceptable baselines, and reproducible results. In these

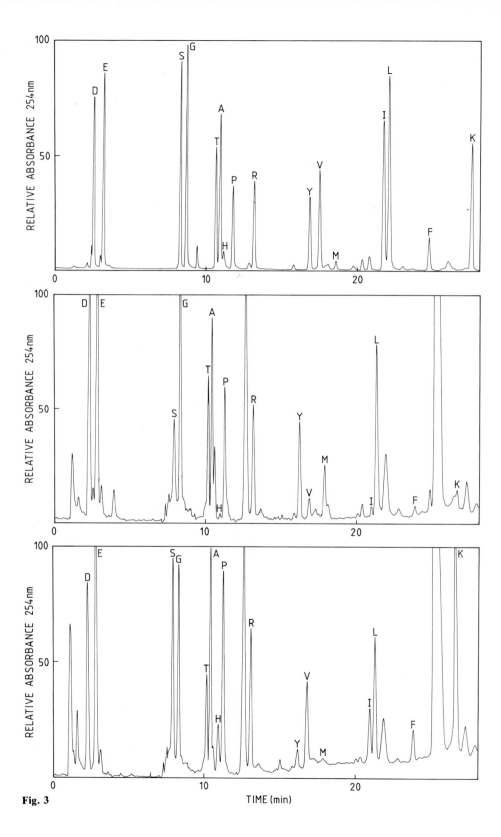

Fig. 3

respects, curves appear better than those previously shown (cf. present Figs. 1 and 3 with figures in [4, 5]). In particular, it should be noticed that one of the standard curves in Fig. 1 is obtained with 5 pmol of each derivative, and that two of the hydrolytic peptide samples in Fig. 3 and Table 4 are obtained with only 3.6 – 4.6 pmol material.

The artifacts visible at the highest sensitivity are probably derived from urea derivatives. They are always obtained in any application using phenylisothiocyanate. However, the gradients can be chosen in such a way that the artifact peaks do not interfere with those of PTC-amino acids, as shown in Fig. 3.

In summary, the present methodology gives excellent separation combined with a high sensitivity. The values obtained from peptide hydrolysates agree with those known before (Table 4). They are also equally good or better than those obtained by conventional ninhydrin-based analysis on ordinary amino acid analyzers (Table 4). The present method using PTC-amino acid analysis on reverse phase HPLC is available for routine use. Analysis at the 3 pmol level increases the sensitivity by a factor of at least 20 – 40 in relation to that from most ordinary amino acid analyzers.

4 Conclusion

The method presented has the following advantages:

- It is easy to use, requiring only the well-known Edman coupling step before analysis. In contrast to methods utilizing ortho-phthalaldehyde, all amino acids are derivatized, without use of oxidizing conditions or other special steps [3]. In relation to methods utilizing still other derivatizations, like for example dimethylaminoazobenzyl sulphonyl chloride [7], the present method is quantitative and therefore gives complete recovery without use of correction factors [8].
- The PTC-analysis method is applicable to use with an ordinary HPLC instrument. The wavelength of detection is 254 nm, and no fluorescence detector or other special device is necessary.
- Essentially, similar results are obtained with both standards and true samples, showing that the method is comparatively insensitive to contamination or the presence of other products.
- Results obtained are identical to those from ninhydrin-based ordinary amino acid analyzers but at least one order of magnitude more sensitive. Values obtained therefore directly reflect the total composition of peptides, proving identical derivatization close to 100% of all amino acids.

Fig. 3. HPLC-separation of PTC-amino acids derived from peptide hydrolysates. *From top to bottom:* the samples are: 55 pmol of a 370-residue viral protein at 0.2 absorbancy unit for full scale (AUFS). 37 pmol of a 19-residue peptide from a plasma protein at 0.005 AUFS (the peptide is also shown in Table 4, peptide 5). 4.6 pmol of a 110-residue hormonal peptide at 0.005 AUFS

Table 4. Comparisons of results obtained by HPLC of PTC-amino acids and by a conventional, nin-

	Peptide 1			Peptide 2			Peptide 3	
	Known	PTC-HPLC (50 pmol)	Ninhydrin-based analysis (1 nmol)	Known	PTC-HPLC (120 pmol)	Ninhydrin-based analysis (1 nmol)	Known	PTC-HPLC (73 pmol)
Asp	1	1.1	1.1	2	1.8	2.1	35	35.2
Glu	2	2.2	2.2	3	3.4	3.3	46	48.4
Ser	–	–	–	2	2.1	2.4	29	27.7
Gly	–	–	–	1	1.6	1.6	30	37.9
Thr	1	1.0	1.0	1	1.3	1.1	23	23.5
Ala	2	2.2	2.0	–	–	–	20	26.2
His	1	0.7	0.9	–	–	–	18	15.1
Pro	–	–	–	1	1.2	1.5	32	23.9
Arg	2	2.0	2.0	–	–	–	32	31.6
Tyr	3	2.5	2.5	1	0.5	0.9	12	13.3
Val	1	1.0	1.1	3	2.9	2.8	21	21.4
Met	–	–	–	–	–	–	11	9.8
Ile	1	1.0	1.1	3	2.7	2.5	23	21.8
Leu	1	1.1	1.0	–	–	–	54	51.7
Phe	–	–	–	1	0.8	0.8	35	30.4
Lys	–	–	–	7	6.7	6.2	24	26.9

Peptide 1 is a 16-residue peptide corresponding to a part of coagulation factor VIII (with one Trp residue which was not detected since the sample was hydrolyzed with HCl).

Peptide 2 is a 28-residue peptide corresponding to a part of an inhibin-like polypeptide (with one Trp and two Cys residues; Trp was not detected since the sample was hydrolyzed with HCl; Cys was not detected because the sample was non-oxidized and non-carboxymethylated).

Peptide 3 is a 452-residue cytochrome P-450 (with Trp and Cys not analyzed, cf. peptide 2, above).

– The temperature for separation is comparatively low. Even in the absence of a thermostatically controlled compartment, the column can easily be held at 36 °C with an ordinary water bath.

In summary, PTC-amino acid derivatization and subsequent reverse phase HPLC has been shown to be a highly reliable, easy to use, and very sensitive analytical method for amino acid analysis. Routine samples down to 2 pmol can be analyzed, and give results equivalent to those obtained by ordinary ninhydrin-based amino acid analyzers but at sensitivities which are higher by a large factor. The present method is recommended for routine use in peptide analysis.

Acknowledgments. This work was supported by the Swedish Medical Research Council (projects 13X-3532 and 13X-1010), Magn. Bergvall's Foundation, the Nordic Insulin Fund, and KabiGen.

hydrin-based amino acid analyzer

	Peptide 4		Peptide 5		Peptide 6		Peptide 7		Peptide 8	
	Known	PTC-HPLC (61 pmol)	Known	PTC-HPLC (37 pmol)	Known	PTC-HPLC (20 pmol)	Known	PTC-HPLC (17.5 pmol)	Known	PTC-HPLC (3.6 pmol)
	1	1.2	3	3.0	1	1.0	24	20.2	3	4.0
	1	1.3	4	3.9	4	4.0	19	19.0	7	6.9
	1	1.6	–	–	3	2.7	11	10.1	7	6.9
	–	–	2	2.6	3	3.1	13	18.0	5	5.1
	1	1.0	1	1.2	2	2.0	9	8.2	4	3.6
	1	1.1	1	1.8	1	1.2	26	23.1	5	4.6
	–	–	–	–	1	1.1	8	7.2	–	–
	–	–	1	1.1	–	–	9	7.6	3	3.3
	–	–	1	1.0	4	3.7	6	7.9	8	8.0
	–	–	1	0.9	–	–	7	10.0	3	2.4
	2	1.2	–	–	1	1.3	10	16.5	4	4.2
	–	–	2	0.5	–	–	2	1.3	–	–
	–	–	–	–	–	–	6	5.5	1	0.8
	1	1.0	2	1.7	6	5.7	21	18.7	6	6.1
	1	0.8	–	–	1	1.0	11	9.5	4	3.9
	1	0.8	–	–	–	–	17	16.3	1	1.0

Peptide 4 is a 10-residue peptide from the plasma protein α_2-macroglobulin.

Peptide 5 is a 19-residue peptide also from α_2-macroglobulin (with one Cys residue not detected because the sample was non-oxidized and non-carboxymethylated).

Peptide 6 is a 27-residue human gastrointestinal hormone (secretin).

Peptide 7 is a 205-residue bacterial enzyme (superoxide dismutase; Trp not analyzed).

Peptide 8 is a 67-residue human growth factor (IGF-2; Cys not analyzed).

References

1. Roth M (1971) Anal Chem 43:880 – 882
2. Klapper DG (1982) In: Elzinga M (ed) Methods in protein sequence analysis. Humana, Clifton, NJ, pp 509 – 515
3. Bohlen P, Mellet M (1979) Anal Biochem 94:313 – 321
4. Bidlingmeyer BA, Cohen SA, Tarvin TL (1984) J Chromatogr 336:93 – 104
5. Heinrikson RL, Meredith SC (1984) Anal Biochem 136:65 – 74
6. Koop DR, Morgan ET, Tarr GE, Coon MJ (1982) J Biol Chem 257:8472 – 8480
7. Chang J-Y, Knecht R, Braun DG (1981) Biochem J 199:547 – 555
8. Jörnvall H, Kalkkinen N, Luka J, Kaiser R, Carlquist M, von Bahr-Lindström H (1983) In: Tschesche H (ed) Modern methods in protein chemistry. de Gruyter, Berlin, pp 1 – 19

1.5 High Sensitivity Amino Acid Analysis Using DABS-Cl Precolumn Derivatization Method

Rene Knecht and Jui-Yoa Chang[1]

Contents

Abbreviations: DABS-Cl = dimethylaminoazobenzene sulfonyl chloride; DABS = dimethylaminoazobenzene sulfonyl; DNS-Cl = dimethylaminonaphthalene sulfonyl chloride; OPA = o-phthalaldehyde; PITC = phenylisothiocyanate; NBD-F = 4-fluoro-7-nitrobenzo-2-oxa-1,3-diazole; HPLC = high performance liquid chromatography.

1 Introduction

Precolumn derivatization of amino acids and subsequent HPLC analysis of amino acid derivatives have provided a viable alternative to the conventional analyzer for high sensitivity amino acid analysis. Several reagents are available for this new technique. All amino acid derivatives of DNS-Cl [1, 2], OPA [3 – 5], NDB-F [6, 7], PITC [8, 9] and DABS-Cl [10 – 17] can be reproducibly prepared and detected at the low picomole level. Among them, the DABS-Cl method is unique in two ways. (a) DABS-amino acids are detected in the visible region. At low picomole level, a stable baseline can be readily obtained with a large variety of solvent and gradient systems. (b) DABS-amino acids, like DNS-amino acids (with their sulfonamide bonds) are comparably the most stable derivatives. They can be left at room temperature for a period of up to two months without any appreciable degradations. This high stability is a prerequisite for the reliable quantitative analysis of amino acid derivatives.

The DABS-Cl method has been routinely used in our laboratory since 1981.

1 Pharmaceuticals Research Laboratories, CIBA-GEIGY Limited Basel, CH-4002 Basel, Switzerland

Advanced Methods in Protein Microsequence Analysis
Ed. by B. Wittmann-Liebold et al.
© Springer-Verlag Berlin Heidelberg 1986

2 Reagents and Equipment

DABS-Cl is commercially available from Fluka (Switzerland), Pierce (USA), Sigma (USA), Aldrich (USA) et al. Double recrystallized DABS-Cl can also be obtained from Pierce. The DABS-Cl should be freshly prepared for each dabsylation. Small portions of DABS-Cl are prepared as follow: A stock solution of DABS-Cl in acetone (2 nmol μl^{-1}) is first prepared. Aliquots (400 μl) are pipeted into Eppendorf tubes, dried under water vacuum pump, the oil vacuum pump, and stored at $-20\,°C$. For dabsylation, each DABS-Cl tube is dissolved in 200 μl of acetonitrile shortly before use. All organic solvents are obtained from Merck with minimum purity of 99.8%. We use the HPLC system from Waters (USA). However, this technique can be applied on any HPLC equipments which include two solvent delivering pumps, a gradient programmer and a detector equipped with visible wavelength filter.

3 DABS-Amino Acid Standards

These can be prepared according to the method described [11] or obtained from commercial source (Pierce, USA). However, it is relevant to mention that those DABS-amino acids standards are used only to establish the chromatographic conditions for DABS-amino acids separation. For quantitative analysis of unknown samples, the DABS-amino acid standards have to be obtained from the dabsylation of an amino acids mixture standard (e.g., Pierce or Hamilton standards, see hydrolysis and dabsylation).

4 Preparation of Peptide and Protein Samples

There are two limitations to the DABS-Cl method which have to be carefully observed during the preparation of protein samples. First, one must know the approximate amount of the protein sample. With the dabsylation conditions described in this paper, the optimal amout of protein sample to be hydrolyzed is approximately $0.1-2$ μg. Second, the protein sample is preferably salt-free. Presence of excess amount of salts like SDS or urea will interfere with the reproducibility of derivatization.

5 Acid Hydrolysis

5.1 Normal Hydrolysis

All glassware used in acid hydrolysis were heated at 500 °C for 6 h. Approximately $0.1-2$ μg of protein was placed in a hydrolysis tube (5 mm i.d. $\times 100$ mm) and dried under vacuum. 25 μl of 6 N HCl (Pierce) were added and the tube was

sealed under vacuum (<0.1 mbar). Hydrolysis was performed at 110 °C for 24 h.
A Pierce amino acid standard (catalog no. 20088) of 500 pmol was also hydrolyzed, together with each batch of unknown samples. After hydrolysis, the acid
was dried, the tube was cut at the point of 40 mm from the bottom, and prepared
for dabsylation.

5.2 Gas-Phase Hydrolysis

Approximately 0.1 – 2 µg of protein was placed in a tube (4 mm i.d. ×50 mm)
and dried in a vacuum centrifuge. Twelve samples can be simultaneously placed
in a Waters gas-phase hydrolysis vessel (no. 07363). At least one Pierce standard
(500 pmol) was always included in each batch of hydrolysis. About 400 µl of 6 N
HCl (Pierce) was then placed in the vessel. The vessel was sealed and attached to
vacuum by opening the valve on the top side, purged with Argon for about 10 s
and placed under vacuum again. This was repeated twice before the vessel was
finally sealed with a good vacuum of lower than 0.1 mbar. Gas-phase hydrolysis
was carried out at 110 °C for 24 h. After hydrolysis, the tubes were dried again
under vacuum and prepared for dabsylation.

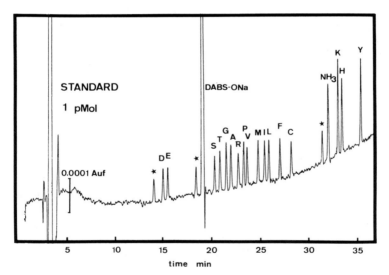

Fig. 1. Original chromatogram for the HPLC separation of 1 pmol of DABS-Cl derivatized amino
acid standard (Pierce, Catolog no. 20088). Five hundred picomole of the standard were dabsylated
according to the method described (see Sect. 6) and 0.2% of the sample was injected. Solvent A was
25 mM sodium acetate, pH 6.5, containing 4% of dimethylformamide (degased with helium). Solvent
B was acetonitrile. The grdient was 15%B – 40%B in 20 min, 40%B – 70%B from 20 – 32 min, kept
at 70%B from 32 – 34 min, returned from 70%B to 15%B from 34 – 36 min. The cycle time from
injection to injection was 44 min. The column was Merck LiChrospher (100, CH-18/2, 5 µm). The
column temperature was 40 °C. The flow rate was 1 ml/min^{-1}. The detector was Kratos 737, variable
wavelength, and set at 436 nm. With each new column or with the aging of the column, DABS-Arg
shifted and usually had longer retention time. This can be corrected by slightly increasing the
concentration of buffer A, for example from 25 mM to 30 mM or 35 mM

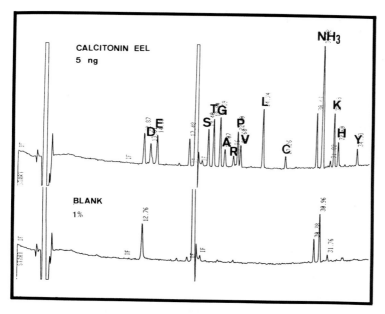

Fig. 2. Original chromatogram for the amino acid composition analysis of 5 ng of calcitonin (eel) by the DABS-Cl method. Five hundred nanogram of the peptide was hydrolyzed by the gas-phase 6 N HCl, dabsylated and 1% of the total sample was injected (top panel). A blank sample was performed to determine the background contamination (bottom panel). The quantity of each amino acid was summarized in Table 1. Chromatographic conditions are as described in the legend of Fig. 1

6 Dabsylation

The DABS-Cl solution should be freshly prepared for each dabsylation (see Sect. 2). Hydrolyzed unknown samples and standards were dissolved in 10 µl of 50 mM sodium bicarbonate, pH 8.1, and added with 20 µl of DABS-Cl solution (4 nmol µl^{-1} in acetonitrile). The mixtures were heated at 70°C for 10–15 min (the mixture become completely soluble after heating). After dabsylation, the samples were directly diluted with the diluting solution (50 mM phosphate, pH 7.0/ethanol, 1:1, v/v) to suitable volumes and used directly for HPLC injection. For instance, the dabsylated standards (500 pmol) was usually diluted to 1 ml and 20 µl was injected for HPLC analysis.

7 HPLC Quantitative Analysis of DABS-Amino Acids

The chromatographic conditions which gives a complete separation of all DABS-amino acids are described in Fig. 1. Several columns, including Merck LiChrospher (100, CH-18/2, 5 µm), Merck LiChrosob RP-18, Vydac C-18, which were tested in our laboratory gave the same separation pattern and it is

Table 1. Amino acid composition of calcitonin (eel) determined by the DABS-Cl method

Amino acids	Amount hydrolyzed (2 µg) Amount analyzed (100 ng)		Amount hydrolyzed (0.2 µg) Amount analyzed (10 ng)		[a]
	Gas-phase hydrolysis	Normal hydrolysis	Gas-phase hydrolysis	Normal hydrolysis	
Asp	1.80	1.85	1.95	1.80	2
Glu	2.85	2.80	2.95	2.90	3
Ser	2.75	2.65	2.85	2.70	3
Thr	3.80	3.80	3.90	4.20	4
Gly	2.85	3.05	2.95	3.30	3
Ala	1.05	1.10	1.20	1.25	1
Arg	1.00	1.10	1.05	1.20	1
Pro	2.00	2.10	2.05	2.30	2
Val	2.15	2.20	2.10	2.00	2
Met	–	–	–	–	0
Ile	–	–	–	–	0
Leu	5.50	5.60	5.00	5.00	5
Phe	–	–	–	–	0
Cys	2.00	1.80	1.80	1.95	2
Lys	2.20	2.20	2.10	1.95	2
His	1.05	0.90	1.05	0.70	1
Tyr	0.95	n.d.	0.90	n.d.	1

[a] Expected value from the amino acid sequence.
n.d. = not determined.

likely that most reversed phase C-18 columns will give similar results. At room temperature, complete separation of all DABS-amino acids can also be achieved [14]. Example for the composition analysis of an unknown sample is shown in Fig. 2 and Table 1.

8 Comments

The data accumulated in our laboratory for the past 5 years has indicated that the reliability of the DABS-Cl method is comparable to that of the conventional post-column detection system. Although having limitations with respect to the preparation of the protein samples (see Sect. 4), the precolumn derivatization technique has the distinct advantages of high sensitivity and efficiency over the conventional amino acid analyzer. The major advantage of the gas-phase hydrolysis over the normal hydrolysis is the lower background of Ser, Gly and Asp. The gas-phase hydrolysis is recommended when less than 1 µg of protein is available.

References

1. Wilkinson JM (1978) J Chromatogr Sci 16:547 – 552
2. De Jong C, Hughes GJ, van Wieringen E, Wilson KJ (1982) J Chromatogr 241:345 – 352
3. Hill DW, Waters FH, Wilson TD, Stuart JD (1979) Anal Chem 51:1338 – 1341
4. Lindroth P, Mopper K (1979) Anal Chem 51:1667 – 1674
5. Ashman K, Bosserhoff A (1985) In: Tschesche H (ed) Modern method in protein chemistry, vol 2. de Gruyter, Berlin, pp 155 – 171
6. Imai K, Watanabe Y (1981) Anal Chim Acta 130:377 – 382
7. Wantanabe Y, Imai K (1981) Anal Biochem 116:471 – 474
8. Heinrikson RL, Meredith SC (1984) Anal Biochem 136:65 – 74
9. Bergman T, Carlquist M, Jörnvall H: This volume, section 4, Chapter 1
10. Lin J-K, Chang J-Y (1975) Anal Chem 47:1634 – 1638
11. Chang J-Y, Knecht R, Braun DG (1983) Methods Enzymol 91:41 – 48
12. Chang J-Y, Knecht R, Braun DG (1982) Biochem J 203:803 – 806
13. Winkler G, Heinz FX, Kunz C (1984) J Chromatgr 297:63 – 70
14. Stocchi V, Gucchiarini L, Piccoli G, Magnani M (1985) J Chromatogr 349:77 – 82
15. Lin J-K, Wang C-H (1980) Clin Chem 26:579 – 583
16. Lammens J, Verzele M (1978) Chromatographia 11:376 – 378
17. Nolan TG, Hart BK, Dovichi NJ (1985) Anal Chem 57:2703 – 2705

Chapter 2
Manual and Solid-Phase Microsequencing Methods

2.1 Modern Manual Microsequencing Methods

CARL C. KUHN[1] and JOHN W. CRABB[2]

Contents

1 Introduction

Today manual Edman sequencing procedures are frequently taken to be too slow, too insensitive or too labor-intensive to be of significant practical, let alone competitive, value to protein structural studies. Over the last 10 years, George Tarr, the man who developed polybrene as a sequencing reagent [1], has enhanced the performance of manual sequencing [2 – 5] to the point that his procedures can be important assets to even the best-equipped protein chemistry laboratories. Tarr's methods are not difficult to master and can produce multiple picomole level analyses more rapidly than most automatic sequenators produce a single analysis. Using Tarr's batchwise procedures, HPLC peptide map fractions may be quickly screened for NH_2-terminal purity and identification of peptides. Quality control sequence analyses of synthetic peptides, which are usually abundant, can also be done efficiently by Tarr's manual methods rather than using valuable automatic instrument time for the task. While automatic sequencing is the method of choice for extended degradations, the manual procedures of Tarr provide an attractive alternative for analyzing short peptides. For example, ten peptides, each of about 6 residues, would in total require about 1 week of automatic gas-phase sequencing time provided two peptides are analyzed per day;

1 Institut für Physiologische Chemie I, Ruhr-Universität Bochum, D-4630 Bochum, FRG
2 W. Alton Jones Cell Science Center, 10 Old Barn Road, Lake Placid, NY 12946, USA

Advanced Methods in Protein Microsequence Analysis
Ed. by B. Wittmann-Liebold et al.
© Springer-Verlag Berlin Heidelberg 1986

all ten could be sequenced in 1 day by using Tarr's batchwise manual strategy. For peptide screening purposes, for quality control analyses, for sequencing short peptides and for investigators on a low budget, the Tarr manual strategies can constitute methods of choice.

2 Methods

2.1 Sensitivity and Cycle Time

The Tarr procedures are similar in sensitivity performance to DABITC manual sequencing [6], and are routinely useful for determining $10-30$ residues with $500-2000$ pmol of peptide or protein. Useful sequence information has, however, been obtained with less than 100 pmol of sample [4]. In contrast to the manual DABITC procedure which requires $120-140$ min per cycle [6], the Tarr strategies are much more rapid, requiring only $15-60$ min per cycle.

2.2 Tarr Sequencing Strategies

Tarr's established sequencing procedures include two batchwise methods, one for small peptides and the other for large peptides and proteins, and another very rapid method useful for analysis of a single large peptide or protein. The batch-wise procedures utilize ethanol or dimethyl formamide coupling mediums, are carried out in 6×50 mm glass culture tubes containing polybrene and have recently been described in detail by Tarr [5]. Repetitive yields are typically about 90% for the large peptide/protein batchwise method and about $80 \pm 10\%$ for the small peptide batchwise procedure. The rapid large peptide/protein method utilizes an aqueous pyridine coupling medium, is carried out in a 1-ml reactivial, exhibits repetitive yields of about 90% and has been described briefly [4, 7] but with subsequent modifications [8]. Examples of the efficacy of these methods include several NH_2-terminal analyses [8 – 10] some of which revealed homologous relationships [11 – 13] or defined the reading frame of a gene [14] as well as complete primary structured analyses of human hypoxanthine-guanine phosphoribosyltransferase [15], rabbit cytochrome P-450 LM_2 [16] and the cellular retinoic acid-binding protein from bovine retina [17]. An outline of each of the Tarr manual sequencing methods is presented in Tables 1, 2 and 3 along with examples of typical results in Figs. $1-3$.

3 Equipment and Supplies

3.1 The Manual Sequencing Station

A work area equipped so that all the steps of the Edman degradation may be carried out from a chair or stool, without moving about the laboratory, is a major

Table 1. Tarr batchwise sequencing of small peptides

1. *Set-up*
   Redry samples (100 pmol+) from TEA in 6×50 mm tubes containing clean polybrene.

2. *Couple*
   ETOH (or DMF): TEA : PITC : H_2O 7:1:1:1.
   10 µl, 8 min 50°C, under N_2. Vacuum dry.

3. *Wash*
   Add 2.5 µl H_2O, centrifuge briefly.
   Add about 250 µl hep: EA 15:1 containing 0.5% TMA.
   Vortex, centrifuge, decant upper phase.
   Wash twice with hep: EA 7:1.
   Vacuum dry.

4. *Cleave and stabilize*
   Add 5 µl conc HCl, 5 min RT or 1.5–2 min 50°C, under N_2.
   Carefully vacuum dry. Add 10 µl MeOH containing an internal standard (MTH-Y), vacuum dry.

5. *Extract*
   Add 4 µl H_2O, 40 µl hep: EA 1:5, vortex and centrifuge.
   Decant into 6×50 mm tube containing 5 µl thiourea (20 nmol µl^{-1}).
   Extract again with 40 µl .02M HFA/.016 TMA (pH 7.2) in hep: EA 1:5, combine extracts and dry.

6. Dry n-1 peptide and recycle.

7. *Convert*
   20 µl 1N HCl/MeOH, 10 min 65°C.

The wash and extraction solvents and the cleavage acid should contain 0.01% ethanethiol. Working with ten samples, one cycle per hour may be achieved with repetitive efficiencies of $80 \pm 10\%$.

Table 2. Tarr rapid sequencing for large peptides and proteins

1. *Set-up*
   Dry salt-free samples (1–50 nmol up to 1 mg) in 1 ml reactivial.
   Precycle then redry from TEA.

2. *Couple*
   Add 40 µl (50% pyridine, 30% H_2O, 20% TEA) and 40 µl (20% PITC in pyridine). Flush with N_2, vortex, incubate 3 min at 50°C.

3. *Wash*
   Gently add 400 µl hep: EA (1:1) containing 0.5% TMA yielding two phases. Slowly rotate nearly horizontal vial, gradually producing one phase in which the sample precipitates as rough film. Decant supernatant, wash two to three times with 500 µl EA, and vacuum dry.

4. *Cleave*
   Add 10 µl TFA, flush with N_2, incubate 4 min at 50°C. Extend cleavage time two to three times for Gy and Pro. Blow off TFA with a N_2 stream, leaving sample as a film. Vacuum dry 20 s and add 10 µl MeOH containing internal standard (MTH-Y), 3 min 50°C, redry.

5. *Extract*
   Extract two to three times with 30 µl benzene: acetonitrile (1:1) containing 0.1% HAc into 6×50 mm tube containing 5 µl thiourea (20 nmol µl^{-1}). Vacuum dry.

6. *Convert*
   Add 20 µl 1N HCl/MeOH, incubate 10 min 65°C, vacuum dry.

The wash and extraction solvents and the cleavage acid should contain 0.01% ethanethiol. Working with one sample, three to four cycles per hour may be achieved with a repetitive efficiency around 90%.

Table 3. Tarr batchwise sequencing of large peptides and proteins

---

1. *Set-up*
   Dry salt-free samples (100 pmol +) in 6 × 50 mm tubes containing clean polybrene. Pre-cycle then redry from TEA.

2. *Couple*
   DMF: TEA: PITC: H_2O 7: 1: 1: 1
   10 µl, 3 min 50°C under N_2. Vacuum dry.

3. *Wash*
   200 µl EA then Hep: EA 15: 1 containing 0.5% TMA then acetone (3 ×)
   Gently rotate tube (do not vortex) and decant each wash.
   Vacuum dry briefly.

4. *Cleave and stabilize*
   Add 8 µl TFA, 6 min 50°C.
   Dry 20 s and add MeOH containing internal standard (MTH-Y) 3 min 50°C. Vacuum dry about 1 min.

5. *Extract*
   2 × with 40 µl benzene: MeCN 1: 1 containing 0.1% HAc.
   Decant into tubes containing 100 nmol thiourea. Vacuum dry.

6. *Convert*
   20 µl 1N HCl/MeOH, 10 min 65°C.

The wash and extraction solvents and cleavage acid should contain 0.01% ethanethiol. Working with ten samples, one cycle per 30 – 40 min may be achieved with repetitive efficiencies around 90%.

prerequisite for rapid and efficient manual sequencing. Preferably, this so-called manual sequencing station should be located in a fume hood for health reasons, as well as to promote inoffensive conditions for co-workers. Locating the sequencing station and an HPLC system close to each other allows on-line identification of PTH amino acids. The sequencing station is composed of a vacuum system, an argon or nitrogen flush system, and a collection of small pieces of equipment and accessory supplies as listed in Table 4. Although the station must be well organized, there is no absolute design and the components may be arranged generally according to individual preference. Figures 4 and 5 show examples of sequencing stations established at the University of Washington, Seattle and at the Ruhr-Universität Bochum, FRG, respectively. Although many of the components in Table 4 are in common laboratory use, the glass vacuum manifold with attached glass-TFE valves must be custom-made by a glassblower. Attach the mininert slide valves to the manifold with vacuum tubing after removing the sliders and enlarging the bore. Keep in mind that the fastest drying times (and therefore most rapid cycle times) will be obtained by using wide bore vacuum hose and keeping the length of hose as short as possible. Incorporate a set of aquarium air valves into the argon or nitrogen system to provide a branch point for multiple use and fine control at the station; this also allows the gas tank to be secured away from the work area. For keeping reagents cool between cycles, a small refrigerator is optional, but more convenient and versatile than an ice bucket and can also be used to support the table top centrifuge. The two heat blocks accommodate Pierce 25 ml and 40-ml screw cap bottles and are set at

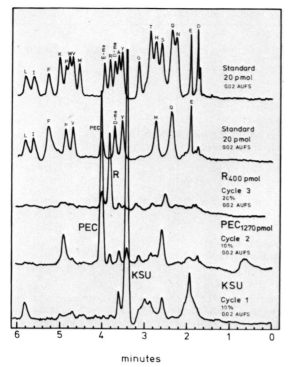

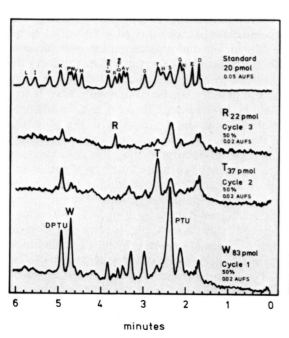

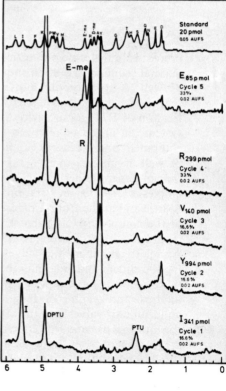

Fig. 1. Representative sequence results using the Tarr batchwise small peptide strategy (Table 1) for analysis of three short peptides from the cellular retinoic acid-binding protein from bovine retina [17]. An isocratic HPLC system [18] was used for PTH amino acid identification at 269 nm on a Beckman Ultrasphere ODS column. The percent of each cycle analyzed and the yield normalized to 100% injection are indicated. Approximately 1600 pmol of KCR, 150 pmol of WTR and 1200 pmol of IYVRE were degraded. *KSU* succinyl lysine; *PEC* pyridylethyl cysteine

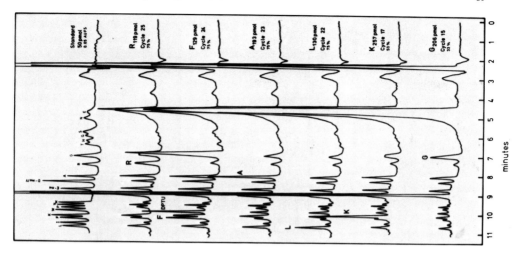

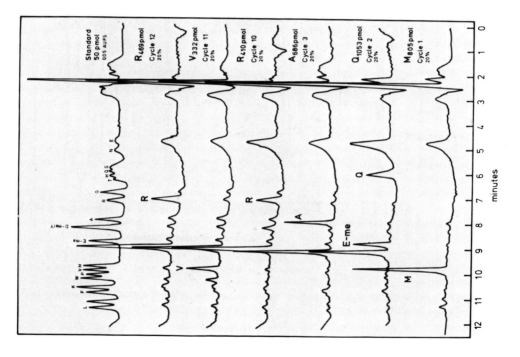

Fig. 2. Representative sequence results using the Tarr large peptide/protein method (Table 2) for analysis of xylose isomerase from *E. coli* (MW, 44,000). Approximately 1500 pmol of protein was degraded with a repetitive yield of about 90%, resulting in the identification of the first 25 residues [14]. A step gradient HPLC system [20] was used for PTH amino acid identification at 269 nm on a Beckman Ultrasphere ODS column

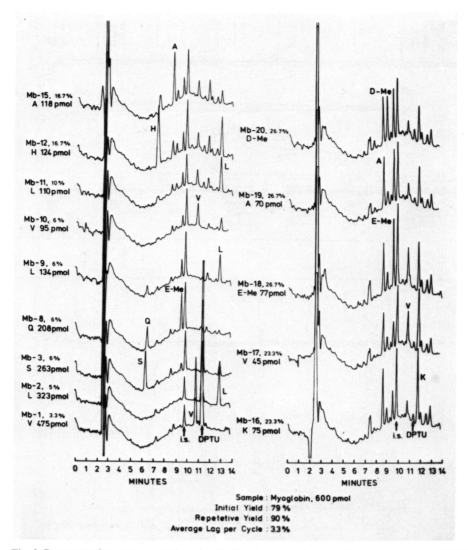

Fig. 3. Representative sequence results using the Tarr batchwise large peptide/protein strategy (Table 3) for analysis of myoglobin. PTH amino acid analysis was as described in Fig. 2

50 °C for coupling and cleavage and at 65 °C for conversion. The Pierce 40-ml screw cap bottles with Teflon/Silicon disc seals serve as solvent reservoirs and should be supported on a ring stand or held in an appropriate rack. Automatic pipets and syringes can be arranged in a test tube rack or also supported on a ring stand.

Table 4. Sequencing station components

I. *Vacuum system*
Vacuum pump (e.g., Leybold-Heraeus Trivac Model D2A)
Vacuum guage (e.g., Televac from the Fredericks Co., Huntingdon Valley, PA 19006)
Vacuum hose and connectors
Cold trap and Dewer
Glass vacuum manifold with glass-TFE Rotoflo valves (Corning, 6 mm bore) and mininert valves (Pierce)

II. *Argon or nitrogen flush system*
Tank of pre-purified gas with regulator
Silicon tubing with an attached hypodermic needle as a nozzle
Aquarium air valves

III. *Small equipment*
2 heat blocks
1 table top centrifuge with swinging bucket rotor
1 variable speed vortex mixer
1 small refirigerator (or ice bucket)
Glass syringes with Teflon-tipped plungers (10, 25 and 100 µl)
Automatic pipet (e.g., Gilson 20, 200 and 1000 µl)

IV. *Accessoires and supplies*
2 timers
Mininert slide valves (Pierce SC-13 and SC-14)
6 × 50 mm glass tubes, appropriate tube rack and dust-free storage box
13 × 100 mm screw cap tubes and appropriate tube rack
25 and 40 ml screw cap bottles (Pierce) with Teflon/silicon discs and appropriate bottle rack
1 ml reactivials (Pierce)
Disposable pipet tips and dust-free storage box
Pasteur pipets and dust-free storage box
100 µl glass capillary pipets
Polypropylene squirt bottle containing ethyl acetate
Ring stand and assorted tube clamps
Clay Adams pipet holder

3.2 Chemicals and Precautions

The chemicals used in the Tarr manual sequencing methods are presented in Table 5. Except for the polybrene, all the chemicals are usually of satisfactory purity in the grade and from the source indicated. However, special precautions must be taken not to introduce contaminants from the glassware (e.g., tubes, bottles, graduated cylinders), syringes, and pipet tips. Rinse apparently clean glassware with glass distilled water, then acetone and ethyl acetate. Clean syringes used for coupling medium and internal standard immediately after use with ethanol or ethyl acetate. Do not touch the disposable pipet tips with bare hands (wear rubber gloves), store tips in a ready-for-loading position in a dust-free box and use only once and discard. Work with small volumes of solvents and reagents and renew frequently rather than going in and out of a larger volume (e.g, about 40 ml of the wash and extraction solvents prepared as required and about 100 µl of PITC, TFA, and TEA fresh daily). About 0.01% ethanethiol should be added to all ethyl acetate-containing solvents and to the cleavage acids

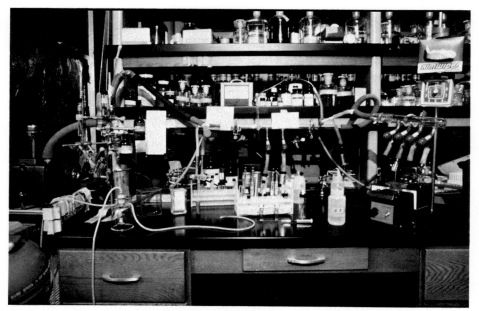

Fig. 4. A one-person manual sequencing station. Note that solvents, reagents, valves, syringes, timer, vortex, and kim wipes are all within easy reach. The rounded object on the *lower left corner* is the table top centrifuge which stands upon a small refrigerator (not shown). A kitchen range hood exhausted into the laboratory fume hood was installed directly over this station

(TFA and HCl). Store the wash and extraction solvents at room temperature, the TEA, TFA, thiourea, and pyridine at 4 °C under N_2 or argon and the stock PITC at −20 °C under an inert atmosphere. The working solutions of PITC, TEA, TFA, conversion medium, and thiourea solution should remain in the refrigerator at 4 °C when not in use.

3.3 Preparation of Tubes for Manual Sequencing

Batchwise Tarr manual sequencing is carried out in 6 × 50 mm glass tubes. Tarr recommends using and reusing toothed 6 × 50 mm sequencing tubes [4, 5]; however, we have not found it essential to put a tooth on the tube. Solution transfers can be successfully made by simply tipping and pressing the lip of one 6 × 50 mm tube into the mouth of another and rolling the tube until the solution drops into the second tube. Accordingly we prepare a supply of polybrene sequencing tubes ahead of time and discard them after they are used once. Polybrene may be cleaned up easily according to Lai [18]. Resuspend Lai purified polybrene to about 25 mg/ml in water and redry 10 µl aliquots (250 µg) in the bottom of each 6 × 50 mm sequencing tube. Wash the dried film with about 400 µl benzene : acetonitrile 1 : 1, repeating the wash two to three times until the solvent remains clear upon contact with the polybrene. Dry the tubes and per-

Table 5. Chemicals

Chemical	Abbreviation	Source
Acetic acid, HPLC grade	HAc	Baker
Acetone		Burdick and Jackson[a]
Acetonitrile	MeCN	Burdick and Jackson
Benzene		Burdick and Jackson
Dimethyl formamide, Gold label	DMF	Aldrich
Ethanethiol	EtSH	Pierce, Aldrich
Ethanol	EtOH	Baker
Ethyl acetate	EA	Applied Biosystems, Burdick and Jackson
Heptane	Hep	Burdick and Jackson
Hexafluoroacetone	HFA	Merck-Schuchardt
Hydrochloric acid	HCl	Merck
Methanol	MeOH	Burdick and Jackson
Methylthiohydantointyrosine	MTH-Y	Sigma
Phenylisothiocyanate	PITC	Beckman, Applied Biosystems
Polybrene		Pierce, Aldrich
PTH amino acid standards		Pierce
Pyridine, sequenal grade	Pyr	Pierce
Thiourea		Sigma
Triethylamine, sequenal grade	TEA	Pierce
Trifluoroacetic acid, sequenal grade	TFA	Pierce, Applied Biosystems
Trimethylamine	TMA	Aldrich
Water	H_2O	glass distilled

[a] Solvents from Burdick and Jackson are UV grade HPLC solvents.

from one to two cycles of Edman degradation (batchwise large peptide strategy) without a sample. Check a few tubes for purity by running the converted extracts on your HPLC PTH amino acid analysis system. Store ready-to-use polybrene tubes in a sealed 25- or 40-ml screw cap bottle.

4 Further Suggestions

The basics of the Tarr manual sequencing methods are not difficult to learn and one of the procedures has even been successfully used in an undergraduate bio-chemistry laboratory course at the University of Washington, Seattle. Neverthe-less, manual dexterity, consistent, well-organized work habits, a sound under-standing of Edman chemistry, and practice are required to develop superior manual sequencing skills. Before attempting any manual microsequencing, establish in the laboratory a HPLC system for identifying PTH amino acids. Also, review the relevant literature, particularly Tarr's most recent advice in the article entitled *Manual Edman Sequencing System* [5]. Within your sequencing station, establish an invariant location for all solvents, reagents, syringes, pipets, tubes, timers, kim wipes and so on to minimize delivery errors and to

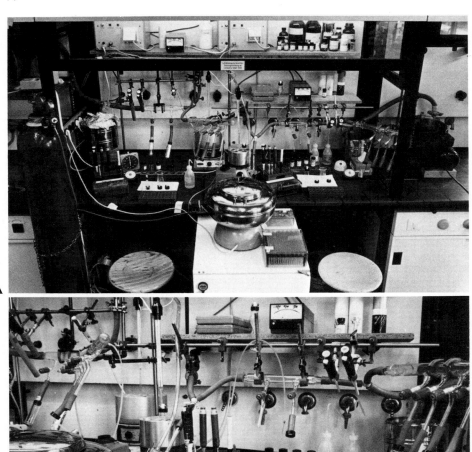

Fig. 5. A Two-person manual sequencing station within a larger fume hood. Except for the refrigerator and table top centrifuge, most components are set up in duplicate. **B** Enlarged view of the work area on the right side of the sequencing station. **C** A schematic of **B**. *1, 2* vacuum manifold; *3* vacuum sensor; *4* vacuum guage; *5* cold trap; *6* dewar; *7, 8* heat blocks; *9* three-way valve connected to the vacuum manifold and the nitrogen line for preparation of vapor HCl hydrolysates [see 5]; *10* lidded-boxes containing pipet tops; *11* lidded box containing 6 × 55 mm tubes; *12* centrifuge; *13* wash and extraction solvents for sequencing; *14* sequencing tubes and rack; *15* test tube rack with pipets and syringes; *16* automatic pipets; *17* refrigerator

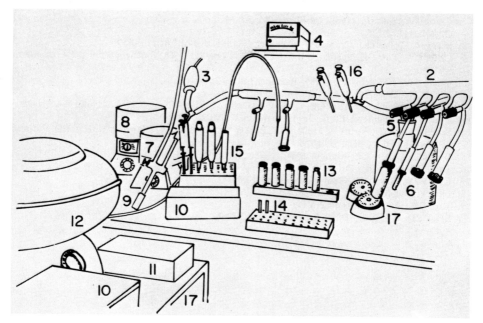

Fig. 5 C

enhance speed and efficiency. As with other microanalysis systems, trace contaminants and reaction by-products including residual vapors can obscure the final sequencing results. To minimize cross contaminations, devote to each particular function dispensing devices, vials, vacuum valves, and mininert slide valves and always use the same combinations. Keep the work area, tubes, and pipet tips dust-free and routinely clean the automatic pipets, syringes, slide valves and vials. Change the vacuum pump oil regularly and occasionally dismantle and clean the vacuum manifold. Include a standard sample such as myoglobin or insulin B chain in all batchwise sequencing. When "junk peaks" are obtained with standard samples which previously yielded clean results, start over with all new solvents and reagents. Finally, an instructional demonstration of the Tarr procedures provides the best introduction to methods; such a demonstration can usually be arranged through the authors.

References

1. Tarr GE, Beecher JF, Bell M, McKean DJ (1978) Anal Biochem 84:622
2. Tarr GE (1975) Anal Biochem 63:361
3. Tarr GE (1977) Methods Enzymol 47:335
4. Tarr GE (1982) In: Elzinga M (ed) Methods in protein sequence analysis. Humana, Clifton, NJ, p 223
5. Tarr GE (1986) In: Shively JE (ed) Microcharacterization of polypeptides: a practical manual. Humana, Clifton, NJ, p 155
6. Chang JY (1983) Methods Enzymol 91:455

7. Crabb JW, Saari JC (1982) In: Elzinga M (ed) Methods in protein sequence analysis, Humana, Clifton, NJ, p 535
8. Saari JC, Teller DC, Crabb JW, Bredberg L (1985) J Biol Chem 260:195
9. Crabb JW, Tarr GE, Yasunobu KT, Iyanagi T, Coon MJ (1981) Biochem Biophys Res Commun 95:1650
10. Crabb JW, Hanstein WG (1985) Biochem Intl 10:385
11. Crabb JW, Saari JC (1981) FEBS Lett 130:15
12. Fujita VS, Black SD, Tarr GE, Koop DR, Coon MJ (1984) Proc Natl Acad Sci USA 81:4260
13. Crabb JW, Heilmeyer LMG, Jr (1984) J Biol Chem 259:6346
14. Schellenberg GD, Sarthy A, Larson AE, Backer MP, Crabb, JW, Lidstrom M, Hall BD, Furlong LE (1984) J Biol Chem 259:6826
15. Wilson JM, Tarr GE, Mahoney WC, Kelley WN (1982) J Biol Chem 257:10978
16. Tarr GE, Black SD, Fujita VS, Coon MJ (1983) Proc Natl Acad Sci USA 80:6552
17. Crabb JW, Saari JC (1986) Biochem Intl 12:391
18. Lai PH (1984) Analyt Chim 163:243
19. Tarr GE (1981) Anal Biochem 111:27
20. Black SD, Coon MJ (1982) Anal Biochem 121:281

2.2 Manual Microsequence Determination of Proteins and Peptides with the DABITC/PITC Method

Brigitte Wittmann-Liebold, Hisashi Hirano, and Makoto Kimura[1]

Contents

1 Introduction

Microsequencing of polypeptides by the Edman degradation technique can be performed either manually or by automated methods [1, 2]. The degradation consists of three parts, the coupling with phenylisothiocyanate (PITC) or an appropriate homolog; the acid cleavage of the first amino acid as phenylthiazolinone from the peptide chain, and the isomerization to the more stable phenylthiohydantoins. This then has to be identified, e.g., by thin-layer techniques [2] or high performance liquid chromatography (HPLC) [3].

The manual methods are of great value if many peptide or protein samples have to be sequenced in parallel or tested for purity by N-terminal sequencing, or they are very useful if sophisticated sequencers or HPLC systems are not available in the laboratory. On the other hand the automated methods can only degrade one sample at a time, but they are beneficial where long stretches of sequence information are needed.

For sequencing in the low nanomole range, different manual and automatic methods have been described, some of which were adapted for picomole quantities. A recent automated method employs delivery of the base at the coupling

1 Max-Planck-Institut für Molekulare Genetik, Abteilung Wittmann, Ihnestr. 63 – 73, D-1000 Berlin 33

Advanced Methods in Protein Microsequence Analysis
Ed. by B. Wittmann-Liebold et al.
© Springer-Verlag Berlin Heidelberg 1986

stage and of the acid at the cleavage stage as vapors [4]. After sequencing, the released PTH-amino acids are identified by HPLC. Usually this is done in a separate HPLC system, but can also be made automatically by an on-line HPLC detection system governed by the same programmer as controls the sequencer [5].

Sensitive manual methods may also employ PITC, but in this case a HPLC detection system for identification of the released phenylthiohydantoins is required. Usually the derivatives are contaminated by UV-positive side products of the reaction, that obscure the identification by thin-layer chromatography.

In the past, another manual sequencing method, the dansyl-Edman technique, has been frequently employed [6]; the N-terminal residue of the peptide chain is detected by reaction with dansyl chloride after normal sequential degradation by PITC. However, the portion of the peptide necessary for dansylation has to be hydrolyzed and is lost for further degradation. Therefore, application of this method to minute peptide amounts is limited.

Alternatively, manual methods make use of isothiocyanate homologs that allow detection of the released amino acids as fluorescent derivatives, e.g., by reaction with fluorescein isothiocyanate [7]. More recently, the application of Boc-aminoalkylphenylisothiocyanates (BAMPITC) for Edman degradation was proposed [8]. This should allow the derivatization of the released phenylthiohydantoin amino acids carrying a free amino group with ortho-phthaldialdehyde or other fluorescence-generating reagents. We examined this reagent, and although degradation of the polypeptide chain occurs at the expected rate, an unusual reaction was observed and it was not possible to isolate specific amino acid derivatives by the reaction with the fluorochrome dabsyl chloride [9, 10].

The reagent DABITC (4-N,N-dimethylaminoazobenzene-4'-isothiocyanate) has been widely applied for manual liquid and solid phase microsequencing in recent years [11]. It is also used in solid phase sequencers [12] (for details see Salnikow, this vol.). This reagent was described by Chang and Creaser in 1976 [13] and has several advantages over the usual degradation with PITC: firstly, it releases the amino acid thiohydantoins as red-colored derivatives which are visible to the eye in picomole quantities on polyamide sheets; and secondly, the side products of the Edman chemistry form blue-colored compounds whose color easily differentiates them from the real red amino acid hydantoins.

However, due to the bulky side group of the DABITC reagent, the coupling yield is lower than that of PITC. Therefore, this reagent can only be applied usefully for microsequencing if the reaction is completed by a second coupling with PITC [14]. Since the introduction of this DABITC/PITC double coupling procedure the method has found wide application; it is simple to perform, requires little sophisticated equipment and can be employed in any laboratory. Students and technicians learn this technique in a few days. The method can easily be used for purity tests of isolated proteins or fragments thereof; it is applicable to small as well as to large hydrophilic or hydrophobic peptides; with short hydrophobic fragments, microsequencing is best done after covalent attachment of the peptide to an aminopropyl glass support [15]. More details of the attachment procedures have recently been compiled [16, 17].

In this article, an updated protocol for the manual DABITC/PITC double coupling method is given; the procedure can be applied to 1 to 5 nanomoles of peptide or protein, and with slight modifications is also suitable for picomole quantities. Even for minute amounts of peptide material the use of a sophisticated HPLC system is not mandatory. Identification can most quickly and simply be performed on stamp-sized polyamide thin-layer sheets [18]. However, HPLC systems for additional confirmation of the identified amino acids and their quantitation have been developed [19, 20].

2 Description of the Manual DABITC/PITC Double Coupling Method

2.1 Equipment

The manual procedures described here need no expensive equipment. However, the following equipment facilitates sensitive sequencing and easy performance. Best suited is a laboratory table, e.g. of 1 m length, within a hood. It should be equipped with the following items:

1. Nitrogen (or argon) tank connected to a pipet by a plastic outlet line closable by a stopcock.
2. Table centrifuge with holders suitable for the Edman tubes, e.g., Labofuge I, purchased from Christ/Heraeus.
3. Vacuum device (pump, cool trap or KOH filter, manifold for 2 to 6 desiccators, vacuum gauge and sensor).
4. One to two aluminum heaters with holes into which the Edman tubes and conversion glasses fit tightly.
5. Edman glass tubes, 4.5 cm long, i.d. 8 mm, with glass stoppers; alternatively: 2.5 cm, i.d. 2 – 4 mm.
6. Conversion glasses, 5 cm, i.d. 4 mm.
7. Plastic holders for 10 to 20 glasses that fit into the desiccators.
8. Automatic pipets (2, 5, 10, 20, 50, 100, and 200 µl).
9. Items for thin-layer chromatography: two to four glass beakers (25 ml, height 5 cm) with polished ends, glass covers with a layer of parafilm to seal the beakers, glass capillaries, polyamide sheets (2.5×2.5 cm), pincets, holders for 10 2.5×2.5 cm sheets, cold fan.
10. HPLC system for the identification of the DABTH derivatives: An isocratic system equipped with one HPLC pump, an analytical C8 reversed phase column with precolumn, column oven, fixed wavelength detector at 436 nm, recorder and injection valve. For gradient HPLC separation of the DABTH-amino acids see [20].

2.2 Reagent and Solvent Purification

1. 4-N,N-dimethylaminoazobenzene-4'-isothiocyanate (DABITC) from Fluka or Pierce, recrystallized from boiling acetone (pro analysis grade, dried over

molecular sieve): 1 g dissolved in 70 ml, passed through a paper filter and allowed to cool slowly; yield 0.7 g of brown needles, m.p. 169° – 170°C.

2. p-Phenylisothiocyanate (PITC), purissimum grade from Fluka, three times redistilled under reduced pressure (oil pump) and nitrogen; stored under nitrogen in small ampules (500 µl to 1 ml aliquots) at – 20°C.

3. Pyridine, pro analysis grade from Merck, distilled successively over KOH pellets, ninhydrin and KOH, b.p. 114° – 116°C; stored in glass-stoppered flasks under nitrogen at 4°C.

4. Trifluoroacetic acid (TFA), purum grade from Fluka, redistilled from $CaSO_4 \cdot 0.5\ H_2O$ (dried at 500°C immediately before use) over a 50 cm column filled with glass rings; b.p. 72° – 73°C; stored in glass-stoppered flasks (100 ml) under nitrogen at – 20°C.

5. Water: twice glass distilled and freshly prepared.

N-heptane and ethylacetate were sequencer grade chemicals (for details see Reimann and Wittmann-Liebold, this vol.). N-butylacetate, acetic acid, toluene, n-hexane and acetone were pro analysis grade from Merck and used without further purification.

2.3 Manual Liquid Phase Method

The reaction scheme for the degradation with DABITC is given in Fig. 1. According to this scheme, the coupling results in the formation of a dimethylamino-azobenzene-thiocarbamyl peptide (DABITC-peptide); under treatment with strong anhydrous acid this releases the first amino acid as dimethylaminoazobenzene-thiazolinone derivative (DABTC-amino acid). After extraction this deriva-

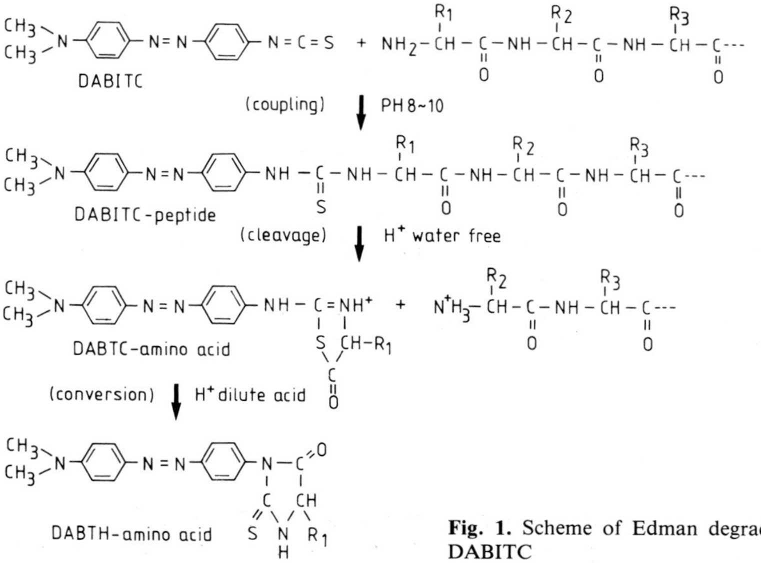

Fig. 1. Scheme of Edman degradation employing DABITC

Table 1. Pretreatment of peptide or protein for microsequencing

I. Polypeptide isolation

a) The sample should be salt-free to guarantee a correct pH at the coupling reaction with the isothiocyanate reagent. HPLC techniques employing volatile buffers are best for the isolation of small quantities of peptides. Desalt on reverse phase precolumn on C8 or C18 support if necessary; dry peptide carefully to remove traces of acids or volatile salts in the sample (repeated drying in vacuo of the peptide with the addition of redistilled water or ethanol).

b) Remove detergents and other impurities by acetone or ethanol precipitation of the protein where necessary.

c) Employ buffers for the polypeptide purifications that contain mercaptoethanol or dithioerythritol (e.g., 0.02%) to preserve the free N-terminal amino groups. In cases where the peptide is isolated from polyacrylamide gels add these scavengers to the electrophoresis buffers and pre-electrophorese the gels before sample application.

d) Store polypeptide frozen in solution (e.g., water, dilute acids); concentrate in Speed Vac centrifuge but do not lyophilize sample (this causes extreme peptide losses if small quantities are lyophilized; further, alterations in solubility are frequently observed).

II. Preparation of sample for degradation

a) Transfer sample to Edman tubes shortly before use; drying is best done in the tubes used for the degradation to avoid transfer losses of sample.

b) Store peptide frozen after completed cycles of the degradation dissolved in a small volume of water.

c) Avoid storage of peptide under vacuum for more than several hours.

d) Remove traces of acids in the sample by repeated drying in vacuo with the addition of redistilled water or ethanol.

tive is converted to the more stable dimethylaminoazobenzene-thiohydantoin, which yields red-colored products upon treatment with HCl vapors.

For completion of the reaction, the coupling is performed twice, firstly with DABITC, and secondly with PITC. Peptides or proteins should be free of traces of acids, and salts have to be removed carefully prior to the degradation. In Table 1, precautions for the isolation of polypeptides subjected to microsequencing are listed.

In Table 2 the protocol of the degradation is described for 2 to 5 nanomole peptide or protein, or alternatively for 500 picomole to 1 nanomole amounts.

2.4 Manual Solid Phase Method [21]

Small hydrophobic peptides are best covalently attached to amino glass prior to manual microsequencing. Different attachment procedures are in use, e.g., attaching the free amino groups of peptides (epsilon-amino groups of lysine and alpha-amino groups of N-terminal residues) to diisothiocyanate-activated aminopropyl glass (APG), coupling carboxyl groups after activation by a water-soluble carbodiimide to amino glass, or by attachment of homoserine lactone peptides to APG. Experimental details of these procedures are given elsewhere ([17], [22] and Salnikow, this vol.).

Here, a protocol for the degradation of 5 – 10 nanomoles of polypeptides attached to aminopropyl glass is described (see Table 3).

Table 2. Manual liquid phase DABITC/PITC double coupling method

The protocol is described for 2 to 5 nanomol peptide or protein and for 500 pmol to 1 nanomol (in brackets).
The degradations are performed in small glass-stoppered tubes, length 4.5 cm, i.d. 8 mm (or in shortened dansyl glass tubes, length 2.5 cm, i.d. 4 mm).
For reagent and solvent purification see Sect. 2.2.

I. Coupling

1. Dry peptide or protein in Edman tube (dansyl glass) in desiccator over KOH pellets at 100 – 200 millitor); repeat drying if purified from acid solution.
2. Add 80 µl (20 µl) of 50% pyridine in redistilled water.
3. Flush with nitrogen or argon.
4. Add 100-fold excess of DABITC, MW 282, dissolved in pyridine, e.g., 150 µg/40 µl (30 µg/10 µl).
5. Flush with inert gas; close tube with glass stopper (parafilm in case of dansyl glass).
6. Keep for 30 to 40 min at 52 °C in heated aluminum block.
7. Add 10 µl (2 µl) conc. PITC.
8. Purge with inert gas, close Edman tube.
9. Keep for 20 min at 52 °C.

II. Wash after coupling

1. Add 3 to 4 times 400 µl (100 µl) n-heptane/ethylacetate, 2:1, v/v.
2. Centrifuge for 2 min, Labofuge I, Heraeus/Christ, (for 2 – 3 s using an Eppendorf tube as holder in the Microfuge B from Beckman) until clear separation of the phases.
3. Discard upper phase by withdrawal with water pump into waste bottle; store in case of small hydrophobic peptides; withdraw carefully in case of precipitation of the polypeptide at the interphase.
4. Dry water phase for 15 to 20 min in desiccator in vacuo at 100 millitorr,
   do not proceed to cleavage reaction if sample is not dried completely;
   repeat drying of the sample in vacuo under the addition of 40 µl (10 µl) ethanol, if necessary.

III. Cleavage

1. Add 50 µl (15 µl) anhydrous TFA.
2. Purge with inert gas, close with stopper.
3. Keep for 10 min at 52 °C.
4. Dry in desiccator over KOH pellets in vacuo at 100 millitorr.
5. Repeat cleavage 1 – 3 times in case of repetitive Val-, Pro- and Ile-residues.
6. Repeat drying with the addition of 50 µl (10 µl) n-butyl acetate if sample is not dried completely.

IV. Extraction of thiazolinone derivative

1. Add 30 µl (30 µl) of water and 100 to 200 µl (30 µl) n-butyl acetate.
2. Purge with inert gas.
3. Vortex and centrifuge for 2 min, Labofuge I (for 3 – 5 s, Microfuge B).
4. Withdraw upper layer containing the thiazolinone derivative into small glass tube, length 5 cm, i.d. 4 mm (2.5 cm, i.d. 4 mm).
5. Repeat extraction and dry the combined extracts in vacuo.
6. Dry water phase containing the residual peptide in vacuo over KOH pellets at 100 millitorr and begin next degradation cycle.

V. Conversion to DABTH-amino acid derivative

1. Add to dried butyl acetate extract 40 µl (10 µl) 40% TFA in water.
2. Purge with inert gas; close with parafilm.
3. Keep for 30 min at 52 °C.
4. Dry in vacuo.

Table 2 (continued)

VI. Identification

1. Dissolve DABTH-amino acid in 2 μl (less than 0.5 μl) of ethanol.
2. Apply approximately 1/20 (1/2 to 1/5) onto 2.5 × 2.5 cm polyamide sheets, spot size of 1 mm; add one droplet of markers, see below.
3. Develop sheets in two dimensions: first dimension: 33% acetic acid in water; second dimension: toluene/n-hexane/acetic acid, 2:1:1, v/v; add the second-dimension solvent to the beaker immediately before use.
4. Repeat sample application with appropriate amounts; comigrate the identified DABTH-amino acid with the corresponding reference DABTH-amino acid.
5. Inject remainder of the DABTH-amino acid into reversed phase C18 HPLC-column for quantitation, or if any doubt exists regarding the identity of the DABTH-amino acid, e.g., differentiation of DABTH-Ile/Leu.

VII. Preparation of markers

1. Pipet 500 μl of 50% pyridine into glass-stoppered test tube.
2. Add 30 μl each of diethylamine and ethanolamine (both redistilled).
3. Flush with nitrogen.
4. Add 250 μl DABITC in 100% pyridine (2.3 g ml^{-1} pyridine).
5. Flush with nitrogen, close with stopper.
6. Keep for 1 h at 52°C.
7. Dry in vacuo.
8. Dissolve in 1 ml ethanol.
9. Apply one small droplet onto polyamide sheet together with sample (the two blue marker spots should be visible after chromatography as very faint spots only).

It is important to note that the first degradation cycle is always carried out with PITC alone to guarantee a complete reaction of the excess of amino groups on the glass support. Then, in the following cycles, the coupling is first performed with DABITC and the reaction completed with PITC. The first amino acid of the peptide is obtained as PTH derivative and can be detected by isocratic HPLC separation as described recently [23]. Even if the coupling to the glass is made through the free amino groups of the peptide, part of the N-terminal residues can be identified as PTH-amino acid due to the incomplete coupling of the N-terminal amino acid to the support.

3 Identification of DABTH-Amino Acid Derivatives

3.1 Thin-Layer Technique

Figure 2 illustrates the resolution of the DABTH-amino acid derivatives by chromatography on stamp-sized polyamide sheets. The procedure for the analysis of the degraded samples and the solvent mixtures used for the two-dimensional chromatography are listed in Table 4. The location of the individual derivatives is determined relative to the migration of two markers, DABITC-reacted diethylamine and ethanolamine (see Table 2, VII). Whereas the released DABTH-amino acid derivatives yield red-colored spots upon exposure to acid vapors, these

Table 3. Manual solid phase DABITC/PITC method

Peptides are covalently attached to glass support (see text). The degradation is performed in glass-stoppered glass tubes (7 cm, i.d. 1 cm) with gentle stirring (by means of a stirring bar; magnetic stirrer placed below the heating block).
Special care has to be taken at the vacuum stages to avoid physical losses of glass-attached peptide. This is prevented by means of a glass adaptor fitted with a G2 sinter glass filter that replaces the stopper during dryings.

First degradation cycle

Employ only PITC (to guarantee complete saturation of excess of amino groups of the glass support) and identify released PTH-amino acid derivative by HPLC (isocratic PTH-amino acid system).

I. Coupling

1. Dry glass-attached peptide in vacuo for 5 min in a Speed Vac Concentrator, then for 20 min in a desiccator.
2. Wash the beads twice with 500 µl of methanol.
3. Dry in vacuo.
4. Add 400 µl of 50% pyridine in water and 50 µl PITC.
5. Purge with nitrogen; close tube with glass stopper.
6. Keep for 40 min with stirring in aluminum block which is set at 52°C.
7. Centrifuge and remove supernatant.
8. Wash the beads twice with 500 µl pyridine and twice with 500 µl of methanol.
9. Dry in vacuo (see above).

II. Cleavage

1. Add 200 µl of anhydrous TFA.
2. Flush with nitrogen, close tube with glass stopper.
3. Incubate for 8 min at 52°C.
4. Dry in vacuo for 20 min.

III. Extraction of thiazolinone

1. Add 400 µl of methanol, stir gently.
2. Centrifuge.
3. Collect methanol into dansyl glass.
4. Repeat extraction with 200 µl of methanol.
5. Dry combined methanol extracts.
6. Repeat cleavage and extraction for bulky residues.
7. Dry glass-attached peptide in vacuo and subject to next cycle.

IV. Conversion to thiohydantoin derivative

1. Add to dried methanol extracts of thiazolinone 80 µl of 50% TFA in water.
2. Purge with nitrogen; then close glass with parafilm.
3. Keep for 30 min at 52°C.
4. Dry in vacuo.
5. Identify PTH-amino acid derivative by HPLC (see text).

Second cycle

I. Coupling with DABITC/PITC

1. Add 400 µl of 50% pyridine in water.
2. Add 200 µl of DABITC in pyridine (2.3 mg ml^{-1}).
3. Purge with nitrogen; close tube with glass stopper.

Table 3 (continued)

4. Incubate for 30 min at 52 °C with gentle stirring.
5. Add 20 µl of PITC.
6. Purge with nitrogen; close tube.
7. Incubate for 30 min at 52 °C.
8. Centrifuge and remove supernatant.
9. Wash beads twice with 500 µl of pyridine and twice with 500 µl of methanol.
10. Dry beads in vacuo.

II. to III. Cleavage and thiazolinone extraction

are performed as described above for the first cycle.

IV. Identification of DABTH derivatives

see Tables 2 and 4.

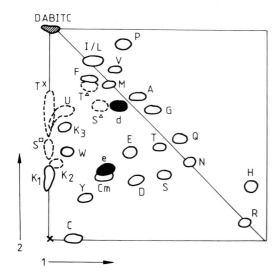

Fig. 2. Separation of DABTH-amino acid derivatives on two-dimensional polyamide thin-layer sheets. For details see Tables 2 (VI and VII), 4 and 5

marker compounds result in blue spots only. Further, all side-products of the Edman chemistry, e.g., DPU (diphenylurea) and DPTU (diphenylthiourea), are not visible on thin-layer sheets or form blue-colored DABITC derivatives. Thus, microsequencing employing DABITC is advantageous and allows an easy differentiation of the amino acid derivatives from the side-products of the reaction. The detection limit for the DABTH-amino acids on micropolyamide sheets is about 20 picomole.

3.2 HPLC Separation of DABTH-Amino Acid Derivatives

For an additional identification of the released DABTH-amino acids, an iso-cratic HPLC system equipped with a laboratory-packed column of C8 (MOS) reversed phase support of 5 µm particle size (purchased from Shandon or Phase

Table 4. DABTH-amino acid identification by TLC

1. Cut polyamide thin-layer sheets (purchased from Schleicher and Schuell) to sizes of 2.5 × 2.5 cm by means of a paper cutter and keep in closed boxes. Use self-made thin glass capillaries for sample application.
2. Add marker samples (see Table 2) as smallest possible spot onto sheet (at 4 mm distance from edges).
3. Dissolve DABTH-derivative in 2 – 5 µl of ethanol and spot portions on top of marker.
4. Develop sheet for about 4 – 5 min in the first dimension (33% acetic acid in water); remove the TLC sheet from the beaker when the solvent front is 2 mm from the top of the sheet; dry with cold fan for 5 min.
5. Develop sheet for 2 min in the second dimension (toluene: n-hexane: acetic acid, 2:1:1, v/v); remove when the front has reached 4 – 5 mm from top; dry by cold fan;
6. Keep sheet over HCl vapors (12 M HCl) under hood and identify red-colored spots in comparison to reference DABTH-amino acid mixture and determine migration relative to blue-colored markers, see Fig. 2.
7. Identification of DABTH-amino acids and additional spots (see also Table 5):
   Symbols denote: DABTH-amino acid derivatives of: A, alanine; C, cysteic acid; Cm, carboxymethyl cysteine; D, aspartic acid; E, glutamic acid; F, phenylalanine; G, glycine; H, histidine; I, isoleucine; K1, alpha-DABTH-epsilon-DABTC-lysine (red); K2, alpha-PTH-epsilon-DABTC-lysine (blue); K3, alpha-DABTH-epsilon-PTC-lysine (blue); L, leucine; M, methionine; N, asparagine; P, proline; Q, glutamine; R, arginine; S, serine; S-delta, dehydro-serine; S□, polymerization product of serine; T, threonine; T-delta, dehydro-threonine; Tx, blue-colored polymerization product of threonine; U, thiourea derivative; V, valine; W, tryptophan; Y, tyrosine; d and e, blue-colored reference markers of DABITC-reacted diethylamine and ethanolamine.

Separations, GB) suffices for resolving all DABTH-derivatives except that of arginine; the latter is eluted using higher concentrations of the organic modifier.

The separation of the DABTH-amino acid derivatives is illustrated in Fig. 3. The separation shown was performed using 50% acetonitrile/0.5% 1,2-dichloroethane/ in 12 mM sodium acetate buffer at pH 5.0. Flow rate was 1.2 ml min^{-1} at 45 °C; the recorder speed was set at 5 mm min^{-1}. The amino acid derivatives were monitored with a variable wavelength detector (Beckman) at 436 nm.

The HPLC buffer can be continuously recycled for one to several weeks; the column is rinsed with 80% acetonitrile/ 0.5% 1,2-dichloroethane/ in 12 mM sodium acetate buffer after 10 to 20 injections (this solvent mixture is also employed for the identification of the DABTH-Arg steps).

Depending on the source of the C8 support used, the percentage of the organic solvent of the HPLC system has to be adjusted between 48 and 51%.

4 Problem-Solving Section

In Table 5, problems frequently observed when employing the DABITC/PITC method for the first time are listed. In this table possibilities for circumventing such difficulties are described. Since salts interfere with an optimal resolution of the amino acid derivatives on thin-layer sheets, mainly in the area of the basic DABTH-derivatives DABTH-Arg and -His, it is necessary to start with salt-free

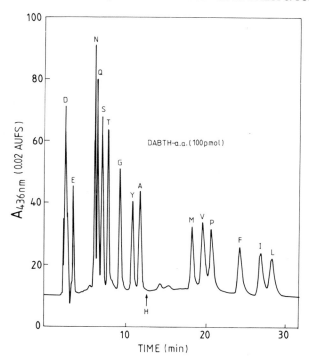

Fig. 3. Isocratic HPLC separation of DABTH-amino acid derivatives. For details see Sect. 3.2

polypeptide preparations for the degradation and to ensure complete removal of the base and the acid after the coupling and cleavage stage of the reaction.

Salts also lower the coupling yields of the polypeptides with isothiocyanates (the pH of the coupling reaction should be adjusted to pH 9.0). A lower pH hinders sufficient coupling yields mainly for histidine and lysine residues. Further, salt contaminations of the peptides cause slower drying of the sample after the coupling and cleavage stages.

By polyamide thin-layer chromatography, the derivatives of isoleucine and leucine cannot be differentiated. However, they separate well by isocratic HPLC, as described above.

Due to the drying stages of the Edman degradation, the derivatives of threonine and serine are dehydrated and undergo further destruction or polymerization. Therefore, in addition to the authentic DABTH-Ser- and Thr-derivatives, other spots are visible on thin-layer (as explained in Table 4) and give characteristic peaks in the HPLC trace. To a certain extent the occurrence of these additional products can be diminished by oxygen-free atmosphere at the thiazolinone extraction and conversion and by the choice of an appropriate conversion medium. However, no conversion medium is available which is optimal for all different thiazolinones. Whereas 1 M HCl is quite appropriate for the conversion of serine and threonine thiazolinones, dilute trifluoroacetic acid is more suitable for the isomerization of the asparagine and glutamine derivatives. These compounds would yield considerable amounts of the free acid thiohydantoins by conversion in HCl. It may be advisable to change from one conversion

Table 5. Trouble shooting

Problems with the manual sequencing procedure that may occur are listed; reasons and precautions to avoiding these are given.

  I. Contents of Edman tube after coupling and/or cleavage dry slowly or remain oily (the sample should be taken to complete dryness in a 15-min period):
  a) Applied vacuum is not efficient; change oil of pump; check cryostat; read vacuum at the pump outlet and at the connections.
  b) Peptide sample contains salt; this may be result from the purification procedures or from the preceding Edman cycles due to incomplete dryings after coupling or cleavage; try to remove salts by repeated evacuation of the sample with the addition of ethanol.

 II. Presence of salts causes spot broadening in the areas of DABTH-Arg and -His:
    Dry DABTH-amino acid under nitrogen, add 50 μl water and reextract the derivative with butyl acetate: the salts will stay with the water phase.

III. Presence of a strong red spot which migrates with the solvent front in the first dimension chromatography (this is due to excess of DABITC reagent):
    Extraction of the excess of reagent after the coupling was not complete; remove upper layer more thoroughly, e.g., by additional extraction; recrystallize DABITC.

 IV. Peptide losses at the extractions:
  a) Watch peptide layer between organic and water phase due to precipitation of hydrophobic peptides: remove upper layer with care; do not touch the precipitate with pipet.
  b) Small hydrophobic peptides are readily dissolved in the organic layer: apply n-heptane/ethyl acetate in the ratio 7:3, or 3:1, respectively; attach peptide covalently to glass support and degrade by manual or automated solid phase sequencing.

  V. Occurrence of "double spots" in the polyamide chromatography (conversion of thiazolinone to the thiohydantoin derivative is not complete):
    Check temperature of the aluminum heating block; check the tight fit of the reagent tube into the block; use aluminum foil for a better fit of the test tube into the block holes; prolong time of conversion or raise temperature; change conversion medium (20% TFA in water; 40% TFA in water; 1 M HCl; 40% TFA in ethyl acetate; methanolic HCl; acetic acid saturated with HCl gas).

 VI. Difficulties with the identification of the hydrophobic amino acid derivatives (DABTH-Ile/Leu; -Val/Phe):
    Use HPLC system for additional identification of these degradation steps; spot sample together with authentic DABTH-amino acid or DABTH-derivatives of previous cycles which contain the same residues.

VII. Identification problems with DABTH-Ser and DABTH-Thr:
    Apply 1 M HCl or HCL-saturated acetic acid as conversion medium.

VIII. Double spots in case of DABTH-Asn and DABTH-Gln, caused by deamidation:
    Apply 20–40% TFA in water as conversion medium to avoid formation of the acids.

 IX. Identification of extra spots, see Table 4.

medium to another for different peptides, depending on their amino acid composition, or to perform a second degradation of the same peptide employing another conversion medium.

Further, the temperature used for the cleavage and conversion reaction may result in the destruction of labile thiazolinones, such as that of arginine and histidine. As the temperature within the glass tubes varies with diameter, the volume of the reaction medium, and tight fit into the holes of the aluminum heater, the appropriate temperature may be evaluated with synthetic peptides that contain such residues. We notice extra spots on thin-layer sheets especially for hydrophobic residues. These are caused by incomplete conversion (too short isomerization time or low temperature).

Differentiation of the hydrophobic residues is difficult if the development in the second-dimension chromatography is excessive. In this case it is recommended to spot the authentic hydrophobic derivatives together with the sample under question on the same sheet and to develop for a shorter time. Comigration in one spot occurs if both samples derive from the same amino acid; non-overlapping spots indicate a different amino acid derivative.

5 Conclusions

Microsequencing employing the DABITC/PITC double coupling method offers several advantages:

1. The technique is sensitive, e.g., 20 picomoles of the DABTH-amino acid derivatives are visible on polyamide thin-layer sheets.
2. The derivatives released are red-colored upon treatment with acid vapors; thus they can easily be differentiated from side products of the reaction.
3. The method needs no sophisticated equipment and can therefore be performed in any laboratory.
4. In the manual mode (liquid and solid phase technique) many polypeptide samples can be degraded simultaneously.
5. In contrast to the dansyl-Edman technique, the method is applicable to large proteins as well as small peptides; thus purity checks of isolated protein and peptides, e.g., after HPLC purification, is possible.
6. The manual method performed in the liquid mode is limited by extraction losses of small-sized hydrophobic peptides; here, covalent attachment of the peptide to a glass support is possible.
7. Attachment of the peptides and proteins to glass can be effected in the presence of SDS and/or salts; these preparations can then be sequenced with DABITC.
8. The DABITC/PITC double coupling method can be applied automatically in a solid phase sequencer.
9. Isocratic HPLC separation of the DABTH derivatives allows an additional identification of the amino acids and confirmation of the sequence derived by polyamide thin-layer chromatography.

References

1. Edman P, Begg G (1967) Eur J Biochem 1:80–91
2. Edman P, Henschen A (1975) In: Needleman SB (ed) Protein sequence determination. Springer, Berlin Heidelberg New York, pp 232–279
3. Zimmermann CL, Apella E, Pisano JJ (1977) Anal Biochem 77:569–573

4. Hewick RM, Hunkapiller MW, Hood LE, Dreyer WJ (1981) J Biol Chem 15:7990 – 8005
5. Wittmann-Liebold B, Ashman K (1985) In: Tschesche H (ed) Modern methods in protein sequence analysis, vol 2. de Gruyter, Berlin New York, pp 303 – 327
6. Gray WR, Hartley BS (1963) Biochem J 89:379 – 380
7. Maeda H, Kawauchi H (1968) Biochem Biophys Res Commun 31:188 – 192
8. L'Italien JJ, Kent SBH (1984) J Chromatogr 283:149 – 156
9. Jin S-W, Wittmann-Liebold B, Palacz Z, Salnikow J (submitted)
10. Wittmann-Liebold B, Jin S-W, Salnikow J In: Bhown AS (ed) Peptide/protein sequencing: Current methodologies. CRC, Boca Raton, FL (in press)
11. Wittmann-Liebold B, Kimura M (1984) In: Walker JM (ed) Methods in molecular biology, vol 1. Humana, Clifton, NJ, pp 221 – 242
12. Laursen RA (1971) Eur J Biochem 20:89 – 102
13. Chang JY, Creaser EH, Bentley KW (1976) Biochem J 153:607 – 611
14. Chang JY, Brauer D, Wittmann-Liebold B (1978) FEBS Lett 93:205 – 214
15. Machleidt W, Wachter E, Scheulen M, Otto J (1973) FEBS Lett 37:217 – 220
16. Laursen RA (1977) Methods Enzymol 47:277 – 288
17. Wittmann-Liebold B (1986) In: Dabre A (ed) Practical protein chemistry – a handbook. Wiley, New York, pp 375 – 409
18. Chang JY, Creaser EH (1977) J Chromatogr 132:303 – 307
19. Lehmann A, Wittmann-Liebold B (1984) FEBS Lett 176:360 – 364
20. Chang JY, Lehmann A, Wittmann-Liebold B (1980) Anal Biochem 102:380 – 383
21. Chang JY (1979) BBA Libr 578:188 – 195
22. Salnikow J, Lehmann A, Wittmann-Liebold B (1981) Anal Biochem 117:433 – 442
23. Ashman K, Wittmann-Liebold B (1985) FEBS Lett 190:129 – 132

2.3 Solid-Phase Microsequencing: Procedures and Their Potential for Practical Sequence Analysis

Werner Machleidt[1], Ursula Borchart[1], and Anka Ritonja[2]

Contents

1 Introduction

The chemistry of stepwise N-terminal degradation of peptides and proteins introduced by Edman in 1949 [1] has remained essentially unchanged. The tremendous progress in speed and sensitivity of protein sequencing is due to improvements of instrumentation and procedures performing these chemical reactions automatically and, probably most important, to the use of high performance liquid chromatography (HPLC) for the identification of the released phenylthiohydantoin (PTH)-amino acid derivatives.

The central problem of automated sequencing based on Edman chemistry is to wash out phenyliosthiocyanate (PITC) and its reaction byproducts, and, after the cleavage reaction, the released amino acid anilinothiazolinone (ATZ) without losing the peptide or protein being degraded. The difference in solubility of the protein and the reaction products was utilized for this separation in Edman and Begg's ingenious spinning-cup sequenator [2]. The difference in solubility was increased by the use of a synthetic "carrier" like polybrene [3], and, recently, most

1 Institut für Physiologische Chemie, Physikalische Biochemie und Zellbiologie der Universität München, Goethestrasse 33, D-8000 München, FRG
2 Department of Biochemistry, J. Stefan Institute, Jamova 39, YU-61000 Ljubljana, Yugoslavia

Advanced Methods in Protein Microsequence Analysis
Ed. by B. Wittmann-Liebold et al.
© Springer-Verlag Berlin Heidelberg 1986

rigorously exploited in the gas-phase sequenator, where minute amounts of peptide are deposited within a polybrene film on a porous glass fiber disc and reacted with vapors of the polar reagents to prevent wash-out [4].

An alternative approach to solve the extraction problem was made by Laursen, who coupled the peptide covalently to a solid support prior to Edman degradation [5]. Theoretically, "covalent solid-phase sequencing" should provide a perfect solution of the extraction problem, because, once covalently bound, the peptide can be treated freely with all kinds of reagents and solvents without any risk of losing it. For similar reasons the solid-phase approach has proven extremely valuable in various fields of biochemistry, such as affinity chromatography and automated synthesis of peptides and DNA fragments. Laursen's solid-phase sequencer [6] was a much simpler and cheaper instrument than the spinning-cup sequenator of that time.

The inherent drawback of covalent solid-phase sequencing is the need for covalent attachment. Unfortunately, there is no general reaction for covalent coupling of a peptide to a solid support exclusively at its C-terminal end. This ideal immobilization of a peptide to be sequenced is restricted to cyanogen bromide fragments carrying a C-terminal homoserine residue. All other peptides have either to be coupled via certain amino acid side-chains (lysine, cysteine) or via their C-terminal and side-chain carboxyls. Due to inherent chemical problems, the latter reaction, though generally applicable, suffers from incomplete yields. Side-chain attachment of certain residues can lead to losses of sequence information ("blanks" where residues remain bound to the support) and to losses of sequenceable peptide whenever a side-chain anchor point is passed.

These problems of covalent attachment have discouraged most investigators from using covalent solid-phase sequencing. Furthermore, considerable experience is needed to obtain good results with chemical coupling of peptides and proteins. Considering the impressive results that can be reached with much less effort using the gas-phase sequenator, the question arises whether the covalent solid-phase approach is a deadlock or still a promising alternative that should be further developed in the future.

Since 1979 we have been working with a self-constructed solid-phase sequencer including on-line injection of PTH-amino acids [7]. Besides continuously improving the methodology, we have been using the covalent solid-phase method as a single tool for sequence analysis of various unknown proteins (e.g. [8] – [15]). This article attempts to summarize our experience, to communicate approved procedures and to provide a sound basis for the evaluation of future methodological development.

2 Supports for Solid-Phase Edman Degradation of Peptides and Proteins

A variety of materials has been proposed and more or less successfully used as supports for solid-phase Edman degradation of peptides and proteins after-

Table 1. Preparation of aminopropyl glass (APG)

1. Suspend approx. 6 g of controlled-pore glass (Fluka or Serva) in a freshly prepared solution of 1.5 ml of 3-aminopropyl triethoxysilane (Serva) in 30 ml of acetone (Merck, p. a.). It is important to open a fresh vial of the silane for each preparation.
2. Degas the suspension for a few minutes by applying a slight vacuum to the flask under careful shaking.
3. Remove excess solution until the soaked support is just covered.
4. Incubate for 15 h at 40 °C without shaking.
5. Wash several times with acetone and methanol (both Merck, p.a.) on a fritted glass filter.
6. Dry the support in a desiccator and store it desiccated in a freezer.

covalent attachment (see [16] for review). The supports used almost exclusively today are derivatives of porous glass. As confirmed by recent results of solid-phase DNA synthesis, porous glass seems to be an ideal matrix for performing automated chemical reactions with covalently bound ligand molecules. Controlled-pore glass (CPG) is commercially available with mean pore sizes ranging from approximately 4 to 300 nm. The large surface of the inner pore space is freely accessible to all kinds of reagents and solvents, and, selecting the proper pore size, to peptides and proteins of all sizes. CPG is rigid and chemically inert with the exception of its lability above pH 8. The glass is available in form of beads of 200 – 400 mesh particle size, allowing them to be packed into small columns producing negligible back pressure.

The functional amino groups required for covalent coupling of peptides are introduced by reaction with a suitable silane. In routine sequencing we use aminopropyl glass (APG) which is prepared easily and reproducibly following the protocol outlined in Table 1. The obtained capacity is approx. 300 nmol mg^{-1}, 200 nmol mg^{-1} and 100 nmol mg^{-1} for CPG of 7.5 nm, 17 nm and 55 nm mean pore size, respectively, thus offering a high excess of functional groups over the nanomole and subnanomole amounts of peptide to be coupled in microsequencing experiments. APG of comparable quality is commercially available (e.g., Fluka, Pierce).

CPG of 7.5 nm pore size is used for coupling of fragment peptides and small proteins with a Mr below 25,000. Larger pore sizes (24 nm) have to be selected for efficient coupling of larger peptides or proteins loaded with detergent (cf. Fig. 1).

Diisothiocyanate-activated aminopropyl glass (DITC-glass) is prepared from APG by reaction with p-phenylene diisothiocyanate (DITC) following the protocol of Table 2. Not more than approximately 20% of the aminopropyl groups is activated, even when a high excess of DITC is used. Obviously the majority of aminopropyl groups is crosslinked via the bifunctional isothiocyanate, giving rise to interfering background peaks when using these supports in microsequencing. For this reason we activate APG with a relatively low molar excess of DITC.

Expecting the immobilized isothiocyanato groups to be sensitive to the same side reactions as PITC used in the degradation (e.g, oxidative desulfurization), fresh batches of DITC glass should be frequently prepared from APG. DITC glass that has been stored for weeks in the freezer should be washed with an-

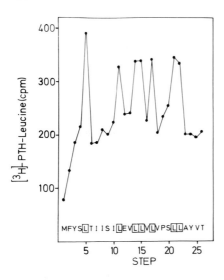

Fig. 1. Partial radiosequencing of the immunopre-cipitated URF-1 protein from *Neurospora crassa* labeled with [³H] leucine [8]. The whole immunopre-cipitate was coupled to 100 mg of DITC glass of 25 nm mean pore size in phosphate buffer containing 1.75% SDS for 12 h at 25 °C. The coupling yield of the labeled protein was 85%. After deformylation (1.5 M HCl for 30 min at 25 °C), the immobilized protein was degraded for 26 steps. Radioactivity was determined in the PTH-leucine fraction separated by HPLC. Seven positions with [³H] leucine were found perfectly matching the positions predicted from the DNA-sequence of the unidentified reading frame (boxes)

Table 2. Preparation of DITC glass

1. Perform the reaction in a flask under a slight stream of nitrogen.
2. Use p-phenylene diisothiocyanate (Fluka) freshly recrystallized from acetone (Merck, p.a.).
3. Prepare a solution of 1.8 g of DITC in 30 ml of anhydrous DMF (Fluka, spectral grade).
4. Suspend, under shaking, 6.0 g of APG and add 0.18 ml triethylamine (Fluka, puriss. p.a.).
5. React 2 h at 45 °C.
6. Wash several times with DMF (Fluka, puriss, p.a.), toluene (Merck, p.a.) and methanol (Merck, p.a.) on a fritted glass filter under a stream of nitrogen.
7. Dry in a desiccator and store desiccated in a freezer.

hydrous dimethylformamide (DMF) immediately before coupling to remove released isothiocyanato groups that will compete with the immobilized groups for the peptide.

3 Reactions for Covalent Coupling of Peptides and Proteins to Sequencing Supports

3.1 Lysine Coupling

The most effective way to bind proteins and large peptides covalently to a sequencing support is lysine coupling to DITC glass (Table 3). Provided the peptide or protein is soluble, coupling yields are usually better than 80%. The coupling reaction is insensitive to high concentrations of salts, detergents, denaturants and even to the dyes commonly used for staining of electrophoretically separated protein bands. However, ammonia as well as primary and secondary amines, competes with the peptide for the immobilized isothiocyanato groups

Table 3. Aqueous coupling of proteins to DITC glass

---

1. Dissolve the protein (0.1 – 5 nmol) in 0.2 – 0.8 ml of 50 mM sodium bicarbonate pH 8.0. If not soluble, add SDS (1 – 2%) or guanidine hydrochloride (1 – 6 M). Readjust with solid bicarbonate to pH 8 – 9 if necessary.
2. Add 10 – 100 mg DITC-glass of 7.5 nm pore size. Use 24 nm pore size or larger for proteins in SDS solution.
3. React 2 – 4 h, mildly shaking at 40 °C.
4. Add 0.02 ml of n-propylamine (Fluka, puriss. p.a.) and react for another 30 min if blocking of excess isothiocyanato groups is needed (not necessary if n-propylamine is used in the sequencer).
5. Centrifuge and save supernatant for amino acid analysis or recovery of nonbound peptide.
6. After resuspension in bicarbonate buffer transfer glass beads into a small funnel with a fritted glass filter.
7. Wash 5 times with bicarbonate and 5 times with methanol (containing traces of triethylamine) applying mild suction by a water aspirator.
8. Stopper the funnel and dry the support in vacuo on the glass filter.
9. Use loaden support immediately for sequencing or store it desiccated in a freezer.

---

and can diminish the coupling yields drastically. It is often possible to reduce the content of reactive amines sufficiently by repeated lyophilization or evaporation of the sample in the presence of 1% triethylamine. Otherwise the sample has to be subjected to dialysis, gel filtration or reverse-phase chromatography prior to coupling.

Coupling to DITC glass in the presence of SDS is well suited for microsequencing of proteins eluted from stained SDS gels and seems to be an alternative approach to electroblotting of proteins onto glass fiber sheets inserted into the cartridge of a gas-phase sequenator [18]. After elution from gels or after column chromatography, the proteins can be extracted from several ml of rather dilute solutions by recycling them through the reaction column filled with DITC glass using a peristaltic pump or a similar device (cf. Salnikow, this vol.). On the other hand, the binding capacity of the solid support can be made high enough to accommodate large amounts of nonspecific protein not interfering with sequencing of trace quantities of a radiolabeled protein. Figure 1 shows an N-terminal radiosequencing experiment with femtomole quantities of a protein (371 residues long) which was isolated by immunoprecipitation and coupled to DITC glass in the presence of a high excess of immunoglobulin (approx. 100 nmol) that would have far exceeded the capacity of a gas-phase sequenator glass fiber filter.

Once covalently bound to the solid support, the proteins can be washed free from all contaminants and effectively sequenced over many steps. Mean repetitive yields of 94 – 96% and an averaged overlap below 2% per step are frequently obtained in these degradations (Fig. 2). Disadvantages of DITC coupling are the more or less complete loss of the N-terminal residue (which is cleaved forming a thiazolinone but remains bound to the support) and the presence of gaps in the sequence at the positions of side-chain-attached lysyl residues. However, complete binding of individual lysyl side chains occurs rarely when native proteins are coupled to DITC glass. This facilitates positive identification of lysines (cf. Fig. 2), but may also be responsible for low coupling yields that can be significantly increased by the addition of SDS or guanidine hydrochloride.

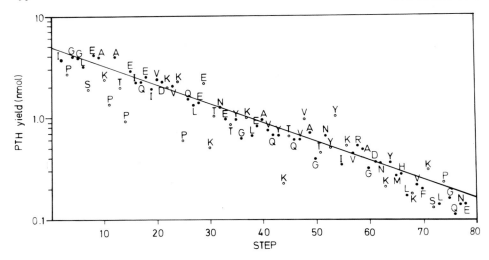

Fig. 2. Solid-phase Edman degradation of stefin, a small protein inhibitor of cysteine proteinases [10]. Starting from 11.5 nmol (100%) of protein, 9.7 nmol (84%) were coupled to 100 mg of DITC glass (7.5 nm) and degraded with an initial yield of 4.9 nmol (43%). A continuous sequence of 78 residues was identified, with the exception of the N-terminal residue, but including all lysines (see text for explanation). The yields determined from the peak areas were corrected for background by subtraction of the averaged yield of several preceding steps. The slope of the regression line calculated from the data represented by *solid circles* corresponds to a mean repetitive yield of 95.8%

Table 4. Nonaqueous coupling of peptides to DITC glass

1. Perform TFA-pretreatment according to Table 6.
2. Dissolve the peptide immediately in 0.2 – 0.5 ml of water-free DMF (Fluka, spectral grade).
3. Suspend 10 – 100 mg of DITC glass and add 0.05 ml of triethylamine.
4. React 1 – 2 h at 40 °C in a thermostatted shaker.
5. React with n-propylamine as in Table 3 if needed.
6. Wash 5 times with DMF (Fluka, puriss. p.a.) and 5 times with methanol (Merck, p.a., containing traces of triethylamine) on the fritted glass filter.
7. Dry in vacuo and store as in Table 3.

Small and medium-sized lysine-containing peptides that have been isolated by reverse-phase HPLC are frequently better soluble in nonaqueous systems and should be coupled to DITC glass following the protocol of Table 4.

3.2 Carboxyl Coupling

The only generally applicable method for covalent attachment of peptides to aminated sequencing supports is carboxyl coupling after activation with a carbodiimide. Due to several side reactions (see [17] for review), carboxyl coupling is never complete but, following the protocol of Table 5, coupling yields greater than 20%, typically around 50%, are usually obtained. It has been confirmed by

Table 5. Carboxyl coupling

1. Perform TFA-pretreatment as in Table 6.
2. Dissolve dry peptide in 0.2 ml of anhydrous DMF (Fluka, spectral grade).
3. Suspend 10 – 50 mg of APG (7.5 nm pore size).
4. Add a freshly prepared solution containing 5 mg of 1-ethyl-3-dimethylaminopropylcarbodiimide (EDC, Fluka) and 2 mg of 1-hydroxybenzotriazole (HBTA, Fluka) in 0.3 ml of anhydrous DMF.
5. React 1 – 2 h at 40 °C in a thermostatted shaker.
6. Proceed as in Table 4, steps 6 and 7.

Table 6. TFA-pretreatment for nonaqueous coupling

1. Dissolve dry peptide in 0.2 ml of anhydrous TFA (Fluka, for sequence analysis).
2. Incubate 30 min at room temperature.
3. Dry in vacuo over KOH pelletes first using a water aspirator until the sample appears to be dry, then using an oil pump for 10 min.
4. Open the desiccator immediately before starting the coupling reaction.

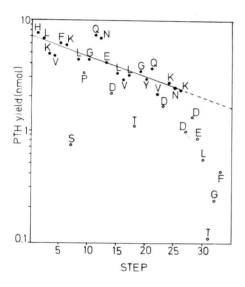

Fig. 3. Solid-phase Edman degradation of the cyanogen bromide fragment 66 – 98 of stefin. Starting from 22.6 nmol (100%) of peptide, 9.9 nmol (44%) were coupled to APG using the nonaqueous carbodiimide procedure (cf. Table 5). The immobilized peptide was degraded with an initial yield of 7 nmol (31%) and a mean repetitive yield of 95.9% to the C-terminal residue. All residues were identified including the side-chain attached dicarboxylic amino acids. However, a pronounced drop of yield is observed when the degradation passes the side-chain-anchored Asp-27, Asp-28 and Glu-29 residues

other investigators that the coupling yields are higher in anhydrous systems [16] than in the various aqueous systems that have been preferably used in the past ([19], [20], and Salnikow, this vol.). Surprisingly, most small to medium-sized peptides and even small proteins are soluble in anhydrous DMF after preincubation with anhydrous trifluoroacetic acid (Table 6). We have found only few exceptions, mainly very polar peptides containing several cysteic acid residues and larger hydrophilic proteins.

Incomplete reaction of the individual carboxyl groups seems to be responsible for the fact that total attachment of aspartyl and glutamyl side chains rarely occurs. In our experience at least 20% of each of these residues remains free, thus permitting its unequivocal identification even at later steps of the degradation. Frequently those side-chain carboxyls located in the N-terminal or C-terminal parts of the peptide are preferentially coupled to the support. This may lead to appreciable losses of sequenceable peptide when the degradation passes the side-chain anchor points. Eventually only that portion of the peptide is being sequenced which is bound via its C-terminal carboxyl group (Fig. 3).

3.3 Lactone Coupling

Coupling of cyanogen bromide fragments via their C-terminal homoserine lactone residue (Table 7) can be considered to be the ideal way to immobilize a peptide for sequencing. Provided the peptide is soluble in anhydrous or almost anhydrous media, high coupling yields (usually better than 80%) are common and the peptide can be degraded to the C-terminal homoserine residue without interference by side-chain attachment. For this reason all strategies for complete sequencing of proteins are based on cyanogen bromide fragments whenever methionines are present. Extended degradations of homoserine-coupled peptides are limited, however, by the increased lability of the homoserine amide bond lowering the average repetitive yield by 2 – 3%. Problems arise not so much from the lower yield, which can be compensated by higher sensitivity of detection, but rather from the decreasing signal-to-background ratio when sequencing cyanogen bromide fragments of more than 50 residues. Figure 4 documents a typical degradation of a 60-residue cyanogen bromide fragment which has been sequenced completely with a mean repetitive yield of 92.8%. As has been proposed earlier, the C-terminal sequence of a long cyanogen bromide fragment can be determined from a subfragment obtained by enzymatic cleavage of the original peptide followed by specific lactone coupling without prior separation [17]. Regarding the non-PTH background, degradations of homoserine-bound peptides are significantly "cleaner" than comparable degradations of peptides coupled to DITC glass.

3.4 Mixed Coupling

The drawbacks of the individual coupling procedures may become critical when the complete sequence of a protein has to be determined from a limited amount of material. To reduce the risk of total failure with unknown peptides, we frequently use a mixed-coupling approach. Two thirds of the material are carboxyl-coupled to APG, one third is attached to DITC glass. After performing

Table 7. Lactone coupling

---

1. Perform TFA-pretreatment as in Table 6.
2. Dissolve dry peptide immediately in 0.2 – 0.5 ml of anhydrous DMF.
3. Suspend 10 – 50 mg of APG.
4. Add 0.05 ml of triethylamine.
5. React 1 – 2 h at 40 °C in a thermostatted shaker.
6. Proceed as in Table 4, steps 6 and 7.

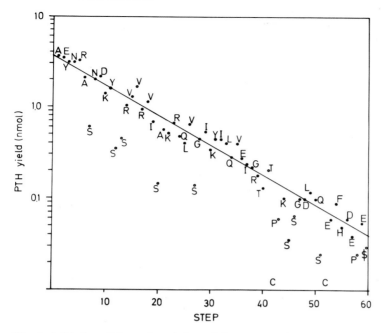

Fig. 4. Solid-phase Edman degradation of the cyanogen bromide fragment 22–81 from chicken cystatin, a small protein inhibitor of cysteine proteinases [11]. Starting from 10 nmol (100%) of the S-carboxymethylated peptide, 6.9 nmol (69%) was coupled to 100 mg APG via its C-terminal homoserine lactone. The attached peptide was degraded with an initial yield of 3.6 nmol (36%) and a mean repetitive yield of 92.8% for 60 steps. All residues were determined including the C-terminal homoserine lactone. The two cysteine residues were identified by radioactivity of their S-[^{14}C]carboxymethyl derivatives

both coupling procedures separately, the loaded glass beads are combined in the reaction column. In the case of an unusually low yield of carboxyl coupling, more extended sequences are obtained due to lysyl anchors in the C-terminal part of the peptide. On the other hand, carboxyl coupling permits identification of the N-terminal residue and the internal lysyl residues (Fig. 5).

4 The Solid-Phase Sequencer

In principle the construction of an automated solid-phase sequencer is simple and does not require much sophisticated technology. Our noncommercial sequencer has been described in detail ([7], [21]). The central part of the sequencer is the thermostated reactor holding the covalently immobilized peptide or protein. Several types of reactor have been proposed, revealing a trend toward miniaturization when used for microsequencing in the subnanomole range [22]. In our experience, a packed column provides the most efficient exchange of reagents and solvents with the inner pore space of the glass support. We use columns made

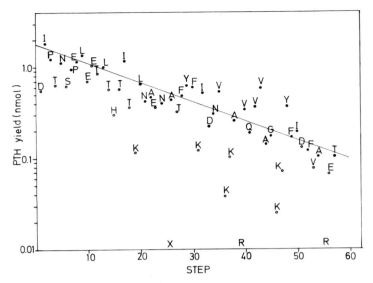

Fig. 5. Solid-phase Edman degradation of a 13-kDa fragment of human low-Mr kininogen which is an inhibitor of cysteine proteinases [14]. A total of approx. 5 nmol was immobilized separately on 50 mg of DITC glass (1/3) using aqueous coupling and on 50 mg of APG (2/3) using the nonaqueous carbodiimide procedure. A partial N-terminal sequence of 57 residues was determined (mean repetitive yield: 94.5%). Due to mixed coupling, the N-terminal Asp residue and all internal lysyl residues were identified. The arginyl residues (*R*) were identified but not quantitated, X denotes a glycosylated Asn residue

from glass tubes with ground surfaces on both sides which are fitted between a Teflon disc on the top and a Teflon disc containing a stainless steel frit plus an adapter of variable length at the bottom. Glass tubes from 0.5 – 2 mm internal diameter and 5 – 700 mm length can adapted. The columns are filled with loaded support (5 – 100 mg) via a small funnel screwed into the column holder instead of the top tube fitting. Fifty to 100 mg of support were used in most experiments.

Considerably reduced overlap (0.5 – 2% per step, depending on the sequence of the peptide) has been obtained when the reagents and solvents are pressurized to 0.5 – 1 bar within the reaction column, thus preventing the formation of gas bubbles and channeling. This approach is not feasible with the usual low-pressure rated solvent delivery systems used in gas-phase or liquid-phase sequenators. In the Munich solid-phase sequencer the reagents and solvents are delivered by syringe pumps [23]. The back pressure to flowing solvents is maintained by a Nupro valve or a coil of capillary tubing in the waste line which can be bypassed via a slider valve whenever a high flow of nitrogen is needed to dry the reaction column. Reagents standing in the reaction column are pressurized with nitrogen against a closed valve after the column. In the course of the degradation cycle (Table 8), the reagents are applied in the flowing mode long enough to wet the support evenly and then allowed to react in the standing mode. A flush of nitrogen is used to deliver the TFA from the column to the fraction collector at the end of the first cleavage step and to dry the reaction column before and after cleavage.

Table 8. Degradation cycle (64 min)

Step	Duration (min)	Functions	
		Degradation unit	Sample processing unit
1	0.5	Rest	Rest
2	5.0	S1 delivery	Sample drying
3	1.0	S1 + PITC delivery	Pressure relief
4	1.5	S1 + PITC delivery	HCl/MeOH delivery
5	0.2	S1 + PITC delivery	Clear sample line
6	6.0	S1 + PITC delivery	Conversion
7	15.0	S1 + PITC standing	HCl/MeOH drying
8	0.5	Pressure relief	Fraction collector advance/start injection
9	15.0	S2 delivery	
10	0.1	S2 delivery	Collect S2
11	3.5	Dry column via bypass to waste	
12	1.0	Pressure relief	Collect
13	1.0	TFA delivery	Collect
14	0.2	TFA standing	Clear sample line
15	4.0	TFA standing	
16	1.0	Purge column	Collect
17	0.5	Pressure relief	Collect
18	1.0	TFA delivery	Sample drying
19	4.0	TFA standing	Sample drying
20	3.0	Dry column via bypass to waste	Sample drying

S1: 5 mM n-propylamine (Fluka, puriss. p.a.) in methanol (Merck, p.a.), 0.6 ml min^{-1}.
S2: methanol (Merck, p.a.), 0.8 ml min^{-1}.
PITC: phenylisothiocyanate (Fluka, puriss. p.a.), 0.03 ml min^{-1}.
TFA: trifluoroacetic acid (Fluka, for sequence analysis), 0.6 ml min^{-1}.
HCl/MeOH: 1.5 − 2.0 M HCl in methanol (Merck, p.a., dried) prepared from acetyl chloride (Fluka), 0.6 ml min^{-1}.
Temperatures: reaction column 50 °C, sample drying ambient, conversion 60 °C, HCl/MeOH drying 60 °C.

5 Reagents and Solvents for Solid-Phase Edman Degradation

Microsequencing in the subnanomole range requires highly purified reagents and solvents. Trace impurities, most of which are chemically not defined, have to be removed if they interfere with the Edman reactions or with subsequent high-sensitivity HPLC detection. Additionally, insoluble scavengers (e.g., aminopropyl glass) have been used to neutralize aminoreactive compounds that may be generated within scrupulously purified reagents during the course of one sequencer experiment [24, 25].

In solid-phase sequencing, quite reasonable repetitive yields of 92 − 94% have been obtained using normal analytical grade reagents and omitting any precautions to exclude oxygen. The high excess of immobilized phenylthiocarbamyl groups on the solid support probably scavenges oxygen and contaminants that would otherwise react with the phenylthiocarbamyl peptide. However, when using highly purified reagent and solvents (according to the standards of liquid-phase sequencing), we were not able to raise the mean repetitive yields above approximately 95%. This could be explained as the effect of amino-reactive intermediates being formed on the porous glass surface after each TFA cleavage step (perhaps mixed anhydrides between silic acid and TFA) that block part of the peptide amino groups as they become deprotonated during the alkaline conditions of the subsequent coupling step. It was indeed observed that repetitive yields increased by 1 – 2% when a primary amine was delivered after the cleavage reaction and before the next coupling step [26].

Surprisingly high repetitive yields were obtained in radiosequencing of femto-moles of a labeled protein in the presence of nanomole amounts of a "carrier" protein (cf. Fig. 1). This observation stimulated us to introduce a scavenger for amino-reactive intermediates into the coupling reaction itself [27]. Five mM n-propylamine in methanol is used instead of the usual coupling buffer. The primary amine acts both as a base effecting deprotonation of the peptide amino group and as a scavenger for amino-reactive intermediates on the solid support and within the coupling medium. The amine is present in a 50 – 1000-fold molar excess over the peptide amino group and both are reacted with a 40-fold excess of PITC. A high amount of phenylthiocarbamyl-propylamine is formed, most of which is removed during the subsequent methanol wash. A small portion of this reaction product, however, is obviously adsorbed to the polyethylene tubes and/or to the porous glass support and elutes as a separated peak before PTH-methionine in the PTH chromatogram. Several minor reaction products of n-propylamine interfere with PTH identification in the subnanomole range when an isocratic system is used, but can be easily separated with a suitable gradient (Fig. 6). Preliminary results indicate that a much lower excess of n-propylamine and PITC is equally effective in microsequencing of subnanomole amounts of peptides then leading to significantly reduced amounts of reaction products.

n-Propylamine raises the repetitive yields by 2 – 3% when compared to triethylamine under otherwise identical conditions. As discussed elsewhere [27], the mean repetitive yield of an extended degradation depends on various sequence-specific factors, such as length of the peptide, distribution of anchor residues, and occurrence of residues which are sensitive to so-called nonspecific cleavage of peptide bonds. The oxidized insulin B chain coupled to DITC glass via its single Lys-29 residue can serve, in good approximation, as a model for the determination of the true "chemical" degradation yield which has been found to be close to 97% (Fig. 7).

The initial sequencing yield, i.e., the percentage of coupled peptide recovered as a PTH-amino acid in the first degradation step, is quite variable in solid-phase sequencing. As in all versions of Edman degradation, the initial yield is naturally dependent on the "history" of the sample prior to sequencing. In our degradation cycle, a certain portion of the released ATZ-amino acid is not utilized for identification, as only the first delivery of TFA is collected being simply purged out of the reaction column by nitrogen without any additional washing of the support (cf. Table 8). Taking into account these factors, the data presented in the legends to Figs. 2 – 4 indicate that quite reasonable initial yields can be obtained

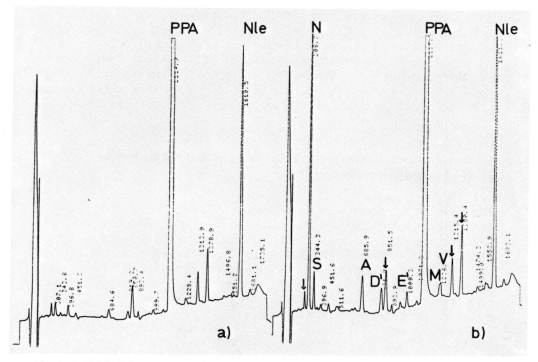

Fig. 6a, b. Original chromatograms from HPLC operating on-line to the sequencer. **a** Blank cycle, 5 steps after complete degradation of a peptide lactone-coupled to APG. 1.0 nmol of PTH-norleucine (Nle) was added to each reaction vessel as an internal standard which is subjected to all steps of automated conversion and on-line injection (1/2 of the sample). The large peak is the reaction product of PITC with n-propylamine, phenylthiocarbamylpropylamine (PPA), the smaller peaks represent the thioureas and minor reaction products of the scavenger. **b** Step 25 of the degradation of a cyanogen bromide fragment of human cystatin C (27 residues long) lactone-coupled to APG. The main signal, PTH-Asn (N), corresponds to 0.37 nmol, the background and overlap signals of PTH-Ser (S), PTH-Ala (A), PTH-Asp methyl ester (D'), and PTH-Glu methyl ester (E') to 50, 78, 50, and 28 pmol, respectively. Non-PTH peaks identical to those appearing in the blank cycle are marked by *arrows*

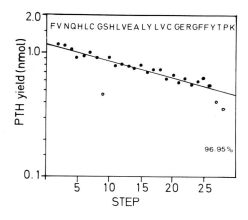

Fig. 7. Solid-phase Edman degradation of oxidized insulin B-chain. Two nmol were reacted with 100 mg of DITC-glass under nonaqueous conditions (cf. Table 4). The immobilized peptide was degraded with an initial yield of 1.2 nmol (60%). The N-terminal residue (F) was completely lost by attachment to the solid support. Yields were determined from the peak areas in the chromatogram without subtraction of back-ground which was negligible in this degradation. The regression line corresponds to an average repetitive yield of 97%

in the Munich sequencer. Most sensitive to blocking reactions prior to sequencing are peptides coupled to DITC glass if coupling is not performed immediately before degradation. Therefore initial yields may be significantly increased by on-column immobilization (cf. Salnikow, this vol.). DITC-coupled peptides are susceptible to premature cleavage leading to "preview" in the degradation. Even washing of the support with analytical-grade methanol effects partial cleavage which can be avoided by the addition of traces of triethylamine to the methanol.

6 Conversion of Amino Acid Thiazolinones and Identification of the Phenylthiohydantoins

The main body of progress in all versions of automated Edman degradation is due to the development of HPLC systems allowing a rapid, sensitive, and quantitative determination of all PTH-amino acids released in the sequencer. The amino acid anilinothiazolinones are instable and have to be converted to the more stable phenylthiohydantoins. When working with subnanomole amounts, the conversion reaction should be performed immediately after each degradation cycle in order to avoid severe losses and the generation of multiple degradation products. Several devices for automated conversion have been used in connection with a solid-phase sequencer (see [21] for review). Most sensitive to destruction are ATZ-Ser and ATZ-Thr which form partially polymerizing dehydro-derivatives mainly in the presence of anhydrous TFA [28]. We have found that reasonable yields of undestroyed PTH-Ser and PTH-Thr can be recovered when the TFA is rapidly evaporated at room temperature after addition of approximately 10% of methanol. This is done automatically in the Munich sequencer (step 10 in Table 8), followed by automated conversion using 1.5 to 2 M methanolic HCl for 8 – 10 min at 60 °C. Decreasing molarity of the HCl solution is compensated by prolonging the conversion time until the acid is renewed when lower than 1 M. The amount of the intermediate phenylthiocarbamyl-glycine is an indicator for the completeness of the conversion reaction. Yields of PTH-Ser and PTH-Thr are usually better than 40% of that of PTH-Leu. PTH-Asp and PTH-Glu are identified in form of their methyl esters (cf. Fig. 6).

An alternative approach frequently used is to perform the conversion under conditions yielding mainly the dehydroderivatives which can be detected with comparable sensitivity by monitoring the HPLC chromatogram at 313 nm [29].

Even after automated conversion, the resulting PTH amino acids are prone to rapid destruction when stored in a fraction collector in subnanomole amounts. An essential step toward extreme microsequencing was the introduction of devices for automated injection of the PTH-amino acids into a HPLC operating on-line to the sequencer. The Munich solid-phase sequencer was the first instrument using on-line injection [7] and, with slight modifications [21], the system has been working permanently for more than 5 years. In the meantime several alternative on-line systems have been constructed in connection with liquid-phase, gas-phase and solid-phase sequencers (e.g., [29]), some of which are commercially available. Using an internal standard, reliable quantitative

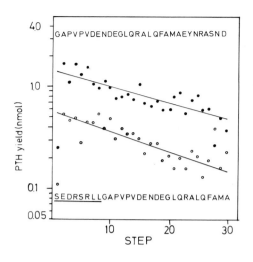

Fig. 8. Solid-phase Edman degradation of a mixture of two forms of chicken cystatin [11] differing by an N-terminal extension of 8 residues (*underlined*). The peptides were coupled to DITC glass as an unresolved mixture. Two sets of data clearly corresponding to the sequences indicated on the top and on the bottom of the panel were obtained. Unequivocal alignment was possible because the sequence of the pure major component (*upper sequence, solid circles*) was known from a separate experiment

results can be obtained that improve the interpretation of extended degradations [30] and even allow the analysis of simple mixtures of sequences (Fig. 8).

A great variety of systems have been proposed and successfully used for reverse-phase HPLC separation of PTH-amino acids. As highly efficient columns have become commercially available there is a strong trend toward isocratic systems which provide a stable baseline, reproducible retention times and the convenience to recycle the elution buffer for more than one week (cf. Ashman, this vol.). Our attempts to adapt isocratic systems to our solid-phase sequencer were not fully satisfying because the minor reaction products resulting from the n-propylamine scavenger were not resolved from the PTH-amino acids, and the separation of undestroyed PTH-Ser and PTH-Thr in the first part of the chromatogram was incomplete. For these reasons we are still using a gradient system having the advantage of equal peak shape and sensitivity over the whole chromatogram. In its present form, the limit of quantitative determination is around 5 pmol. All non-PTH background peaks are separated from the PTH-amino acids (cf. Fig. 6).

7 Concluding Remarks

Our results indicate that a solid-phase sequencer with a packed reaction column, automated conversion and on-line injection is well suited for microsequencing of peptides and proteins in the 5 – 0.1 nmol range. Increasing the sensitivity to the 100 – 10 pmol range seems feasible with some minor modifications. We have been able to sequence all fragments needed for complete sequence analysis of various unknown proteins. The same performance may be not obtainable with some of the few commercially available solid-phase sequencers in other laboratories. However, using a commercial sequencer, L'Italien and Laursen were able to sequence peptides that had been isolated from 1 nmol of a tryptic digest [31], and another commercially available instrument was successfully modified for microsequencing with DABITC ([22], and Salnikow, this vol.). An inexpensive noncommercial solid-phase sequencer working in the nanomole range has been assembled recently from common medical and chromatography equipment parts [32].

When comparing the performance of the solid-phase sequencer with that of the gas-phase sequenator, reports of extreme microsequencing in the low picomole range are rare for the former instrument. However, the results of radio-sequencing experiments (cf. Fig. 1) suggest that microsequencing of proteins covalently bound to DITC glass might become a technique equally effective as direct polybrene-mediated degradation in the gas-phase sequenator. Coupling yields of subnanomole amounts of lysine-containing proteins to DITC glass are essentially quantitative. Once covalently bound, the proteins can be degraded with high repetitive yield. Missing the first residue and some internal lysyl residues (which remain bound to the support) seems not very important when N-terminal protein sequencing is envisaged for the design of oligonucleotide probes or identification of gene products.

Whenever complete protein sequencing is required, high repetitive yields favor the covalent solid-phase approach, preferably in all cases where long cyanogen bromide fragments can be obtained. As shown in Figs. 2 and 3, the complete amino acid sequence of a 98-residue protein was derived from two automated sequencer runs including the information needed to overlap the two fragments.

As Wittmann-Liebold has pointed out [20], none of the three technical versions of automated Edman degradation should be completely dismissed at the moment. It rather seems that all three approaches − even including the technically complicated spinning-cup version − may have advantages of their own that should be utilized for the solution of individual sequencing problems. The automated sequencer of the future will be a multi-purpose sequencer using common units for storage and delivery of reagents and solvents and for automated conversion and on-line injection, but differing in the kind of reactor (cup, cartridge, packed reaction column) designed as an exchangeable module. Within this modular sequencer, the covalent solid-phase version should offer the greatest flexibility for new approches to an improved chemistry of degradation or increased sensitivity using fluorescent or chemiluminescent reagents.

Acknowledgments. The skillful and enthusiastic work of Helmut Hofner, Anna Esterl, and Karin Wiedenmann is gratefully acknowledged. Parts of the reviewed investigations were supported by the Sonderforschungsbereich 207 der Universität München (grant C-2), by the Kernforschungsanlage Jülich, Internationales Büro, and by the Bundesministerium für Forschung und Technologie, GFR (grants PTB 8600 and PTB 8651).

References

1. Edman P (1949) Arch Biochem 22:475
2. Edman P, Begg G (1967) Eur J Biochem 1:80
3. Tarr GE, Beecher JF, Bell M, McKean DJ (1978) Anal Biochem 84:622
4. Hewick RM, Hunkapiller MW, Hood LE, Dreyer WJ (1981) J Biol Chem 256:7990
5. Laursen RA, Bonner AG (1970) Fed Proc 29:727

6. Laursen RA (1971) Eur J Biochem 20:89
7. Machleidt W, Hofner H (1980) In: Birr Ch (ed) Methods in peptide and protein sequence analysis. Elsevier/North Holland, Amsterdam, p 35
8. Burger G, Scriven C, Machleidt W, Werner S (1982) EMBO J 1:1385
9. Schägger H, von Jagow G, Borchart U, Machleidt W (1983) Hoppe-Seyler's Z Physiol Chem 364:307
10. Machleidt W, Borchardt U, Fritz H, Brzin J, Ritonja A, Turk V (1983) Hoppe-Seyler's Z Physiol Chem 364:1481
11. Turk V, Brzin J, Longer M, Ritonja A, Eropkin M, Borchart U, Machleidt W (1983) Hoppe-Seyler's Z Physiol Chem 364:1487
12. Ritonja A, Popovic T, Turk V, Wiedenmann K, Machleidt W (1985) FEBS Lett 181:169
13. Borchardt U, Machleidt W, Schägger H, Link TA, von Jagow G (1985) FEBS Lett 191:125
14. Müller-Esterl W, Fritz H, Machleidt W, Ritonja A, Brzin J, Kotnik M, Turk V, Kellermann J, Lottspeich F (1985) FEBS Lett 182:310
15. Ritonja A, Machleidt W, Barrett A (1985) Biochem Biophys Res Commun 131:1187
16. Machleidt W, Wachter E (1977) Method Enzymol 47:263
17. Laursen RA, Machleidt W (1980) In: Glick D (ed) Methods of biochemical analysis, vol 26. Wiley Interscience, New York, p 201
18. Vandekerckhove N, Bauw G, Puype M, Van Damme J, Van Montagu M (1985) Eur J Biochem 152:9
19. Wittmann-Liebold B, Kimura M (1984) In: Walker JM (ed) Methods in molecular biology, vol 1, Proteins. Humana, Clifton, NJ, p 221
20. Wittmann-Liebold B (1983) In: Tschesche H (ed) Modern methods in protein chemistry. de Gruyter, Berlin New York, p 229
21. Machleidt W (1983) In: Tschesche H (ed) Modern methods in protein chemistry. de Gruyter, Berlin New York, p 267
22. Salnikow J, Lehmann A, Wittmann-Liebold B (1981) Anal Biochem 117:433
23. Machleidt W, Hofner H, Wachter E (1975) In: Laursen RA (ed) Solid-Phase methods in protein sequence analysis. Pierce Chemical Comp, Rockford, IL, p 17
24. Frank G (1982) In: Elzinga M (ed) Methods in protein sequence analysis. Humana, Clifton, NJ, p 91
25. Bhown AS, Mole JE, Bennet JC (1981) Anal Biochem 110:355
26. Horn MJ, Bonner AG (1982) In: Elzinga M (ed) Methods in protein sequence analysis. Humana, Clifton, NJ, p 159
27. Machleidt W, Hofner H (1982) In: Elzinga M (ed) Methods in protein sequence analysis. Humana, Clifton, NJ, p 173
28. Tarr GE (1977) Methods Enzymol 47:335
29. Wittmann-Liebold B, Ashman K (1985) In: Tschesche H (ed) Modern methods in protein chemistry, vol 2. de Gruyter, Berlin New York, p 303
30. Machleidt W, Hofner H (1981) In: Lottspeich F, Henschen A, Hupe K-P (eds) High performance liquid chromatography in protein and peptide chemistry. de Gruyter, Berlin New York, p 245
31. L'Italien JJ, Laursen RA (1982) In: Elzinga M (ed) Methods in protein sequence analysis. Humana, Clifton, NJ, p 383
32. Amiri I, Neimark J, Waltz A, Lentz J, Reinbolt J, Boulanger Y (1985) J Biochem Biophys Meth 10:329

2.4 Automated Solid-Phase Microsequencing Using DABITC, On-Column Immobilization of Proteins

Johann Salnikow[1]

Contents

1 Introduction

The iterative nature of the stepwise Edman degradation of proteins early lead to the development of automated machines of the spinning-cup type; proteins to be analyzed are contained in the reaction chamber exclusively by physical forces. One major problem encountered soon, however, is loss of sample material by extraction specifically when shorter peptides are sequenced. Although the use of polybrene as a physical "glue" proved a valuable remedy permitting the construction of the modern gas-phase instruments, the introduction of additional complex substances into the aggressive milieu of the Edman chemistry often creates new problems, in particular when the sample to be analyzed is scarce and high sensitivity is desired.

The application of the solid-phase technology to the Edman degradation involving covalent immobilization of peptides, and its automation as conceived originally by Laursen and Bonner [1, 2] constitutes nowadays one of the major techniques for peptide sequence determination. The main advantages of this methodology are prevention of extractive peptide losses and less complex instrumentation requirements compared to commercial liquid-phase and gas-phase sequencers. In fact, automated solid-phase sequencers with nanomole sensitivity detection can be assembled from common chromatography equipment parts [3].

---

1 Institut für Biochemie und Molekukare Biologie, Technische Universität Berlin, Franklinstraße 29, D-1000 Berlin 10

Advanced Methods in Protein Microsequence Analysis
Ed. by B. Wittmann-Liebold et al.
© Springer-Verlag Berlin Heidelberg 1986

Adaptation of the successful double coupling method [4] using 4-N,N-di-methylaminoazobenzene 4'-isothiocyanate (DABITC) and phenyl isothiocyanate (PITC) to automated solid-phase sequencing [5, 6] permits reliable sequence determinations in the low nanomole and subnanomole range; the dual identification methodology – two-dimensional micro thin-layer chromatography [4] as well as high-performance liquid chromatography [7, 8] – provides, in particular, high confidence paired with satisfactory sensitivity.

A major drawback of the solid-phase technology concerns the specificity and yield of covalent peptide and protein attachment.

Since the amino terminus has to be unmodified and reactive toward the Edman reagents, useful handles for the chemical immobilization of polypeptides are very limited; in fact, only three methods are of general practical use:

1. *Carboxyl Coupling.* Peptide attachment to aminopropyl resin via the terminal (and glutamic/aspartic side chain) carboxyl after carbodiimide activation (peptides, some proteins).
2. *DITC Coupling.* Peptide attachment to p-phenylene diisothiocyanate (DITC)-activated amino resin via the amino terminus and lysine-derived side chains (lysine peptides and proteins).
3. *Lactone Coupling.* Attachment of peptides possessing carboxyl terminal lactones to amino resins (methionine peptides after cyanogen bromide cleavage; tryptophan peptides after chemical fragmentation [9, 10]).

Whereas the first method is generally applicable with reasonable coupling yields (except for peptides with carboxyl terminal lysine), the other two techniques require the occurrence of specific residues in certain positions (→tryptic lysine peptides; methionine and tryptophan-containing proteins yield respective fragments). The amino attachment to DITC resin allows the Edman degradation to proceed after cleavage of the first residue, with the polypeptide remaining anchored by lysine side chains. This method is in particular effective for the immobilization of proteins.

A further limitation in solid-phase sequencing using DABITC is the choice of the resin. Since DABITC absorbs very strongly to amino polystyrene yielding high background, only silica-based supports can be employed. A suitable support with high capacity and excellent flow rates is amino-propylated controlled-pore glass (APG) and the derivative thereof modified with p-phenylene diisothio-cyanate (DITC-glass) [11].

2 Materials and Methods

2.1 Chemicals and Solutions

APG was prepared by reaction of controlled-pore glass CPG 10/75, 200 – 400 mesh (Serva, Heidelberg, FRG) with 3-aminoethyl triethoxysilane (Pierce, distributed in the FRG by Karl OHG, Geisenheim). APG with 75 Å pore diameter can also be obtained from Pierce directly (CPG/3-Aminopropyl, Cat. No. 23538). EDC was purchased from Serva (Heidelberg, FRG) and DITC from Pierce (Cat. No. 26942) or Fluka (Neu-Ulm, FRG; Cat. No. 78480), the latter product was recrystallized from acetone. DMF of analytical grade was purified by the following procedure: a mixture of DMF, benzene, and water (263 : 34 : 12 v/v/v) was distilled and the fraction, boiling below 150 °C, discarded. The translucent condensating bulk amount was redistilled over P_2O_5 under reduced pressure and stored over 3-Å molecular sieve pellets (Merck, Darmstadt, FRG). Pyridine was distilled twice over KOH pellets and ninhydrin and N-ethylmorpholine was redistilled over $NaBH_4$ and ninhydrin. Hexane and benzene were of analytical grade.

Sequencing Reagents. Methanol of analytical grade was from Merck, as an aldehyde scavenger n-propylamine (0.2% v/v) was added as suggested by Machleidt and Hofner [12]. DABITC was purchased either from Pierce or Fluka and recrystallized from acetone only, if required; a 0.5% solution in redistilled DMF (v/v) was used which was prepared fresh every 24 h. PITC was from Merck and redistilled in vacuo and stored in ampules sealed under nitrogen. Prior to use, PITC was filtered through a small Al_2O_3-column under nitrogen (aluminium oxide, neutral, from Merck, Cat. No. 1077) in order to remove yellowish contaminants. A 5% PITC solution in redistilled DMF (v/v) was used. The buffer was prepared by mixing 20 parts of 0.4 M N-ethyl morpholine titrated with TFA to pH 9.0 with 40 parts redistilled pyridine and 40 parts redistilled DMF. To 100 ml of the buffer 10 µl n-propylamine were added [12]. TFA was distilled over freshly dehydrated $CaSO_4 - 1/2\ H_2O$.

The buffer for on-column immobilization of proteins consisted of 0.1 M $NaHCO_3$ containing 1% sodium dodecyl sulfate (w/v) and 10% isopropanol (v/v).

2.2 Peptide Attachment to Amino Propyl Glass (APG) by Carboxyl Activation with 1-Ethyl-3-Dimethylaminopropyl Carbodiimide (EDC)

The most useful general attachment method for smaller peptides employs carbodiimide activation according to Wittmann-Liebold and Lehmann [13], as pointed out by Laursen et al. [14]. The procedure often results in lower yields, but it permits identification of all amino terminal amino acids including aspartic and glutamic acid. The carboxyl side chains of the latter react only partially with the support, thus permitting the identification of these dicarboxylic amino acids. Peptides containing carboxyl terminal lysine show less satisfactory attachment yields [6]. Peptides are treated prior to the immobilizing reaction for solubilization with trifluoroacetic acid (TFA). This treatment serves, in addition, to protonate the amino terminus and to prevent peptide polymerization during carboxyl activation, since no reversible amino group blocking, as common in chemical peptide synthesis, is used. Peptides of small and large size can be immobilized with yields from 30 – 75% [6], even some smaller proteins can be coupled to APG provided they are soluble in dimethyl formamide (DMF); for longer pro-

Table 1. Carboxyl coupling of peptides to APG with EDC

---

1. Dry the peptide (50 nmol) in a small test tube (the peptide has to be free of *salt, acids* (formic or acetic acid) and sodium dodecyl sulfate.
2. Dissolve the peptide in 100 μl anhydrous TFA and incubate at room temperature for 15 min.
3. Dry in the rotary evaporator and thereafter in vacuo over KOH pellets for 25 to 30 min (prolonged drying that might render peptides insoluble has to be avoided).
4. Add 200 μl EDC-solution (2 mg/200 μl DMF, freshly prepared) and sonicate for 1 min.
5. Add 50 mg APG and deaerate the mixture shortly.
6. Incubate at 40 °C for 60 min with gentle stirring with aid of a magnetic microbar.
7. Remove the magnetic bar and wash the peptidyl-APG with 2×1 ml methanol.
8. Wash the peptidyl-APG with 1 ml ethyl ether, centrifuge and dry in vacuo over KOH pellets overnight.

---

teins, however, probably due to their insolubility in the reaction solvent in general, no good results have been obtained. Evaluation of coupling yields as a function of reaction time and carbodiimide concentration [6] yielded the attachment protocol as outlined in Table 1.

Comments. The amount of peptide can be reduced to 5 – 10 nM with a parallel reduction of APG (5 – 20 mg). A mean pore size of APG of 75 Å proved sufficient for peptides at different lengths, increasing the pore size to 200 Å for longer peptides and proteins showed no improvement. In fact, amino propyl glass support lacking inner pores as utilized in high performance liquid chromatography (LiChroprep NH_2, 25 – 40 μm, from Merck) can be used as well with satisfactory results (A. Lehmann, pers. comm.); this support is distinguished by the appearance of an unspecific by-product in the chromatography system used for identification.

Successful carboxyl coupling of a protein of 238 residues under slightly modified conditions has been reported recently [15].

2.3 Assembly and Filling of the Reaction Column

The reaction column consists of a 4-cm piece of Teflon tubing (1.5 mm I.D., 3 mm O.D.) which is flanged on both sides and equipped with the appropriate fittings to permit facile and tight joining to the Teflon lines of the sequencer with the help of suitable polypropylene connectors. For improved tightness at the connection interfaces and in order to minimize dead space at both ends of the reaction column, short flanged pieces of small-bore Teflon tubing are inserted. The microcolumn displays good flow characteristics and has moreover the advantage of an inexpensive one-way item. In order to secure the glass support in the column a porous Telfon disc (Zitex H 622-123, product of Chemplast, USA) is placed between the bottom connectors. A modified version of this microcolumn with a constricted bottom end has been described [16].

Table 2. Filling of the reaction column

1. Pour with the help of a micropipet tip as funnel a few inert glass beads into the reaction column to cover the filter disc at the bottom.
2. Then pour the peptidyl-APG in small portions; packing of the column will be facilitated by gentle tapping.
3. Rinse the test tube that contained the peptidyl-APG with small portions of inert glass beads, pack the reaction column up to the top and close the open end with the flanged small-bore tubing piece.
4. Insert the filled reaction column into the thermostated jacket and test for air-tightness by pressurizing the column with nitrogen. Leaks will be indicated by small gas bubbles escaping into the surrounding water.

2.4 On-Column Immobilization of Proteins on DITC Glass

DITC glass is at present undoubtedly the optimal support for the coupling of proteins and lysine-containing peptides. Attachment yields are often complete, resulting in a sequence pattern with blanks in positions occupied by the amino terminus and lysine residues, thus necessitating a separate end-group determination (although sometimes spurious traces of the first amino acid sufficient for identification are recovered). Proteins and peptides can be immobilized in aqueous solutions which may contain salts, denaturants, detergents, or even dyes, provided they will not react with the support (ammonium ions as competitive reactants are not tolerated). In particular, proteins eluted electrophoretically from stained SDS gel can be attached to DITC glass, yielding valuable sequence information for the interpretation of DNA sequence data [17] or the construction of suitable DNA probes.

Reaction of APG with DITC and protein coupling to DITC glass are usually operations performed separately from the sequencing process. It was observed in this laboratory, however, that freshly prepared DITC glass shows better reactivity, and packing DITC glass charged with protein sometimes tends to generate column back-pressure, probably due to the appearance of small glass bead fragments after prolonged stirring during the diverse chemical reactions. Therefore, following Machleidt [15], an on-column immobilization procedure was devised whereby activation of APG *and* protein attachment are carried out on the same APG column by recycling the respective chemicals through the prepacked support (Fig. 1).

The apparatus for on-column immobilization consists of a four-way Teflon slide valve (0.8-mm bore) connected in line with an air-pressure-operated membrane micropump (specially manufactured by Fa. Roesler, Berlin) and the Teflon microcolumn filled with APG as outlined in Section 2.3 and accommodated in the thermostated water jacket. Connecting lines are made of Teflon tubing (0.8 mm I.D., 1.5 mm O.D.), which are kept as short as possible in order to minimize the inner volume.

Comments. Although originally, DITC-activation of APG is carried out in DMF [11], benzene is preferred because of its higher volatility in the subsequent washing step. On-column derivatization will lead to some cross-linking of amino

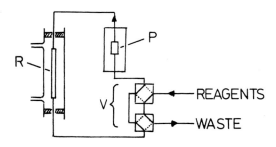

Fig. 1. Apparatus for on-column immobilization of proteins on DITC glass. *R* thermostated reaction column; *P* pump; *V* four-way slide valve. The valve position for recycling is shown. Flow rate: $1.0 - 1.8$ ml min^{-1}; temperature: 50°C

Table 3. On-column immobilization of proteins on DITC-glass

1. Connect the APG-column to the pump and the four-way valve.
2. Fill the system with benzene with the four-way valve in open position.
3. Prepare a fresh solution of 5% DITC (w/v) in benzene, add to 200 µl of this solution 20 µl triethylamine and draw the sample into the system by pumping.
4. Slide the four-way valve into the recycling position and let the column react for 1/2 h.
5. Open the valve and pump through the system ca. $3 - 5$ ml benzene, followed by the same amount of hexane.
6. Dry the whole system by connecting the inlet to a nitrogen line.
7. Fill the system with 0.1 NaHCO$_3$ containing SDS and isopropanol (see Sect. 2.1).
8. Dissolve the dry protein sample in 200 µl of the same buffer, add 20 µl triethylamine and draw the sample into the system (alternatively: replace the inlet and outlet fittings by a sample loop which will be filled with the protein sample).
9. Slide the four-way valve back into the recycling position and let the column react for 1 h. Thereafter, the column is ready for sequencing.

groups of the APG and, consequently, to a reduced coupling capacity of the DITC-glass; nevertheless, the capacity proved sufficient for microsequencing of $1 - 5$ nanomoles of sample.

The membrane micropump can be replaced by any piston pump provided the seals are solvent-resistant. Small bubbles introduced into the recycling system can be ignored as long as the flow rate is not impaired. In fact, they may serve as visible indicators for correct flow.

2.5 Automated Solid-Phase Sequencing

The solid-phase instrument used in this laboratory is a commercial sequencer (LKB, Model 4020) with a two-column system which was adapted for microsequencing, and modified for nitrogen purge of the reaction column and automated conversion of the thiazolinones as additional functions [6]. Automated conversion is accomplished by programmed in-line dilution of the TFA-effluent with distilled water to about 30% and incubation in the second column at 50°C for about 70 min (approx. 1 sequence cycle period). The thiohydantoins are collected in the built-in fraction collector as diluted solutions and dried in a Savant Speed Vac concentrator for analysis.

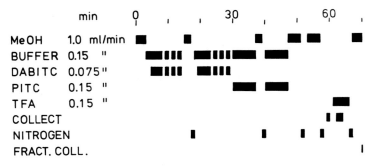

Fig. 2. Solid-phase microsequencing program

The degradation program of a total of 70 min/cycle has been devised to provide efficient coupling (Fig. 2). During the first 30 min the peptide is reacted twice with DABITC in a double-coupling procedure with an intermittent methanol wash (after 15 min) and nitrogen drying step. Similarly, the following PITC coupling is performed repeatedly in two separate consecutive steps. The 1-min collect step after 59 min serves to segment the already converted thiazolinone from the last degradation step in the conversion coil from the next sample. Actual cleavage time with TFA is 4 min. The eluate of the first 2 min is collected because the bulk amount of DAB-thiazolinone is released with the front of the TFA zone. After blowing off the TFA with a 1-min nitrogen purge, the sequence cycle is terminated with a methanol wash.

2.6 Identification of the DAB-Thiohydantoins

The colored thiohydantoins are identified by two-dimensional micro thin layer chromatography on polyamide [4], as applied for the manual DABITC technique and, in addition, by high-performance liquid chromatography [7, 8].

3 General Remarks and Outlook

The classical technique of solid-phase sequencing using PITC exclusively was developed at a time when sensitive detection of the phenyl thiohydantoins was not yet required and the available chromatographic methods proved satisfactory. Scarcity of material, however, made this limitation increasingly transparent and stimulated the search for microsequencing methods which led to the development of the elegant and facile DABITC/PITC technique as one of the spinoffs. The advent of sophisticated HPLC equipment now permits the detection of phenyl thiohydantoins in a sensitivity range which parallels that of the DAB-thiohydantoins. In fact, recent publications prove that microsequencing using solely PITC can be performed with high efficiency [18, 19]. However, the problem of UV-absorbent contaminants still persists, necessitating careful elimination of all possible sources. The elegance of the DABITC method – apart from the simple, yet sensitive and rapid micro thin-layer chromatography – resides in the spectral shift of the detection range into a region where UV-absorbent by-products become invisible and can be neglected.

Extended solid-phase sequencing using PITC can approach the power of sophisticated liquid-phase sequencers, as has been documented by Machleidt and co-workers; thus, a partial sequence of 79 out of a total of 81 residues was determined for an ATPase subunit [20] and with 100 nmol of a 555-residue protein a continuous sequence of 64 amino acids with a mean repetitive yield of 95% was still obtained [15]. For 10 nmol of a cyanogen bromide fragment of 54 residues Machleidt also reports a successful sequence determination [15].

With the DABITC solid-phase microsequencing method as outlined above, the sequencing range extends to about 30 – 40 residues for shorter peptides with a repetitive yield of ca. 92 – 94% under optimal conditions and somewhat less for proteins.

Further developments in solid-phase sequencing may involve alternative attachment techniques like thiol alkylation – reaction of iodoacetyl glass with cysteine side chains [21] – or azo coupling with a diazotized glass support [22]; the potential of such immobilization techniques is still largely unexplored. Furthermore, the synthesis of new fluorescent reagents may extend the sequencing range to the subnanomole or even femtomole level; thus, for the thiohydantoins obtained from fluorescein isothiocyanate a sensitivity of less than 0.5 pmol has been reported [23]. Unfortunately, this reagent shows less coupling yields generating considerable overlap, and the thiohydantoins derived thereof appear complex due to the formation of additional by-products (Dr. Jin, personal communication).

References

1. Laursen RA, Bonner AG (1970) Fed Proc 29:727
2. Laursen RA (1971) Eur J Biochem 20:89
3. Amiri I, Neimark J, Waltz A, Lentz J, Reinbolt J, Boulanger Y (1985) J Biochem Biophys Methods 10:329
4. Chang JY, Brauer D, Wittmann-Liebold B (1978) FEBS Lett 93:205
5. Hughes GJ, Winterhalter KH, Lutz H, Wilson KJ (1979) FEBS Lett 108:92
6. Salnikow J, Lehmann A, Wittmann-Liebold B (1981) Anal Biochem 117:433
7. Chang JY, Lehmann A, Wittmann-Liebold B (1980) Anal Biochem 102:380
8. Lehmann A, Wittmann-Liebold B (1984) FEBS Lett 176:360
9. Fontana A, Dalzoppo D, Grandi C, Zambonin M (1982) In: Elzinga M (ed) Methods in protein sequence analysis. Humana, Clifton, NJ, p 325
10. Hermodson MA (1982) In ref 9, p 313
11. Wachter E, Machleidt W, Hofner H, Otto J (1973) FEBS Lett 35:97
12. Machleidt W, Hofner H (1982) In ref 9, p 173
13. Wittmann-Liebold B, Lehmann A (1982) In ref 9, p 81
14. Laursen RA, Obar R, Chin F, Whitrock K (1980) In: Birr C (ed) Methods in peptide and protein sequence analysis. Elsevier/North-Holland, Amsterdam, p 9
15. Machleidt W (1983) In: Tschesche H (ed) Modern methods in protein chemistry. de Gruyter, Berlin New York, p 267

16. Salnikow J, Lehmann A, Wittmann-Liebold B (1982) In ref 9, p 181
17. Walker JE, Auffret AD, Carne A, Gurnett A, Hanisch P, Hill D, Saraste M (1982) Eur J Biochem 123:253
18. L'Italien JJ, Strickler JE (1982) Anal Biochem 127:198
19. Lu HS, Gracy RW (1984) Arch Biochem Biophys 235:48
20. Sebald W, Machleidt W, Wachter E (1980) Proc Natl Acad Sci USA 77:785
21. Chang JY, Creaser EH, Hughes GJ (1977) FEBS Lett 78:147
22. Chang JY, Creaser EH, Hughes GJ (1977) FEBS Lett 84:187
23. Muramato K, Kamiya H, Kawauchi H (1984) Anal Biochem 141:446

Chapter 3
Gas-Phase and Radio-Sequence Analysis

3.1 Gas-Phase Sequencing of Peptides and Proteins

FRANK REIMANN and BRIGITTE WITTMANN-LIEBOLD[1]

Contents

1 Introduction

Since 1967 Edman degradation has been used in an automated sequencer for determining the primary structure of proteins [1]. For the degradation of peptides which were covalently attached to a solid support, the solid-phase sequencer was constructed [2]. These first automatic degradations included only the coupling of the polypeptide with phenylisothiocyanate (PITC), the cleavage reaction, and the collection of the released thiazolinones. Later, isomerization to the more stable phenylthiohydantoin derivatives was also automated for the modified liquid-phase Beckman sequencer [3].

Automatic conversion devices were also incorporated into solid-phase machines [4 – 6] and the recently designed gas-phase sequencer [7]. The next step for complete automation of the process was the on-line detection of the phenylthiohydantoin derivatives by HPLC employing an isocratic solvent mixture [6, 8 – 10] or a gradient system [11]. On-line detection of the released phenylthiohydantoin derivatives in the isocratic mode gives very reproducible separation in conjunction with a stable baseline over long periods of use.

---

1 Max-Planck-Institut für Molekulare Genetik, Abteilung Wittmann, Ihnestraße 63 – 73, D-1000 Berlin 33

Advanced Methods in Protein Microsequence Analysis
Ed. by B. Wittmann-Liebold et al.
© Springer-Verlag Berlin Heidelberg 1986

2 Technical Devices of the Berlin Sequencer

The main parts of this sequencer were described recently [10]. The Berlin Sequencer has been designed by the assembly of different units, most of which can also be incorporated into a solid-, liquid- or gas-phase machine. This construction principle permits easy rearrangement of individual units. Table 1 shows the description of the different units for all these types of machine (see Fig. 1).

Table 1. Sequencer units

---

1. Nitrogen supply system

This contains the nitrogen supply system for the reaction chamber (nitrogen line N1 and N2), for the conversion flask (nitrogen line N3 and N4), for the fraction collector (nitrogen line N5) and for the solvent and reagent delivery system (pressurizing the bottles).

The nitrogen pressure has to be reduced to a constant pressure of 30 – 150 mbar, depending on the chemistry being performed in the machine.

2. Solvent and reagent bottles

Normal use requires ten bottles for reagents and solvents. Each bottle has one line for pressurizing with nitrogen, one as vent line to exhaust and one as delivery line. Usually, the press and the delivery line reach to the bottom of the bottle. In the gas-phase machine some lines have to be modified. The delivery line for the gaseous delivery of the reagents trimethylamine and trifluoroacetic acid should be cut off above the liquid level. The press line of trimethylamine is also cut off above the liquid level, otherwise the nitrogen will be saturated with water, which would condense in the delivery line and stop the gaseous stream. The bottle of trimethylamine is cooled to 16° – 18 °C; this is important for a continuous and reproducible stream of the base. For delivery of gases the vent line is closed.

3. Delivery valve system

According to Fig. 1, the delivery of each reagent and solvent is made by operating individual membrane valve ports mounted in series in one delivery block containing the central delivery line which is emptied by the flush of nitrogen (for details see [10]). The nitrogen, collect, and waste lines are mounted similarly and all these valves closed by pressed air and opened by vacuum.

4. Valves for governing the delivery system

The delivery valve system is controlled by three-way valves (Festo valves) which supply pressed air and vacuum, respectively [10].

5. Reaction chamber for Edman degradation

This unit depends on the type of degradation selected, e.g., the cup device, the solid-phase column or cartridge, and whether acid and base are delivered in the liquid or gaseous mode. All units need a temperature-control device.

a) The liquid-phase unit contains a spinning cup with appropriate drive, and a tight cup compartment, and must contain a vacuum system. In addition there is one nitrogen delivery line necessary for pressurizing the cup (line N1).

b) The solid-phase unit is somewhat simpler, it needs only a column filled with the covalently bonded protein or peptide.

c) The gas-phase contains a cartridge for the reaction, and the vacuum system is not mandatory.

6. The converter

The converter is used for the isomerization of thiazolinones to phenylthiohydantoins. The size of the conversion flask depends on the size of the reaction chamber and the volume of the solvent (S3) nec-

Table 1 (continued)

essary to collect the thiazolinone. This unit is adapted to the different reaction chambers. It also needs a temperature control.

It is preferable to construct one unit with the reaction chamber, the converter, and the temperature control for both (two different temperature settings are desirable).

7. One-line detection system

The next unit is the on-line detection system which in our sequencers is an isocratic HPLC-system. This unit is described in a separate chapter of this book.

8. The fraction collector unit

A sample of every cycle will be collected for additional identification or further investigation, e.g., to identify a modified amino acid.

9. The microprocessor

The microprocessor governs all functions, in the reaction chamber, conversion flask, the HPLC-system and the fraction collector. This microprocessor has to be freely programmable, so that all kinds of degradation can be governed by an individual appropriate program with the same programmer.

Table 2. Test reactions

A. Tollens-test for all solvents [1].
   1. 0.25 ml of a 10% silver nitrate solution are mixed with 0.25 ml of 10% sodium hydroxide solution.
   2. To this mixture a diluted aqueous ammonia solution (10%) is added dropwise, until the precipitate of silver oxide is redissolved.
   3. 0.5 ml of the solvent are mixed with 0.5 ml of the reagent and kept dark at ambient temperature. No discoloration or cloudiness should appear within 1 h.

B. Test for water with sodium.
   This test is applicable only to inert solvents like benzene or heptane, not for alcohols or esters. A small piece of sodium is cut with a knife, the crust has to be removed. This piece of sodium is washed three times with the solvent, thereafter the surface of the metal has to stay clear, and no hydrogen should be developed.

C. Water test for all solvents [12].
   To 1 ml of the solvent 20 µl tetrapropylorthotitanate (p.a. grade, Merck) is added. If the water content is higher than 0.005 to 0.025%, the solution becomes opalescent; with more water the titanate would precipitate. The solution has to remain clear for 10 minutes. The detectable water amount depends on the solvent as listed in Table 3.

3 Purification of Solvents

Most of the solvents are specially purified for sequencer use. They are first stored over aluminum oxide (neutral, activity I, dried at 100°C, from Woelm, Eschwege) overnight. Then they are refluxed over a hydride compound and distilled. The second time they are distilled without any additive over a 1-m column filled with glass rings and a reflux ratio of 1:5.

Table 3. Solvents

1. Ethylacetate (p.a. grade, Merck); b.p. 77.06 °C.
   The hydride compound used is calcium hydride (p.a. grade, Fluka, $2 \, g \, l^{-1}$), the water content after the purification is less than 0.025%.
2. Chlorobutane (p.a. grade, Merck); b.p. 78.6 °C.
   The same procedure and the same reagent as given for ethylacetate; the resultant water content is less than 0.019%.
3. Benzene (Uvasol grade, Merck); b.p. 80.2 °C.
   The hydride compound is lithium aluminum hydride (Serva $1 \, g \, l^{-1}$); the water content is less than 0.005%.
4. n-Heptane (Uvasol grade, Merck); b.p. 98.34 °C.
   The hydride compound is lithium aluminum hydride ($1 \, g \, l^{-1}$); the water content is less than 0.008%.
5. Methanol is purchased from Merck (Uvasol grade) and used without further purification (b.p. 64.7 °C).
6. Tetrahydrofuran, Uvasol grade, Merck, is used without further purification (b.p. 65.5 °C).
7. 2-Propanol (Uvasol grade, Merck) is refluxed for 5 h over calcium hydride ($5 \, l^{-1}$) and distilled over a 30-cm column filled with glass rings (b.p. 82.4 °C).

Table 4. Reagents

1. Phenylisothiocyanate (PITC); b.p. 221 °C.
   The PITC is purissimum grade (Fluka); it is redistilled under reduced pressure (oil pump) three times and stored under nitrogen in small flasks at -20 °C. It is used as a solution in n-heptane of 1 to 5%.
   The purity may be checked by filtration through aluminum oxide (neutral), where no yellowish band should appear. Otherwise the reagent should be purified by this filtration (under nitrogen).
2. Quadrol.
   The quadrol is purchased as a 1-M solution from Beckman. Two hundred ml of quartz-distilled water are flushed with nitrogen, then the content of one ampule (50 ml) is added and mixed under nitrogen. This solution is 0.2 M and used without further purification.
3. Heptafluorobutyric acid (HFB).
   The HFB is technical grade from Flourad 3 M, Neuss. It is first refluxed over barium oxide or chromium trioxide (p.a. grade, Fluka, 1 g/100 ml) and distilled. It is kept overnight over aluminum oxide (alumina N activity I, purchased from Woelm Pharma, Eschwege) under nitrogen and redistilled from dithioerythritol (10 mg/100 ml); the boiling point is 120 °C. The HFB is redistilled for a third time without any additive. The distillate is stored in 50-ml aliquots in bottles under nitrogen at -20 °C.
4. Trifluoroacetic acid (TFA).
   The TFA is purum grade and purchased from Fluka. It is distilled from calcium sulfate (dried at 500 °C immediately before use) over a 30-cm column filled with glass rings, the boiling point is $72-73$ °C. The TFA is stored in 50-ml portions in sealed glass-stoppered flasks under nitrogen at -20 °C. TFA is used as gas for the cleavage and in 40% aqueous solution as conversion medium.
5. Trimethylamine (TMA).
   The trimethylamine is liberated from its hydrochloride (for synthesis, Merck) with sodium hydroxide (for synthesis, Merck, double molar quantity). It is redistilled from the anhydride of phthalic acid (for synthesis, Merck). The pure TMA is then stored for 14 days over the anhydride of phthalic acid at -20 °C. It is diluted to a 15% aqueous solution with quartz-distilled water immediately before use.

122

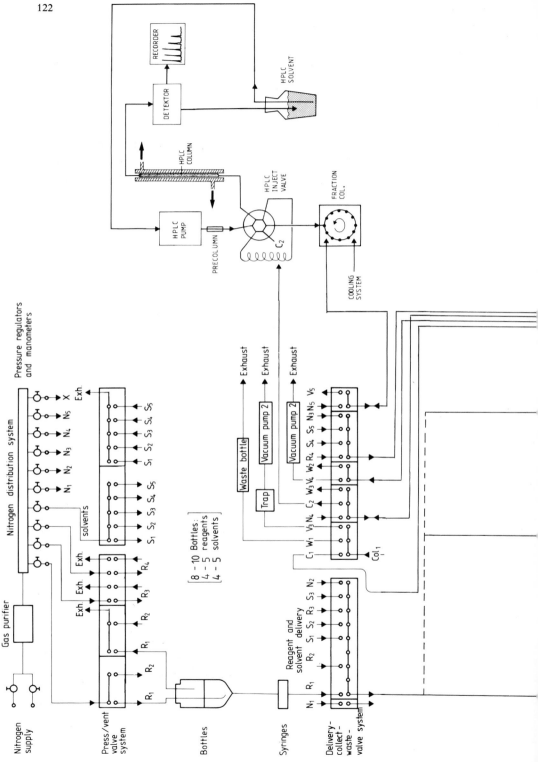

All solvents are tested for the absence of aldehyde and water with special tests as described in Table 2. For purification of the solvents see Table 3.

The solvents are stored under nitrogen, and the bottles closed with glass stoppers and with parafilm. They are stored at 10°C.

4 Purification of Reagents

Most of the reagents are purified specially for use in the sequencer (see Table 4). The best test of purity is a sequener run without any sample. The on-line HPLC-detection should produce a straight baseline, i.e., a chromatogram with no peaks except the injection peak.

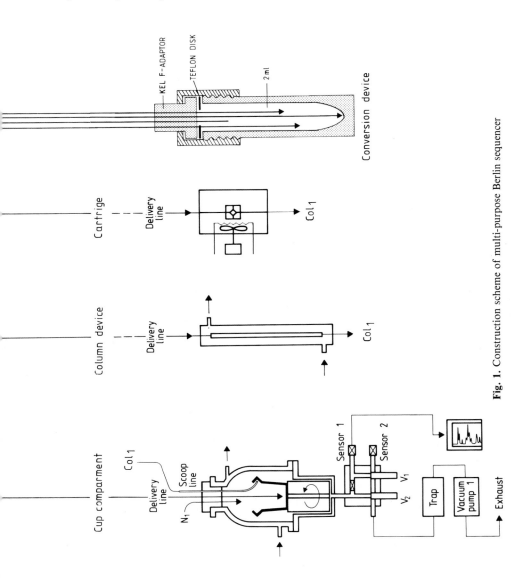

Fig. 1. Construction scheme of multi-purpose Berlin sequencer

Table 5. Degradation program

Block	Time	Cartridge	Volume	Converter	Volume
1	300 s	Cleavage TFA (gas)	7 ml min^{-1}	Dry	
2		delay		Deliver methanol	80 μl
3		Collect chlorobutane	200 μl		
4	300 s	Cleavage TFA (gas)	7 ml min^{-1}	Dry	
5		delay		Deliver methanol	80 μl
6		Collect chlorobutane	200 μl		
7	200 s	Cleavage TFA (gas)	7 ml min^{-1}	Dry	
8	150 s	Cleavage TFA (gas)	7 ml min^{-1}	Deliver 25% TFA	50 μl
9		wash chlorobutane	120 μl	Conversion	
10	80 s	Prepare coupling TMA (gas)	7 ml min^{-1}	Conversion	
11		deliver 1% PITC	50 μl	Conversion	
12	600 s	Coupling TMA (gas)	7 ml/min^{-1}	Conversion	
13		deliver 1% PITC	50 μl	Conversion	
14	600 s	Coupling TMA (gas)	7 ml min^{-1}	Dry	
15		deliver 1% PITC	50 μl	Dry	
16	600 s	Coupling TMA (gas)	7 ml min^{-1}	Deliver HPLC-solvent Inject to HPLC-column	
17		Wash ethylacetate	600 μl	Wash methanol	600 μl

5 Sample Handling

The carrier for the sample is a glass fiber filter (Whatman, Maidstone, GF/C, 12 mm), which is supported by means of a Teflon membrane filter (coarse, Thomapor, Reichelt Chemie Technik, Heidelberg, 25 mm).

The glass fiber filter is kept in 50% aqueous solution of trifluoroacetic acid for at least 24 h, then it is dried under vacuum for 20 min and the protein (0.5 to 2 nM) is applied and dried.

The test proteins for the sequencer are myoglobin, RNase, and insulin B. These proteins are dissolved in water with a concentration of 1 nM in 0.01 ml and stored frozen at $-20\,°C$. At most 0.04-ml solutions can be applied onto the glass fiber filter; if the concentration of the protein is too low, several aliquots have to be applied with drying after each application.

The protein is redissolved on the filter with 0.04 ml TFA and dried for 1 h in vacuum. This is very important for obtaining a homogeneous film of the protein on the filter, so that all reagents and solvents can come in close contact. For larger proteins the application of biobrene is not necessary; the advantage is that no precycles are necessary and the first cycles are much cleaner.

For smaller peptides like insulin B the washout can be decreased to one half by the use of biobrene (1 mg nmol^{-1}).

The pretreatment of TFA increases the initial yield by a factor of 5 to 10.

6 Program

The gas-phase program has a cycle time of about 80 min, the reaction temperature is 45 °C for the coupling, the cleavage, and the conversion. The nitrogen pressure for the solvents and the reagents is about 80 mbar, the pressure for the delivery of gases is 30 mbar. It is essential that the gas stream is slow (7 ml min^{-1}); it has to be as laminar as possible. The solvents for collecting and washing should be as gentle as possible to decrease the wash-out rate of protein.

It is preferable to first deliver a small amount of the solvent, so that the protein can precipitate, and then to wash with a larger amount of solvent. The nitrogen pressure to blow out the delivery line is 50 to 150 mbar, depending on the solvent.

Table 5 contains a short description of the usual degradation program. The program is designed in blocks, which combine several steps, e.g., to pressurize and deliver one reagent and blow out the delivery line.

References

1. Edman P, Begg G (1967) Eur J Biochem 1:80 – 91
2. Laursen RA (1971) Eur J Biochem 20:89 – 102
3. Wittmann-Liebold B, Graffunder H, Kohls H (1976) Anal Biochem 75:621 – 633
4. Birr C (1975) In: Solid phase methods in protein sequence analysis. Proc 1st Int Conf Pierce, Rockford 111, pp 115 – 129
5. Bridgen J (1977) IN: Hirs CHW, Timasheff SN (eds) Methods in enzymology, vol 47. Academic Press, London New York, pp 385 – 391
6. Machleidt W, Hofer H (1980) In: Birr C (ed) Methods in peptide and protein sequence analysis. Elsevier/North-Holland, Amsterdam, pp 35 – 47
7. Hewick RM, Hunkapiller MW, Hood LE, Dreyer WJ (1981) J Biol Chem 15:7990 – 8005
8. Wittmann-Liebold B (1982) In: Elzinga M (ed) Methods in protein sequence analysis. Humana, Clifton, NJ, pp 27 – 63
9. Wittmann-Liebold B, Ashman K (1985) In: Tschesche H (ed) Modern methods in protein chemistry. de Gruyter, Berlin New York, pp 303 – 327
10. Wittmann-Liebold B (1983) In: Tschesche H (ed) Modern methods in protein chemistry. de Gruyter, Berlin New York, pp 267 – 302
11. Rodriguez H, Kohr WJ, Harking RN (1984) Anal Biochem 140:538 – 547
12. Merck product information: "Trocknen im Labor"

3.2 Water Contents and Quality Criteria of Microsequencing Chemicals. Preliminary Results of a Reevaluation

LOTHAR MEINECKE and HARALD TSCHESCHE[1]

Contents

1 Introduction

The sequential degradation of peptides and proteins by the manual technique of Edman [1] and the automated procedure of Edman and Begg [2] requires several reagents and solvents of high quality. The general principle of this technique and the kind of chemicals needed have scarcely been changed since then. This is rather surprising, because since its introduction the protein quantity to be sequenced has decreased from 300 – 1000 nmol to less than 200 pmol of material

---

1 Universität Bielefeld, Fakultät für Chemie, Lehrstuhl für Biochemie, Universitätsstraße, D-4800 Bielefeld, FRG

Advanced Methods in Protein Microsequence Analysis
Ed. by B. Wittmann-Liebold et al.
© Springer-Verlag Berlin Heidelberg 1986

that allow determination of up to 40 amino acid residues of an unknown sequence.

A great number of developments have led to this increase in sensitivity, e.g., the HPLC identification of the PTH-amino acids [3 – 5], the on-line identification [6, 7], the auto-conversion device [7, 8], the improvements of the Beckman machine [9], the introduction of polybrene [10, 11], the immobilization of proteins and peptides by fixation to a solid support [12, 13] and the late concept of a gas-phase sequencer [14, 15].

With the increase of sensitivity the quality of chemicals used for sub-5 nmol sequencing becomes an important factor. The success of an extended sequence analysis, expressed by a high repetitive yield, is strongly dependent on the quality of the chemicals used. The criteria of quality in automated Edman degradation are that:

1. all reagents for the coupling and cleavage steps should allow a quantitative or almost quantitative reaction,
2. all solvents should be inert against protein, the Edman reagent, and cleavage products.

During automated Edman degradation all reagents and especially solvents are used in high molar excess over the protein. Therefore any interfering impurity, however small, leads to a decrease in the repetitive yield.

In 1970 Edman [16] described aldehydes and oxidants as major impurities in reagents and solvents. The present article will deal with the water content of reagents and solvents that could be of considerable importance. Published quality criteria and purification procedures of sequencer chemicals are compared and the results for water determination and purity are presented and discussed.

2 Methods

2.1 Quantitative Water Determination at 1900 nm

A Beckman spectrophotometer, model Acta M IV, was used.

Pathlength: a) water contents $>0.1\%$: 0.2 cm,
           b) water contents $<0.1\%$: 1.0 cm.

The reference cell is filled under nitrogen with the standard sample (e.g., Pierce, sequanal grade, as given in the figures). To 10 ml of this standard is added $0.1 - 10\ \mu l$ water (to water contents from 0.001 to 0.1%) under nitrogen and the spectrum is taken from 2100 to 1800 nm.

2.2 Infrared Spectroscopy

A Perkin-Elmer spectrophotometer, model 1310, was used. The spectra were taken from capillary layers of the samples between sodium chloride plates.

2.3 Ultraviolet Spectroscopy

A Shimadzu double-beam spectrophotometer, model UV-210 A, was used; path-length 1 cm.

Quartz distilled water was the reference solvent for ethylacetate, butylchloride and benzene; Uvasol grade heptane, Merck, was the reference solvent for phenylisothiocyanate.

2.4 Tollens Test for Aldehydes [18]

a) 1 ml (0.1 ml) of a 10% aqueous silvernitrate and 1 ml (0.1 ml) of a 10% sodium hydroxide solution are mixed.
b) A 10% aqueous ammonia solution is added dropwise until the precipitate is redissolved.
c) 2 ml (0.2 ml) of this solution is mixed with 2 ml (0.2 ml) of the sample. In a negative test no discoloration or cloudiness should occur within 1 h. If two phases are formed, the tube is shaken at frequent intervals.

2.5 Iodine-Starch Reaction for Oxidants [18]

a) To 10 ml (0.2 ml) of solvent in a test tube is added 1 ml (0.1 ml) of a 1% aqueous KJ solution.
b) The organic solvent is completely evaporated in a stream of nitrogen. In a negative test no blue color should occur on the addition of a 2% starch solution within 1 h.

2.6 Test for Water with Tetrapropylorthotitanate [28]

To 20 ml of the solvent 2 drops of tetrapropylorthotitanate (p.a. Merck) are added in a dry Erlenmeyer flask. After short shaking no opalescence or precipitation should occur within 5 min when the water content of the solvent is lower than $0.005 - 0.025\%$ (depending on the solvent).

2.7 Drying of Aluminum Oxide (Al_2O_3)

Al_2O_3 (neutral, activity I, Woelm, Eschwege, or Merck) is dried for 4 h at 150 °C immediately before use and always used freshly.

2.8 Activating of Molecular Sieve 4A (MS4)

MS4 (Merck) is activated by drying at 300° – 350 °C in oil-pump vacuum. For this purpose MS4 and a desiccator are heated in parallel (300 °C), then the MS4 is poured into the desiccator and an oil pump with cooling trap is joined to it; when the MS4 has reached room temperature (4 – 5 h), the desiccator is flushed with dry nitrogen. Regeneration of MS4: When the MS4 is saturated with water and organic solvent it is poured into a large excess of distilled water that expels the organic solvent (that is subsequently decanted together with the water).

2.9 Regeneration of Calcium Sulfate

For drying of heptafluoro butyric- and trifluoro acetic acid calcium sulfate-semi hydrate ($CaSO_4 \cdot 0.5\ H_2O$, Merck) is used directly. After use it may be generated at $190° – 230 °C$. This is not always necessary because it is a cheap chemical.

2.10 Use of MS4 and Al_2O_3

A column (Teflon cock, fine frit at the bottom, $2 – 3 \times 80$ cm for MS4, $2 – 3 \times 50$ cm for Al_2O_3) is filled with Al_2O_3 or MS4 and a separatory funnel (1000 – 2000 ml, Teflon cock) is connected to it. The dropping speed is regulated in that 2 l of solvent pass the column within 2 h. The eluent is slightly flushed with nitrogen. The use of plastic tubes for the delivery of solvents to the column is not recommended because they often contain soluble impurities with UV absorbance.

3 Results

3.1 Phenylisothiocyanate (see Table 1)

The most effective purification procedure for phenylisothiocyanate is the fractional distillation in oil-pump vacuum (9 torr $\triangleq$ 12 mbar, 90 °C; 2 torr $\triangleq$ 2.7 mbar, 60°C). The boiling points of the main impurities of water (100°C), aniline (184.4 °C) and diphenylthiourea (decomposes with higher temperatures) are lower than that of the mustard oil (221 °C) and may therefore be removed by distillation. Published purification procedures differ mainly in the number of subsequent distillations. Braunitzer [19] crystallizes the isothiocyanate from absolute ethanol before he distills it on a Vigreux column as described by Edman and Henschen [18]. This procedure has some disadvantages:

1. Ethanol has to be used as an additional solvent.
2. The impurities aniline and diphenylthiourea are partly co-crystallized with isothiocyanate.
3. Water may be trapped at $- 25 °C$.

Control of purity by UV spectroscopy is not feasible (Fig. 1). IR spectroscopy (Fig. 2) makes it possible to follow the success of purification by fractional distillations of phenylisothiocyanate with disappearance of the water absorption below 3.0 μ.

Table 1. Phenylisothiocyanate, PITC, boiling point: 221 °C

I.   Purification procedures:

Number of distillations in oil-pump vacuum	Reference	Starting material
3	Wittmann-Liebold and Ashman [7]	Fluka, purissimum grade
2	Shively et al. [17]	Eastman product
1 [a]	Edman and Henschen [18]	Fluka, purissimum grade
1 [a]	Braunitzer [19]	Fluka, purissimum grade
1 or 0	Hunkapiller et al. [15]	Beckman, sequencer grade

II. Distillation:
    1. Column: Vigreux, optimal: split tube column
    2. Vacuum: oil-pump vacuum (2 torr $\triangleq$ 2.7 mbar, 60 °C or 9 torr $\triangleq$ 12 mbar, 90 °C), important: vacuum adjusted constant

III. Purity criteria:
    1. Useless: a) UV spectroscopy (Fig. 1)
              b) ^{1}H-NMR spectroscopy
    2. Useful: IR spectroscopy (Fig. 2)

IV. Storage at -25 °C in small ampules (see text)

[a] After crystallization from absolute ethanol.

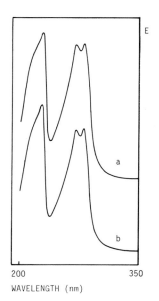

E

a

b

200   350
WAVELENGTH (nm)

Fig. 1. UV spectra of **a** purissimum grade PITC; **b** three times distilled PITC in oil-pump vacuum. Dilution: 1 : 200,000 in spectral grade heptane (Merck, Uvasol). Reference: spectral grade heptane (Merck, Uvasol). Pathlength: 1 cm

    Fractional distillation is best achieved with a well-proportioned Vigreux column; split-tube or spinning-band columns are optimal but expensive.
    The purified phenylisothiocyanate is suitably stored in small ampules at -25 °C where it crystallizes after some time. Before use it is partly thawed and the superseding liquid discarded [20].

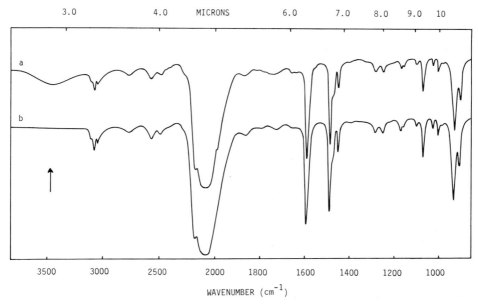

Fig. 2. IR spectra of *a* purissimum grade PITC; *b* three times distilled PITC in oil-pump vacuum. *Arrow* indicates the water absorption band below 3.0 μ

3.2 Quadrol

The purification procedure for Quadrol, as described by Edman and Henschen [18], is time-consuming and requires pure chemicals and a great number of handling steps. Most of the authors engaged in microsequencing use Quadrol buffer from Pierce or Beckman without further purification [7, 19, 22, 23]. Some authors use pretreated Quadrol (Table 2); these procedures are either not useful for further purification or the chemicals used (aminoethyl cellulose) are not pure enough [20].

Table 2. Quadrol

Purification principle	Starting material	Reference
Degasing and standing under vacuum at 50°C	Wyandotte chemicals	Shively et al. [17]
Adding of purified aminoethylcellulose	Pierce or Fluka	Frank [19]

3.3 Heptafluoro Butyric Acid (see Table 3)

The main differences between the methods of various authors in the purification for heptafluoro butyric acid, HFBA, are the drying agents (BaO, Al_2O_3, CrO_3) used. Drying with Al_2O_3 and BaO is not very effective (Table 3). Oxidation with

Table 3. Heptafluoro butyric acid, HFBA, boiling point: $120° - 121°C$

I.   Purification procedures:

Purification principle	Starting material	Reference
Refluxing over BaO, drying on Al_2O_3, once dist. over DTE, once without any addition	Fluorad, 3M, Neuss	Reimann and Wittmann-Liebold [27]
Two distillations from CrO_3, drying with $CaSO_4 \cdot 0.5\ H_2O$, dist.	Minnesota Mining	Edman and Henschen [18]
As in [18], see above	Fluorad, 3M, Neuss	Braunitzer [19]
Distillation from CrO_3 and Al_2O_3	Pierce, sequanal grade	Hunkapiller and Hood [22]

II.   Water contents:
   1. According to citation [25, 26]: $0.01 - 0.04\%$ after drying as in [18]
   2. Own measurements at 1900 nm (see Fig. 4):
      a) 3M, Neuss without drying: $0.48 - 1.0\%$, differing from batch to batch, dependent from the time the bottle was opened
      b) Beckman, sequanal grade: 1.4%
   3. Relative and qualitative determination at 1900 nm (as in Fig. 4, reference: Merck, sequanal grade HFBA after 14 days digestion with excess TFA-anhydride):
      Beckman, sequanal grade $\gg$ 3M, Neuss > purified as in [27] $\gg$ Pierce, sequanal grade > Merck, sequanal grade Merck, sequanal grade incl. excess TFA-anhydride
   4. According to specification:
      a) Pierce, sequanal grade: $<0.1\%$
      b) Merck, sequanal grade: $<0.2\%$

III.   Fractional distillation:
   1. Column: a) useful with filling material Wilson spirals (3 – 4 mm) or small glass balls
                    (1.5 – 3 mm)
               b) with lower fractionating efficiency: Vigreux or Widmer columns, glass rings as filling material
   2. Reflux ratio: to be held constant

IV.   Methods for the determination of water in HFBA:
   Qualitatively: IR spectroscopy (Fig. 3)
   Quantitatively: Near-infrared spectroscopy (Fig. 4, [25, 26])

CrO_3 offers the advantage of a visual color change from red to green that indicates whether the oxidation has been exhaustive. Additional drying with $CaSO_4 \cdot 0.5\ H_2O$ leads to a water content of $0.01 - 0.04\%$ [25, 26]. This low water content should be achieved. Otherwise random cleavage of the protein increases during extended sequencer runs. Drying by addition of acid anhydrides should be avoided because residual traces in the acid prevent the Edman degradation by blocking the N-terminus. (When TFA-anhydride is used in HFBA the preparation should subsequently be fractionated with great care: Boiling point of TFA-anhydride: $40°C$, that of HFBA: $120° - 121°C$.)

For fractional distillations columns filled with Wilson spirals or small glass balls are more effective than columns with glass rings, Vigreux or Widmer columns (compensatable with greater length).

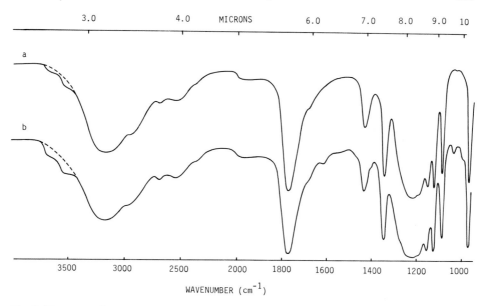

Fig. 3. IR spectra of *a* Pierce sequanal grade HFBA; *b* Beckman sequanal grade HFBA. *Dotted line* indicates the curve of a standard spectrum [23]

3.4 Trifluoro Acetic Acid (see Table 4)

It is recommended to purify TFA by the same procedure as given for HFBA. (Experiments are in progress for the quantification of water contents in TFA preparations from different purification procedures.)

3.5 Benzene (see Table 5)

There are different purification procedures for benzene described in the literature (Table 5). The purification of benzene by fractional freezing at about 4 °C is not sufficient for a decisive removal of water: A tight crystal cake instead of single crystals is formed that encloses the water with the mother liquid.

Drying with molecular sieve 4 Å and subsequent treatment with Al_2O_3 (see Table 10) for the removal of traces of oxidants (that escape the iodine-starch reaction) with final fractional distillation on an effective column is the best purification procedure for Uvasol grade benzene, Merck, in our hands.

The use of $LiAlH_4$ may be recommended if the solvent is distilled off with great care subsequently, but the procedure is not as effective as the treatment with Al_2O_3 and molecular sieve for drying. In one batch of sequanal grade benzene (e.g., Fluka) examined no opalescence occurred with tetrapropylorthotitanate, indicating (after Merck specification for tetrapropylorthotitanate usage)

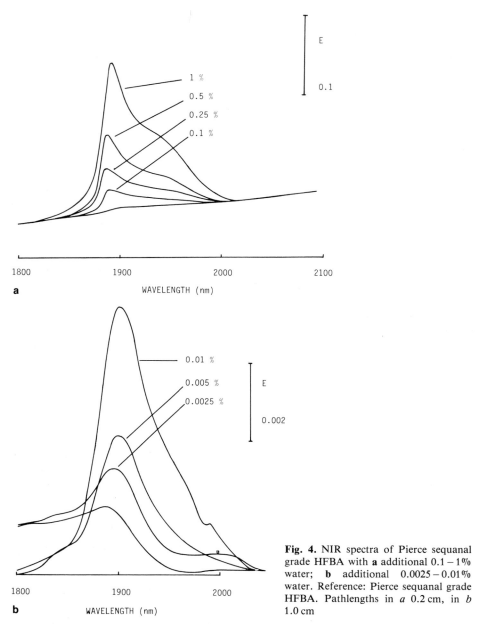

Fig. 4. NIR spectra of Pierce sequanal grade HFBA with **a** additional 0.1 – 1% water; **b** additional 0.0025 – 0.01% water. Reference: Pierce sequanal grade HFBA. Pathlengths in *a* 0.2 cm, in *b* 1.0 cm

that the water content was below 0.005% although with NIR measurements a value of 0.008% was found (Table 5, Fig. 5). Wittmann-Liebold [20] assays the water content in benzene with sodium and tetrapropylorthotitanate. With the relatively low water content Fluka sequanal grade benzene is a recommendable starting material for the purification procedure.

Table 4. Trifluoro acetic acid, TFA, boiling point: $72° - 73 °C$

I. Purification procedures:

Purification principle	Starting material	Water content after purification	Reference
Drying with $CaSO_4$, distillation	Fluka, purum grade	Considerably higher than Pierce TFA, sequanal grade[a]	Wittmann-Liebold and Ashman [7][b]
Two distillations from CrO_3, drying with $CaSO_4$, distillation	not described	–	Edman and Henschen [18]
Distillation from CrO_3 and Al_2O_3	Pierce, sequanal grade (0.1%, Pierce specification[c])	–	Hunkapiller and Hood [21], Shively et al. [17]

II. Fractional distillation:
  a) Useful with filling material Wilson spirals (3 – 4 mm) or small glass balls (1.5 – 3 mm)
  b) With lower fractionating efficiency, compensatable with greater column length: Vigreux or Widmer columns, glass rings as filling material
  c) Reflux ratio to be held constant

III. Methods for the determination of water:
  a) Useful: NIR spectroscopy
  b) Useless: IR spectroscopy

[a] Determined according to NIR spectrum, this chapter; preliminary results.
[b] Used as 40% TFA in water as conversion medium.
[c] No measurements of our own.

3.6 Ethylacetate (see Table 6)

In all published purification procedures for ethylacetate, the initial treatment with charcoal is common and useful to reduce aromatic impurities. However, in some cases batches were found that led to a contamination of the solvent by UV-absorbing impurities released from the charcoal during static or dynamic drying. Traces of oxidants are most often removed by Al_2O_3. The final step is a fractional distillation on various kinds of columns. Drying on molecular sieve 4 Å (see Table 10, this chapter) as described by two authors is recommended because it leads to a final water content of 0.004% (Merck specification, personal experience, reference solvent for NIR measurements). Ethylacetate of p.a. quality from several sources requires treatment with Al_2O_3 to remove oxidants. These oxidants are not present in sequanal grade solvent from Fluka. Traces of aldehydes were found in ethylacetate from Merck p.a. and university batches (Hochschulliefe-rungen), however not in the sequanal grade solvents from Fluka or Applied Biosystems (very expensive). Therefore final distillation is recommended not only to remove traces of dust from charcoal, molecular sieve, and Al_2O_3. Optimal for this purpose are split-tube or spinning-band columns.

In the solvent from Applied Biosystems the relatively high water content of 0.014% was found.

Table 5. Benzene, boiling point: 80.2 °C

I. Purification procedures:

Purification principle	Starting material	Water content		Oxidants and aldehydes	Reference
		before purification	after purification		
Fractional freezing at 4 °C, distillation	Analar grd. Hopkins and Williams	–	–	–	Edman and Henschen [18]
As in [18], without distillation	Merck, Uvasol	0.01 – 0.04% [a]	between 0.04 and 0.005% [b]	Not found	Braunitzer [19]
Drying over Al_2O_3 dist. over $LiAlH_4$ redist. over 1 m column with glass rings	Merck, Uvasol	0.01 – 0.04% [a]	0.006% [b]	Not found	Reimann and Wittmann-Liebold [27]
Stirring with Al_2O_3, filtering and destillation in spinning band still	Baker, HPLC grade	<0.05% [c] (Baker specification)	–	–	Hunkapiller et al. [15]
Storage on Al_2O_3	Fluka, analytical grade	–	–	–	Frank [20]
Drying on molecular sieve 4 Å and Al_2O_3, dist.	Merck, Uvasol	0.01 – 0.04% [a]	<0.005% [d]	Not found	This chapter (a)
As described above	Fluka, sequanal grade	0.008% [b]	<0.005% [d]	Not found	This chapter (b)

II. Distillation:
   1. Column: a) Optimal: spinning-band or split-tube column
      b) Useful: filling materials like Wilson spirals (3 – 4 mm) or small glass balls (1.5 – 3 mm)
      c) With lower fractionating efficiency, compensatable with greater column length: glass rings as filling material
   2. Reflux ratio to be held constant

III. Purity criteria:
   1. Negative Tollens test (see methods)
   2. Negative iodine-starch reaction (see methods)
   3. Water content below 0.003% (see Merck specifications for molecular sieve drying)
   4. No opalescence with tetrapropylorthotitanate (see methods)

IV. Useful qualitative method for the examination of purity:
   UV spectroscopy (Fig. 6)

V. Useless methods for the examination of purity:
   1. IR spectroscopy
   2. ^{1}H-NMR spectroscopy

[a] Determined according to NIR spectrometry as given in Fig. 5. Different batches of the supplier gave different values with this method.
[b] Determined according to NIR spectrometry as given in Fig. 5.
[c] No measurements of our own.
[d] Determined with tetrapropylorthotitanate; according to specification for molecular sieve drying; reference solvent for NIR measurements.

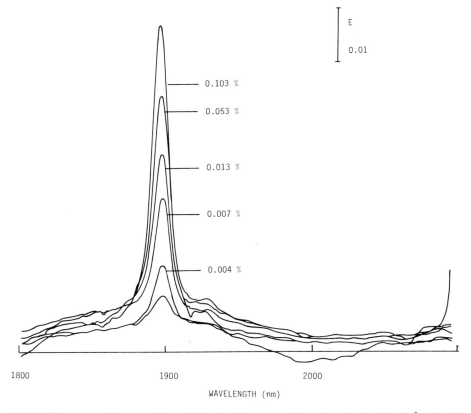

E
0.01

0.103 %

0.053 %

0.013 %

0.007 %

0.004 %

1800   1900   2000

WAVELENGTH (nm)

Fig. 5. NIR spectra of benzene (Uvasol, Merck, after passage through molecular sieve 4 Å and Al_2O_3 columns) with additional $0.001 - 0.1\%$ water. Reference: Benzene, Uvasol, Merck, dried as described above. Pathlength: 1 cm

The UV absorption maxima at 260 and 265 nm, as given by Edman, are only observed from ethylacetate that was purified by the method of Braunitzer (Fig. 7 Ib).

Useless methods for the purity control of ethylacetate from different sources are IR spectroscopy and ^{1}H-NMR spectroscopy.

3.7 Butylchloride (see Table 7)

Butylchloride is commonly purified by treatment with charcoal, Al_2O_3 and final distillation. However, in some cases batches were found that led to contamination of the solvent by UV-absorbing impurities released from the charcoal during static or dynamic drying. Edman and Braunitzer described the additional drying on molecular sieve while Reimann and Wittmann redistill the solvent over CaH_2. Although no oxidants could be detected in butylchloride of different sources by the iodine-starch reaction, the treatment with Al_2O_3 is recommended to remove

Table 6. Ethylacetate, boiling point: 77.06 °C

I. Purification procedures:

Purification principle	Starting material	Water content		Oxidants		Aldehydes		Fulfilment of UV-criteria after purification	Reference
		before purific.	after purific.	before purific.	after purific.	before purific.	after purific.		
Passage through Al_2O_3 and charcoal, distillat. and molecular sieve	Commerc. grade	–	–	n.d.	n.d.	n.d.	n.d.	+ [18]	Edman and Henschen [18]
Passage through charcoal, distill. passage through Al_2O_3 and molec. sieve	Industrial technical grade	–	0.004%[a]	+	–	+	–	+ (Fig. 7)	Braunitzer [19]
Drying over Al_2O_3 distill. over CaH_2 redist. over 1 m column with glass rings	Merck, p.a. grade	–[b]	Below 0.005%[b]	+	–	+	–	– (Fig. 7)	Reimann and Wittmann-Liebold [27]
Stirring with charcoal and Al_2O_3, distill. on spinning b. still	Baker, HPLC grade	0.05% (Baker specif.)[c]	–	n.d.	n.d.	n.d.	n.d.	n.d.	Hunkapiller et al. [15]

II. Water contents in ethylacetate from different sources:
   a) Applied Biosystems: 0.014%
   b) University batches: 0.008%
   c) Fluka, sequanal: 0.007%

III. Distillation:
   1. Column: a) Optimal: spinning-band or split-tube column
             b) Useful: filling materials like Wilson spirals (3 – 4 mm) or small glass balls
                         (1.5 – 3 mm)
             c) With lower fractionating efficiency, compensatable by greater column length:
                glass rings as filling material, Vigreux columns
   2. Reflux ratio to be held constant

IV. Purity criteria:
   1. Negative Tollens test (see Sect. 2)
   2. Negative iodine-starch reaction (see Sect. 2)
   3. Water content below 0.004% (see above, after drying with molecular sieve)
   4. UV absorption below 0.1 above 260 nm and below 0.01 above 265 nm ([18], Fig. 7)

V. Useless methods for the determination of purity:
   1. IR spectroscopy
   2. ^{1}H-NMR spectroscopy

n.d. = not determined.
[a] After Merck specification, personal experience, reference solvent for NIR measurements.
[b] Determined according to NIR spectrum, see Sect. 2, in the p.a. solvent, Merck, different water contents in different batches were found.
[c] No measurements of our own.

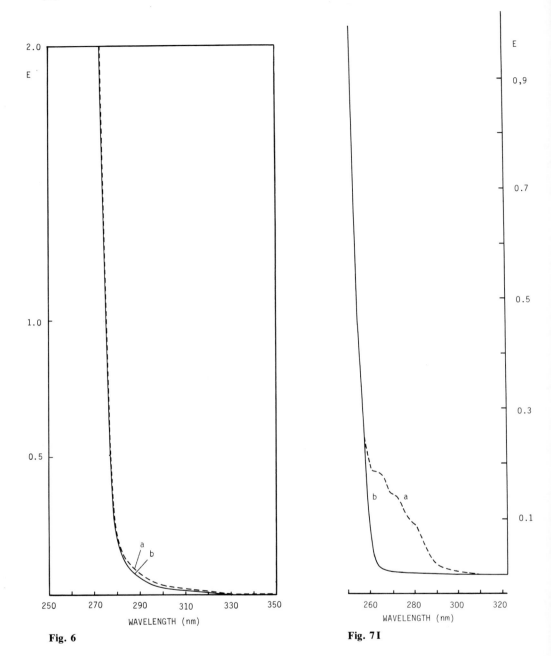

Fig. 6

Fig. 7I

Fig. 6. UV spectra of benzene preparations. Reference: bidistilled water. Pathlength: 1 cm. *a* benzene, purified by Reimann and Wittmann-Liebold [27], *b* benzene, Fluka, sequanal grade

Fig. 7I. UV spectra of ethylacetate preparations. Reference: bidistilled water. Pathlength: 1 cm. *a* Ethylacetate, university batch (Hochschullieferung); *b* ethylacetate, purified by Braunitzer (see text)

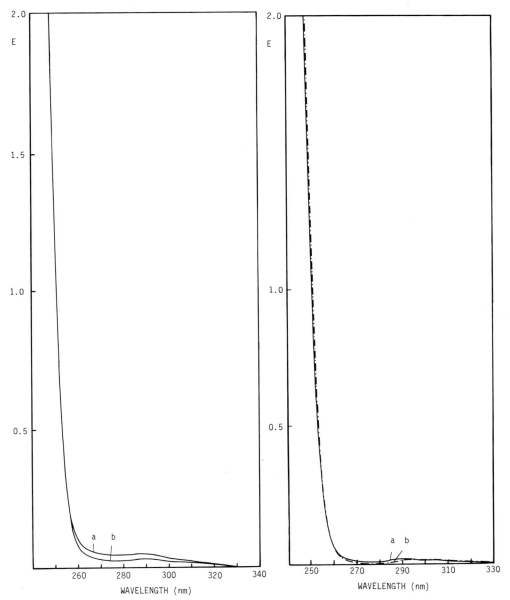

Fig. 7 II

Fig. 7 III

Fig. 7. UV spectra of ethylacetate preparations. Reference: bidistilled water. Pathlength: 1 cm. **II** *a* ethylacetate, Fluka, sequanal grade; *b* ethylacetate, Applied Biosystems. **III** *a* ethylacetate, purified by Reimann and Wittmann-Liebold [27]; *b* ethylacetate, Applied Biosystems

Table 7. Butylchloride, boiling point: 78.6 °C

I. Purification procedure:

Purification principle	Starting material	Water content		Oxidants		Aldehydes		Fulfilment of UV criteria after purification	Reference
		before purific.	after purific.	before purific.	after purific.	before purific.	after purific.		
Passage through charcoal and Al$_2$O$_3$, distill. and molec. sieve	Commercial grade	–	–	n.d.	n.d.	+ [18]	– [18]	+ [18]	Edman and Henschen [18]
As in [18], without Al$_2$O$_3$	Industrial grade, technical	–	0.006% [a]	–	–	+	–	+	Braunitzer [19]
Refluxing over CaH$_2$, dist. over 1 m column with glass rings	Merck, for synthesis	0.01% [b]	0.007% [b]	–	–	+	–	–	Reimann and Wittmann-Liebold [27]
Stirring with Al$_2$O$_3$, filter., distill. on spinning band still	Burdick and Jackson	–	–	n.d.	n.d.	n.d.	n.d.	n.d.	Hunkapiller et al. [15]
Storage over Al$_2$O$_3$	Fluka, p.a.	–	–	n.d.	n.d.	n.d.	n.d.	n.d.	Frank [20]

II. Distillation:
1. Column: a) Optimal: spinning-band or split-tube column
      b) Useful: filling materials like Wilson spirals (3 – 4 mm) or small glass balls
         (1.5 – 3 mm)
      c) With lower fractionating efficiency, compensatable with greater column length:
         glass rings as filling material, Vigreux column
2. Reflux ratio to be held constant

III. Purity criteria:
1. Negative Tollens test (see Sect. 2)
2. Negative iodine-starch reaction (see Sect. 2)
3. Water content below 0.006% (after molecular sieve drying, see Merck specification, own
   experiences)
4. UV absorption below 0.15 above 230 nm and below 0.01 above 240 nm (Fig. 9)

IV. Useless methods for the determination of purity:
1. IR spectroscopy
2. ^{1}H-NMR spectroscopy

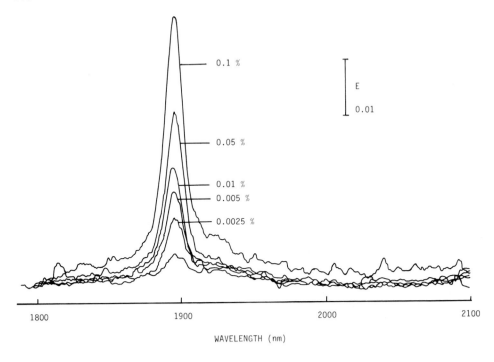

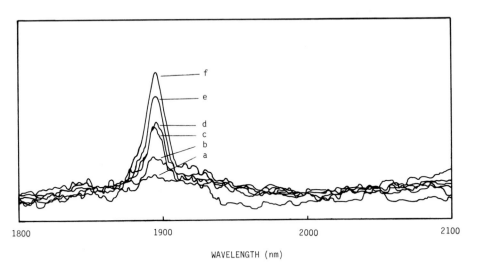

Fig. 8. I NIR spectra of butylchloride (Sequanal grade, Fluka, after passage through molecular sieve 4 Å and Al_2O_3-columns) with additional 0.0025 – 0.1% water. Reference: butylchloride, dried as described above. Pathlength: 1 cm. **II** NIR spectra of different butylchloride preparations. Reference: butylchloride, sequanal grade, dried with molecular sieve 4 Å and Al_2O_3. *a* Reference; *b* butylchloride, purified by Reimann and Wittmann-Liebold [27]; *c* butylchloride, Applied Biosystems; *d* butylchloride, Merck, for synthesis; *e* butylchloride, Fluka, sequanal grade; *f* butylchloride, university batch (Hochschullieferung)

Table 8. Heptane, boiling point: 98.34 °C

I. Purification procedures:

Purification principle	Starting material	Reference
–	*Hexane* of Uvasol grade without purification	Edman and Henschen [18]
–	Merck, Uvasol grade without purification	Braunitzer [19]
Refluxing over charcoal, distillat., redist.	Merck, Uvasol grade	Wittmann-Liebold and Ashman [7]
Stirring with Al_2O_3, filtering, distillat. in twisted Teflon spinning-band column	Burdick and Jackson	Hunkapiller et al. [15]

II. Distillation:
  1. Column: a) Optimal: Spinning-band or split-tube column
          b) Useful: Filling materials like Wilson spirals (3 – 4 mm) or small glass balls (1.5 – 3 mm)
          c) With lower fractionating efficiency, compensatable with greater column length: glass rings as filling material, Vigreux columns
  2. Reflux ratio to be held constant

III. Purity criteria:
  1. Negative Tollens Test (see Sect. 2)
  2. Negative iodine-starch reaction (see Sect. 2)
  3. Water content below 0.005%

IV. Useless methods for the determination of purity:
  1. IR spectroscopy
  2. ^{1}H-NMR spectroscopy
  3. NIR spectroscopy

V. Result from qualitative water determination:
Heptane, Merck, Uvasol grade: no opalescence with tetrapropylorthotitanate (see Sect. 2), water content below 0.008% (see specification)

VI. Result from test for aldehydes and oxidants (see Sect. 2):
Negative for untreated heptane, Merck, Uvasol grade

small traces that escape the test. Drying on a molecular sieve is necessary because no measured batch from different sources had a water content below a batch that was dried with Al_2O_3 and molecular sieve (Fig. 8).

Aldehydes were found in the solvent of synthesis grade, Merck, and university batches (Hochschullieferungen grade). Fractional distillation on an effective column should be used for final purification.

Useless for assay of purity of butylchloride are IR- and ^{1}H-NMR spectroscopy.

The UV absorption maxima at 230 nm and 240 nm, as given by Edman [18], are only observed from butylchloride that was purified according to the method of Braunitzer (charcoal, distillation, and molecular sieve 4 Å; not shown).

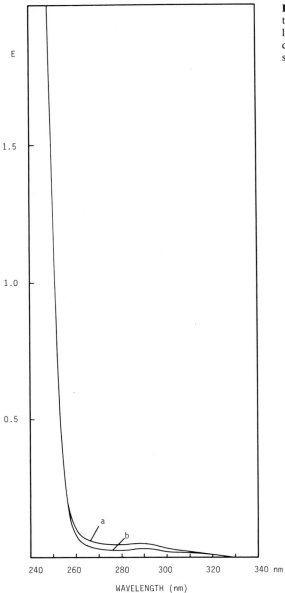

Fig. 9. UV spectra of butylchloride preparations. Reference: bidistilled water. Pathlength: 1 cm. *a* butylchloride, Fluka, sequanal grade; *b* butylchloride, Applied Biosystems

Treatment with calcium hydride is not without problems. In the presence of contaminating traces of water the strong base $Ca(OH)_2$ is formed, which may lead to side reactions with butylchloride and with impurities. These can only be reduced or removed by careful distillation on an effective column.

3.8 Heptane (see Table 8)

Heptane may be purified, e.g., from Uvasol grade, Merck, by the same procedure as described for benzene except that drying on a molecular sieve is not necessary. This solvent is specified with a water content less than 0.005% and no opalescence occurs with tetrapropylorthotitanate. For final fractional distillation effective columns should be used. When $LiAlH_4$ is used, it should be distilled off with great care, as described for benzene.

For the control of the purity of heptane IR spectroscopy and ^{1}H-NMR spectroscopy are not useful.

Aldehydes and oxidants could not be detected in Uvasol grade heptane, Merck.

Acknowledgment. This work was supported by a BMFT-grant Nr. 0385980 from the Federal Minister of Research and Technology, Bonn.

References

1. Edman P (1950) Acta Chem Scand 4:283 – 294
2. Edman P, Begg G (1967) Eur J Biochem 1:80 – 91
3. Zimmermann CL, Appella E, Pisano J (1977) Anal Biochem 77:569 – 573
4. Lottspeich F (1980) Biol Chem Hoppe-Seyler 361:1829 – 1834
5. Ashman K, Wittmann-Liebold B (1985) FEBS Lett 190:129 – 132
6. Machleidt W (1983) In: Tschesche H (ed) Modern methods in protein chemistry, vol 1. de Gruyter, Berlin New York, pp 267 – 302
7. Wittmann-Liebold B, Ashman K (1985) In: Tschesche H (ed) Modern methods in protein chemistry, vol 2. de Gruyter, Berlin New York, pp 303 – 327
8. Wittmann-Liebold B, Graffunder H, Kohls H (1976) Anal Biochem 75:621 – 633
9. Wittmann-Liebold B (1983) In: Tschesche H (ed) Modern methods in protein chemistry, vol 1. de Gruyter, Berlin New York, pp 229 – 266
10. Tarr GE, Beecher JF, Bell M, McKean DJ (1978) Anal Biochem 84:622 – 627
11. Klapper DG, Wilde CE, Capra JB (1978) Anal Biochem 85:126 – 131
12. Laursen RA (1971) Eur J Biochem 20:89 – 102
13. Machleidt W, Wachter E (1977) In: Hirs CHW, Timasheff SN (eds) Methods in enzymology, vol 47. Academic Press, London New York, pp 263 – 277
14. Hewick RM, Hunkapiller MW, Hood LE, Dreyer WJ (1981) J Biol Chem 256:7990 – 8005
15. Hunkapiller MW, Hewick RM, Dreyer WJ, Hood LE (1983) In: Methods in enzymology, vol 91. Academic Press, London New York, pp 399 – 413
16. Edman P (1970) In: Needleman SB (ed) Protein sequence determination. Springer, Berlin Heidelberg New York, pp 211 – 255
17. Shively JE, Hawke D, Jones BN (1982) Anal Biochem 120:312 – 322
18. Edman P, Henschen A (1975) In: Needleman SB (ed) Protein sequence determination. Springer, Berlin Heidelberg New York, pp 232 – 279
19. Braunitzer G (pers. comm.)

20. Wittmann-Liebold B (pers. comm.)
21. Frank G (1979) Biol Chem Hoppe-Seyler 360:997 – 999
22. Hunkapiller MW, Hood LE (1978) Biochemistry 17:2124 – 2133
23. Buse G (pers. comm.)
24. Pouchert CJ (1981) In: The Aldrich library of infrared spectra, 3rd ed. Aldrich Co, Milwaukee, WI, p 297
25. Begg GS, Pepper DS, Chesterman CN, Morgan FJ (1978) Biochemistry 17:1739 – 1744
26. Brandt WF, Henschen A, Holt C (1980) Biol Chem Hoppe-Seyler 361:943 – 952
27. Reimann F, Wittmann-Liebold B (1986) Chap. 3.1, this volume
28. Merck product information: "Trocknen im Labor"

3.3 An Improved Gas-Phase Sequenator Including On-Line Identification of PTH Amino Acids

Heinrich Gausepohl, Marcus Trosin, and Rainer Frank[1]

Contents

1 Introduction

Since 1967, when P. Edman described the first automated liquid-phase sequenator [1] its basic chemistry has remained unchanged. However, many modifications in instrumentation and automation of Edman degradation have led to improvements in sensitivity, and finally to the realization of routine subnanomole sequence analysis in the spinning cup as well as in improved solid-phase instruments. Among others, modifications contributing to this advance are automated conversion of anilinothiazolinones (ATZ-derivatives) to PTH amino acids, miniaturization of the instruments, introduction of Polybrene as carrier [2], extensive purification of reagents and solvents, and, recently, on-line identification of PTH amino acids by HPLC [3].

In 1981, Hood's group presented a new concept, the gas-liquid-phase sequenator [4]. Although this new sequenator type combines many of the previously mentioned improvements, the construction of the instrument is much simpler than the spinning-cup sequenator. In combination with new sample preparation

1 European Molecular Biology Laboratory, Meyerhofstraße 1, D-6900 Heidelberg, FRG

Advanced Methods in Protein Microsequence Analysis
Ed. by B. Wittmann-Liebold et al.
© Springer-Verlag Berlin Heidelberg 1986

Fig. 1. Front view of the gas-liquid phase sequenator

techniques and high sensitivity PTH amino acid analysis by HPLC, the gas-phase sequenator is capable of sequence determinations at the low picomole level.

The construction principle of the sequenator we have developed in our laboratory (Fig. 1) is based on the prototype instrument published by Hewick et al. [4]. The original design has been modified in order to simplify construction and maintenance and to improve operation and efficiency. Among other advantages, these modifications led to an overall miniaturization of the instrument and to a remarkable reduction of costs. To increase the efficiency and to avoid the tedious problems encountered with sample storage and transfer, our sequenator has been equipped with an on-line HPLC system similar to that developed by Wittmann-Liebold group for the spinning-cup sequenator (see Ashman, this vol.).

2 Sequencer Design

2.1 Mechanical Design

The design of our sequenator is shown schematically in Fig. 2. A more detailed description of the components and of the modular construction principle which has also been used to build an oligonucleotide synthesizer and a peptide synthesizer is given in a previous paper [5].

The general scheme differs from that of Hewick et al. [4] in the following points. The instrument operates without a vacuum system, since it turned out that reagents and solvents can be effectively removed from the lines and the cartridge by a stream of dry argon and by using 0.5-mm Teflon tubing instead of 0.3-mm microbore tubing. The argon supply system has been simplified. Only four different argon pressures are used, namely 25 mbar for delivery of the gaseous reagents R2 and R3 resulting in a stream of 4 ml min^{-1} and 50 mbar for delivery of the liquid reagent R1 and R4 and all the solvents giving a flow rate of 0.4 to 0.6 ml min^{-1} depending on the viscosity of the liquid. To blow out the delivery valves and lines and to dry the cartridge and conversion flask, a pressure of 150 mbar is used and 30 mbar for agitating liquids in the conversion flask. The

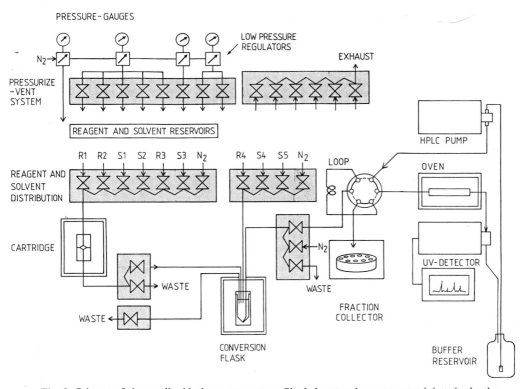

Fig. 2. Scheme of the gas-liquid phase sequenator. *Shaded rectangles* represent miniaturized valve blocks of the Wittmann-Liebold type. The reagent and solvent bottles are not shown

conversion flask has been miniaturized to an internal volume of 1.0 ml (Pierce re-activial). Since the volume of solvent S3 necessary to extract the thiazolinone from the cartridge is only about 250 μl, the elution of the PTH amino acid from the miniaturized conversion flask is effectively done using 50 – 80 μl of solvent S4. This small volume is an essential prerequisite for the on-line identification of the PTH amino acid.

2.2 Electronic Programming Unit

All the functions of the sequenator and of the peripheral instruments such as the fraction collector, HPLC injection valve, UV-detector auto zero and chart re-corder, are controlled by the central electronic programming unit based on an 8-bit microprocessor (Z80). The software is menu-oriented and allows up to 50 dif-ferent degradation cycles to be created and stored on a floppy disc. During a se-quenator run every function or step can be manually modified and the delay time can be shortened or lengthened without interrupting the running program. This programming flexibility proved to be very useful for optimizing the parameters affecting the performance and efficiency of the instrument.

2.3 On-Line HPLC System

The on-line HPLC detection system used in our gas-phase sequenator is based on an isocratic HPLC system. It is almost identical to that developed by the Witt-mann-Liebold group for the liquid-phase sequenator [3]. The connection of the sequenator to the HPLC system is performed via a Rheodyne HPLC-injection valve inserted in the line between the conversion flask and fraction collector (see Fig. 1). During transfer of a sample from flask to collector, the valve is in the load position and the loop (20 – 70 μl) is filled. As soon as it is full, the valve is switched to the inject position and the sample in the loop is injected into the HPLC column. At the same time the UV-detector is set to baseline zero and the recorder is started.

    To guarantee constant injection conditions reproducibility, exact timing, highly constant flow rates, and constant back-pressure are required. To achieve this, drift-free low pressure regulators must be used and the collect line has to be blown completely dry by a prolonged argon flush. The reproducibility of the on-line system has been tested by injecting 50 μl standard PTH-mixture delivered through the S5 port into the conversion flask, drying down and redissolving in 80 μl elution solvent S4 (25% acetonitrile in water). An overall accuracy of ± 5% was obtained. The HPLC elution buffer (2 l) is recirculated and can be used for 2 – 3 weeks when sequencing in the picomol range without any problems with baseline drift or noise even in the high sensitivity detector range (0.002 – 0.005 AUFS).

2.4 PTH Separation System

Presently there are two different columns available for isocratic PTH amino acid separation, the Spherisorb ODSII from Phase Separation Ltd. and the Super-

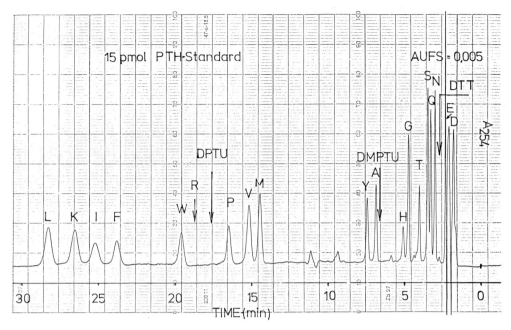

Fig. 3. Isocratic HPLC separation of PTH amino acids on a Supersphere RP8 column. The separation conditions are described in Table 1. The positions of PTH-Arg, DPTU, DMPTU and DTT are depicted by *arrows*

Table 1. Separation conditions for PTH amino acids

Column	Spherisorb ODS II (Bischoff)	Supersphere RP8 (Merck)
Dimension	250 · 4.6	250 · 4.0
Particle size	3 μ	4 μ
Eluent	68% 15 mM sodium acetate, pH 5.2	
	32% acetonitrile (Baker, HPLC grade)	
	+0.5% 1,2 dichloroethane (Merck, Uvasol)	
	+10 – 50 mg DTT/l	
Flow	0.9 ml min^{-1}	1.0 ml min^{-1}
Temperature	62 °C	62 °C
Column pressure	160 bar	150 bar
Detection	254 nm	254 nm
Detection limit	2 – 5 pmol	2 pmol
Analysis time	32 min	30 min
Injection volume	20 – 50 μl	20 – 70 μl
Injection buffer	25% acetonitrile	25% acetonitrile

sphere RP8 from Merck. Both show very similar separation characteristics [6, 7]. In Fig. 3 the separation of PTH amino acids on a Supersphere RP8 column is shown using a sodium acetate/acetonitrile/dichloroethane buffer system as it was published in [7]. In our system we use a slightly higher molarity and higher pH for the aqueous buffer (Table 1). The molarity has to be changed from batch to batch to adjust the position of PTH-His and PTH-Arg in the elution diagram.

The slightly higher pH gives better separation between PTH-Glu and DTT peak. The flow rate has been reduced to elongate column life time, which for Superspher is 3 – 4 months, and for Spherisorb 1 – 4 months. Even when using a quite high injection volume (up to 70 μl) we do not observe any band broadening of the peaks of the hydrophobic PTH amino acids, since the low concentration of organic solvent in the injection buffer gives a concentration effect on top of the column. Further, the use of acetonitrile reduces the injection peak which interferes with PTH-Asp and PTH-Glu identification.

With some of our Spherisorb columns we found a rather spectacular phenomenon. After a few weeks (sometimes days) of operation these columns showed a dramatic decrease in detection sensitivity of all PTH amino acids. The peak areas decreased to 10% of the expected value. Regeneration of the columns could be achieved by adding either PTH-Lys or DTT (up to 50 mg l^{-1}) to the elution buffer. To prevent this phenomenon, which we cannot yet explain, we now add routinely 10 mg DTT l^{-1} to the buffer.

2.5 Solvents and Reagents

The composition of the reagents and solvents and the volume used per cycle are shown in Table 2. Purification procedures have been published by several authors [8, 9] and our first sequenator prototype was operated using self-purified chemicals. However, a consequent calculation of costs for purification and quality control led us to the decision to buy the highest quality of chemicals available. Currently we mostly use chemicals of "gas-phase sequencing" grade from Applied Biosystems. The chemicals are connected in their original containers to the sequenator and immediately pressurized with argon. The reagents R2 (12.5% trimethylamine) and R3 (trifluoroacetic acid) are delivered to the reaction cartridge as vapors transported by a stream of argon (4 ml min^{-1}) blown across the surface of the liquids. To achieve saturation of the argon with reagent it is important not to fill more than 100 ml of TMA solution and 50 ml of TFA into a 200-ml reagent bottle.

Table 2. Composition of reagents and solvents and volume used per cycle

Reagent/Solvent		Usage per cycle
R1	5% PITC in heptane	3 · 30 μl→90 μl
R2	12.5% TMA	100 ml for 120 cycles
R3	TFA	60 μl
R4	A 25% TFA	30 μl + 40 μl→70 μl
R4	B 1 N HCl in methanol	40 μl
S1	Heptane	0.7 ml
S2	Ethyl acetate	1.8 ml
S3	Butyl chloride	250 μl + 150 μl→400 μl
S4	Acetonitrile	1.0 ml
S5	25% acetonitrile	80 μl + 120 μl→200 μl

2.6 Sample Application

A Whatman GF/C glass-fiber disc (12 mm O) which has been washed with TFA is prewetted with 5 µl methanol and aqueous Polybrene (1.8 mg in 20 µl) is spotted onto the disc. The disc is precycled for three cycles using a program which contains a shortened TFA treatment (120 s) and a coupling time of only 300 s. The sample is spotted on the precycled disc and air-dried. Precycling is started at the coupling step and the first degradation cycle is started with a short TFA treatment followed by the normal extraction and coupling procedure.

2.7 Operation

The first programs we used in our prototype sequenator were much more complicated than the program shown in Table 2. They employed double cleavage and

Table 3. Scheme of the sequenator program

Step	Reaction chamber	Conversion flask	Delay
1	Deliver R3	Deliver S5	
		Pause	
		Inject/collect	
		Deliver S4	800
		Collect	
		Deliver S4	
18	Deliver R3	Dry/waste	
19	Dry	Dry	60
	Deliver S3	Transfer	15
	Pause	Deliver R4	15
	Deliver S3	Transfer	50
	Pause	Dry	20
	Dry	Transfer	30
	Deliver S3	Dry	30
30	Dry		120
31	Deliver R2		100
	Deliver R1		4
	Dry	Dry	50
	Dry	Deliver R4	20
	Deliver R2	Wait	400
	Deliver R1		4
	Dry		50
	Deliver R2		400
	Deliver R1		4
	Dry		50
47	Deliver R2	Wait	400
48	Dry	Dry	60
	Deliver S1		20
	Pause		20
	Deliver S1		60
	Dry		30
	Deliver S2		220
	Pause		20
	Dry	Dry	100

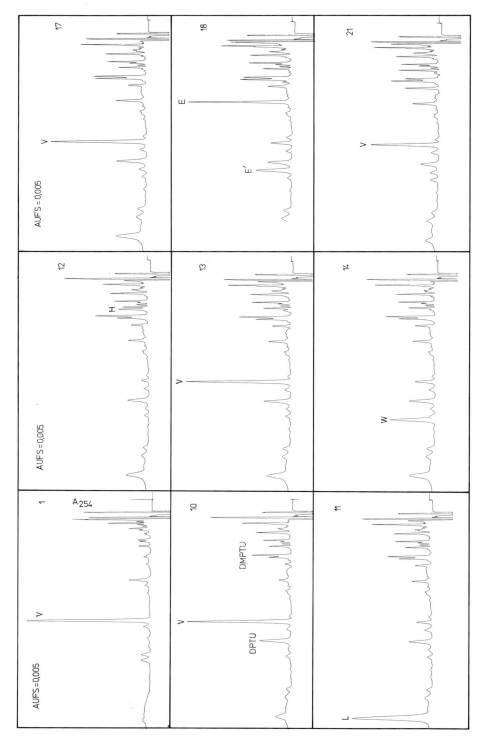

Fig. 4. Sequence analysis of 300 pmol sperm whale apomyoglobin. Fifty percent of each sample was analyzed in the isocratic on-line HPLC system described in Fig. 3 and Table 1

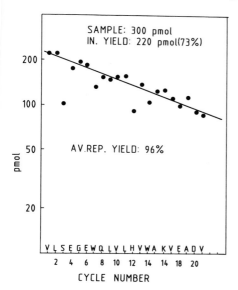

Fig. 5. Yields of PTH amino acids obtained from sequence analysis of sperm whale apomyoglobin. The yields were normalized to 100% injection. The average repetitive yield was calculated from the least squares fit to PTH amino acid yields

coupling steps and contained up to 100 steps with a total duration of 80 min. The programs we presently use have been optimized with respect to sequenator background, repetitive yield, and cycle time. They now contain only a single cleavage and a single coupling step with a triple delivery of PITC. Looking at the parameters which affect the performance of the sequenator, we found an inverse reciprocal correlation between the yield of PTH-Ser and PTH-Thr and the duration of the drying after TFA cleavage and after evaporation of butylchloride in the conversion flask. This indicates that the destruction of the Ser and Thr derivatives takes place during drying steps and not during TFA cleavage. Therefore we shortened drying after TFA delivery to 60 s and prewet the conversion flask, before butylchloride extraction, with 30 µl of 25% TFA. Subsequent evaporation of the butylchloride is then exactly timed to leave behind a few microliters of water, avoiding the destructive exposure of the ATZ-derivatives to the hot dry glass surface. The shortening of drying after TFA-cleavage further improves the extraction yield of the positively charged ATZ-derivatives from the cartridge, since a small amount of TFA is required for their extraction as TFA ion pairs.

When comparing methanol/HCl conversion and water/TFA conversion, we found no remarkable difference beside the fact that identification and quantitation of PTH-Glu and PTH-Asp are more accurate when previously converted to the methyl esters in the methanol/HCl system. Presently we use the water/TFA conversion, since this chemistry is compatible with aqueous prewetting of the conversion flask (see above).

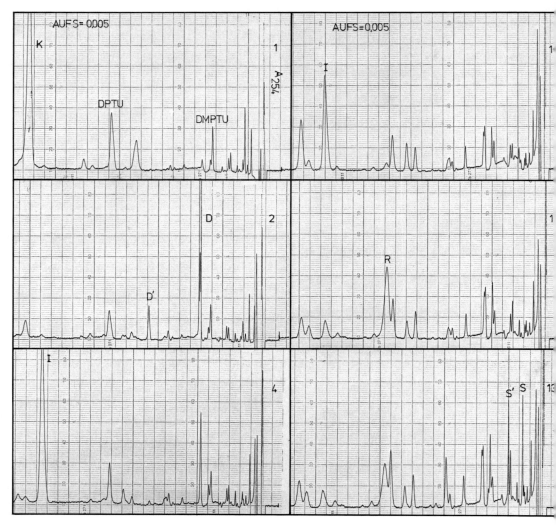

Fig. 6. Sequence analysis of 410 pmol of a 73 residue CNBr fragment from mitochondrial adenylate kinase from bovine heart (AK2)

3 Results

To demonstrate the performance of the instrument, a few representative chromatograms from the degradation of 300 pmol of sperm whale apomyoglobin are shown in Fig. 4. The initial yield is 73% and the average repetitive yield is 96% (Fig. 5). The diagrams show a constant background peak of DPTU and DMPTU which do not interfere with any PTH amino acid. In Fig. 6 sequence analysis of

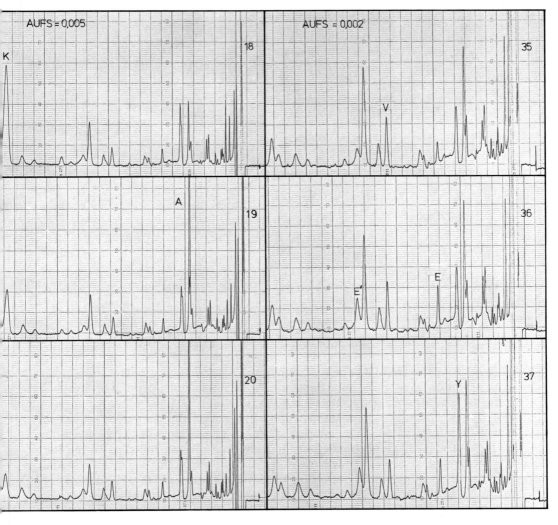

Fig. 6 (continued)

410 pmol of a 73-residue CNBr fragment from AK2 [10] is shown. The initial yield at cycle 1 was 48% and an average repetitive yield of 93% was obtained (Fig. 7). The chromatograms in Fig. 6 show that the signal level of PTH-Tyr in cycle 37 is only 30% higher than the background level in the previous cycle, therefore the unambiguous interpretation of the chromatograms was only possible up to cycle 41. This illustrates one of the major problems of the gas-liquid phase method appearing during sequence analysis of larger peptides and proteins. The sequenator background caused by unspecific cleavage of the protein chain is relatively high if compared to the spinning-cup sequenator. This type of background is peculiar to a given protein and is often the limiting factor in se-

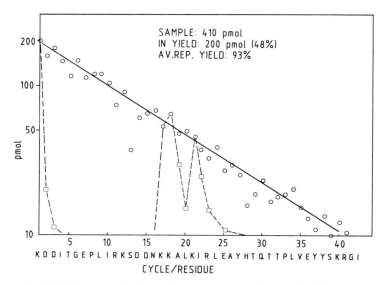

Fig. 7. PTH amino acid obtained from sequence analysis of the CNBr fragment shown in Fig. 6. The yields were normalized to 100% injection. The average repetitive yield was calculated from the least-squares fit of the PTH amino acid yields. *Dotted line* shows the carryover of PTH-Lys (K) in steps 1, 17, 18, and 21 respectively

quence analysis of larger proteins. A more general phenomenon is the increasing carryover of a given PTH amino acid into the following degradation steps. An investigation of coupling and cleavage efficiency has shown that the carryover is mostly caused by incomplete coupling reactions.

Further investigation of the reaction conditions in the gas-phase sequenator may help to overcome these limiting factors in microsequencing. Especially the effect of residual water trapped in the Polybrene film on the glass fiber filter must be studied, since on one hand it is known that traces of water are essential for good coupling yields, but on the other hand water may cause hydrolysis of peptide bonds during cleavage reaction.

References

1. Edman P, Begg G (1967) Eur J Biochem 1:80–91
2. Tarr GE, Beecher JF, Bell M, McKean DJ (1978) Anal Biochem 84:622–627
3. Wittmann-Liebold B (1983) In: Tschesche H (ed) Modern methods in protein chemistry. de Gruyter, Berlin New York, pp 229–266
4. Hewick RM, Hunkapiller MW, Hood LE, Dreyer WJ (1981) J Biol Chem 256:7990–7997
5. Frank R, Trosin M (1985) In: Tschesche H (ed) Modern methods in protein chemistry, vol 2. de Gruyter, Berlin New York, pp 287–302
6. Ashman K, Wittmann-Liebold B (1985) FEBS Lett 190:129–132
7. Lottspeich F (1980) Biol Chem Hoppe-Seyler 361:1829–1834
8. Shiveley JE, Hawke D, Jones BN (1982) Anal Biochem 120:312–322
9. Hunkapiller MW, Hewick RM, Dreyer WJ, Hood LE (1984) Methods Enzymol 91:399–413
10. Frank R, Trosin M, Tomasselli AG, Noda LH, Krauth-Siegel RL, Schirmer RH (1986) Eur J Biochem 154:205–211

3.4 Amino-Acid Composition and Gas-Phase Sequence Analysis of Proteins and Peptides from Glass Fiber and Nitrocellulose Membrane Electro-Blots

Wolf F. Brandt and Claus von Holt[1]

Contents

1 Introduction

For proteins which can only be isolated in very small amounts, gas-phase sequencing [1, 2] has become an essential tool in the field of molecular genetics and biotechnology. This is true for the elucidation of complete structures, as well as for the identification of amino-acid sequence stretches in order to design suitable oligonucleotide probes. In addition, sequencing of small amounts of proteins will become of increasing importance in establishing postsynthetic protein modifications, identification of epitopes, and in general structure function relationship, to name a few. In order to perform the structural characterizations of proteins and peptides, these have to be isolated in pure form, a process which in most cases is the most time-consuming step. Powerful analytical methodology for proteins has been developed in many laboratories over the last decade in the form of the numerous variants of polyacrylamide gel electrophoresis. In this field the arsenal of identification methods for the fractions separated electrophoretically has

---

1 UCT-CSIR Research Centre for Molecular Biology, Department of Biochemistry,
  University of Cape Town, Private Bag, Rondebosch 7700, Rep. of South Africa

Advanced Methods in Protein Microsequence Analysis
Ed. by B. Wittmann-Liebold et al.
© Springer-Verlag Berlin Heidelberg 1986

been considerably enlarged by the introduction of Western blotting techniques [3, 4]. Replacing the nitrocellulose membrane in such methodology with the more inert glass microfiber filter as the adsorbing material in the transfer opens the way to subjecting proteins and peptides separated electrophoretically on an analytical scale directly to amino-acid and sequence analysis.

Practical details for such blotting procedures are described. In these experiments mainly histones have been used as model proteins. Therefore the protocols given may not apply without modifications to other proteins. We have, however, attempted to throw light on some of the underlying principles in order to allow rational methodology development. As the procedures to be described have only recently been developed, improvements in various areas will still have to be made.

2 Protein Blotting

2.1 Adsorbing Membranes

2.1.1 Glass Microfiber Filters

In order to characterize the blotted proteins, high sensitivity methods are employed, e.g., for amino-acid analysis post- or pre-column OPA derivatization. It thus becomes imperative to reduce the amino-acid background, in particular in the materials used in the gel electrophoretic and blotting phases of the procedures. The major contamination with peptides and amino acids arises from handling. Fingerprints may contain up to 100 nmol of glycine and serine per cm^2 [5].

The background amino acids on glass microfiber filters have to be determined before and after hydrolysis as outlined in Section 3 (Fig. 10) in order to estimate free and peptide bonded amino acids. Should the background be unacceptably high, it can be reduced by pyrolysis at 500 °C for a few hours or by acid or base washes. Either one of the following effectively reduces the background: 50 – 100% trifluoroacetic acid at room temperature, 10 M HCl at 50° – 60 °C or 0.25 M NaOH at room temperature. These washings are followed by extensive rinsing with high quality water free of contaminants, until neutrality of the final rinse has been achieved. Scrubbed glass fiber filters must be handled with gloves and tweezers. Untreated or pyrolyzed glass filters were used for all the experiments described here.

2.1.2 Modified Glass Microfiber Filters

Depending on the nature of the proteins or peptides to be electroblotted, it may be advantageous to use either unmodified glass fiber filters or derivatized filters.

a) Covalent Modification. Amino groups are introduced by treatment with 5% 3-aminopropyltriethoxysilane in acetone at 50 °C for 4 h [6]. This is followed by rinsing with acetone and drying at 120 °C for 1 h. Such membranes can not be used for subsequent staining to identify proteins as the substituted filter itself stains darkly with Coomassie Brilliant Blue. Introduction of either quarternary amino or carboxyl groups is also possible with suitable silylation reagents.

b) Physical Adsorption. Glass microfiber filters can be coated with Polybrene as for sequencing purposes. Prior to blotting, the filter is washed briefly with the blotting buffer.

The treatment with Polybrene confers to the filter two properties aiding the binding of proteins to the glass fiber, namely the strong positive charge of a quarternary ammonium group and hydrophobicity of the aliphatic chains in the polymer. Binding studies indicate that $5 \mu g \, cm^{-2}$ remain tightly bound even in the presence of 0.1% SDS (Sect. 2.2.3, Table 1). This corresponds to an ion exchange capacity of $25 \, nmol \, cm^{-2}$. The blotting on Polybrene-charged membranes has been investigated in detail [7].

2.1.3 Adsorption to Nitrocellulose Membranes

[Schleicher and Schuell or Amersham (Hybond)]
This can be used as an intermediate step prior to transfer onto glass fiber filter. The amino-acid background of such filters is acceptably low (Sect. 3).

2.2 Protein Binding Studies

Three simple experiments can be undertaken to quantify the degree of adsorption of proteins or peptides to the blotting filters (modified or unmodified) in the presence of different buffers in order to optimize the transfer conditions.

2.2.1 Determination of Protein Adsorption via Amino-Acid Analysis

$0.5 \, cm^2$ pieces of glass microfiber filter are wetted with a $1 \, mg \, ml^{-1}$ protein reference solution. The filter is rinsed with H_2O or the blotting buffer. The amount of protein bound is determined by amino-acid analysis after hydrolysis, as described in Section 3. Alternatively, radioactive protein can be used to quantify the bound fraction.

Table 1 shows that positively charged peptides and proteins appear to bind preferentially to unmodified glass fiber. This affinity for positive charges can be completely saturated by Polybrene, the equivalent weight of the latter is similar to that of the Arg-Leu peptide (Table 1).

2.2.2 Determination of the Binding Capacity of a Membrane for a Given Protein with Quantification by Spot Size

Increasing amounts of protein or peptide in a constant volume of the transfer solution are applied to the membrane. The solvent is allowed to evaporate and the

Table 1. $0.5\ cm^2$ of glass microfiber membrane were soaked in a solution of 1 mg of the peptide or protein per ml. The membrane was then washed extensively with water and the adsorbed material quantified via amino-acid analysis (Sect. 3). The Polybrene concentration was determined with the diffusion test (Fig. 1, Sect. 2.2.2)

Material	$\mu g\ cm^{-2}$
Boc Glu-Tyr	0
Met-Ala	1
Arg-Leu	5
BSA	16
Ribonuclease (CM)	22
Myoglobin	20
Histone H2B	21
Polybrene	5

dry filter can then be stained with a suitable dye (Sect. 2.3). The diameter of the stained spot (Fig. 1) is proportional to the protein concentration in the solution applied. From such studies it is apparent that 25 μg myoglobin cm^{-2} are bound on both GF/C glass microfiber filter and nitrocellulose filters in distilled H_2O.

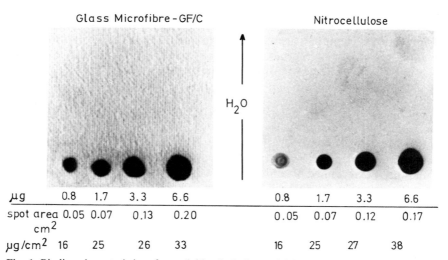

Fig. 1. Binding characteristics of myoglobin. 3 μl of myoglobin solutions of various concentrations were applied to the membranes, developed in the ascending mode in water and stained with Coomassie (Sect. 2.2.2)

2.2.3 Chromatographic Test to Assess Blotting Buffers and Eluants

Protein or peptide is applied to 1×5-cm membrane strips and developed by ascending chromatography in the blotting buffer or transfer solution to be as-

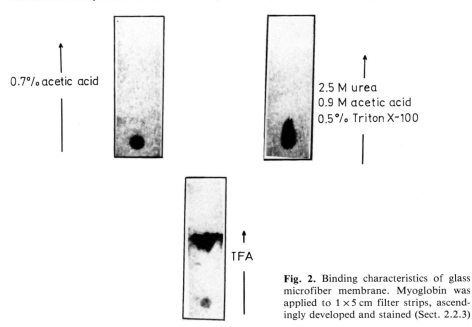

0.7% acetic acid

2.5 M urea
0.9 M acetic acid
0.5% Triton X-100

TFA

Fig. 2. Binding characteristics of glass microfiber membrane. Myoglobin was applied to 1×5 cm filter strips, ascendingly developed and stained (Sect. 2.2.3)

sessed. The strip is then dried and stained (Fig. 2). A good blotting buffer should be a poor chromatographic mobile phase, whereas in a good eluant the sample should exhibit a high mobility. Examples in Fig. 3 show that SDS-containing buffers are not suitable for blotting onto unmodified glass filters. TFA is a very good eluant, and thus suitable to concentrate or remove proteins from a glass fiber filter (even after staining) (Figs. 2 and 3; see also Sects. 3.1 and 3.2). TFA is equally well suited to elute proteins or peptides from nitrocellulose. Although the latter is known to be very unstable in acid, TFA leaves the nitrocellulose membranes tested intact.

2.3 Staining

2.3.1 Glass Microfiber Filters

Glass fiber filters are stained with 0.2% Coomassie Brilliant Blue in 40% methanol 9% acetic acid for $1-5$ min and destained under warm running water ($+30\,^{\circ}$C) until the background is stain-free. Alternatively a solution of 20% (v/v) methanol in distilled water can be used for destaining, followed by washes with warm water. Destaining solutions containing 40% methanol remove stain from the protein bands but leave the latter on the membrane. The stained filter can be stored in 0.7% acetic acid or in the dry state at $-20\,^{\circ}$C. Protein bands tend to fade somewhat on storage, but can be restained. Glass microfiber filters that have been treated with aminopropylethoxysilane or Polybrene cannot be stained, due to high adsorption of the stain. Instead of staining after blotting,

Nitrocellulose

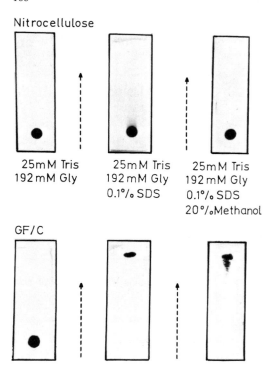

Fig. 3. Effect of buffers on the binding of myoglobin to glass microfiber and nitrocellulose membranes. Filter strips charged with myoglobin were developed ascendingly (Sect. 2.2.3)

protein can be prelabelled with fluorescent groups like dansyl chloride prior to electrophoresis (Fig. 7). Alternatively, Coomassie stained gels can be blotted in the presence of SDS onto modified filters (Sect. 2.1.2), as Coomassie, as well as most proteins, migrates toward the anode and is bound to the modified membrane, thus obviating subsequent staining.

2.3.2 Nitrocellulose

Proteins on nitrocellulose membranes are stained with 0.1% amino black in 20% methanol, 5% acetic acid for 1 min followed by destaining in 20% methanol, 5% acetic acid until background is stain-free. Coomassie Brilliant Blue is strongly adsorbed by the filter and leads to a high background and thus is an unsuitable stain.

2.4 Blotting Apparatus

The blotting apparatus has been described in detail by Bittner et al. ([8], see also [3]). The electrode symmetry is very important in order to obtain an even electric field and thus even blotting.

Upon the completion of the electrophoretic separations of proteins, the slab gel is placed on a scouring (Scotch Brit) pad that has been submerged in the blotting buffer, then the blotting membrane is placed on top of the gel, carefully avoiding the trapping of air bubbles. The filter membrane is then covered by a second scouring pad. This assembly is placed between two plastic grids and then inserted into the slots of the electrophoresis tank previously filled with blotting buffer. The electro blots described have been achieved in an electrical field between $4-7$ V cm^{-1} over 1 to 4 h. The rate of transfer depends inter alia on the molecular weight and charge of the proteins.

2.5 Blotting — General Considerations

In principle any of the large number of polyacrylamide gel electrophoresis methodologies can be applied to achieve the optimal separation of the proteins or peptides under investigation (e.g., [12, 15]). The partial or complete transfer of the separated proteins to an adsorbing membrane again can be achieved by various means, i.e., the slow process of diffusion [9] accelerated by buffer flow or a vacuum [9]. Faster and more complete transfer is achieved by electrophoretic means [3, 4, 11] (Sect. 2.4).

Two important choices have to be made, namely the nature of the adsorbing membrane and a suitable blotting buffer have to be selected (Sects. 2.1 and 2.2). If possible, the running buffer used during the electrophoretic separation should also be used for the blotting procedure. Should the running buffer, however, interfere with the protein binding to the filter (Sect. 2.2) or the subsequent characterization of the protein, it then becomes necessary to condition the polyacrylamide gel post-electrophoretically with a more suitable buffer which is subsequently used for the blotting step. Successful blots have been executed after conditioning periods of up to 12 h, although a certain amount of blurring of the electrophoretic bands occurs after such extended periods of conditioning. During the electrophoretic transfer the gel should not change in size, as this reduces the sharpness of the protein bands. Such swelling or contraction during blotting due to change in the hydration of the polyacrylamide gel is prevented by conditioning the gel for about 1 h in the blotting buffer.

It is important to pay attention to the correct polarity to the electrical field in order to ascertain the desired direction of migration of the proteins after the assembly of the blotting apparatus (Sect. 2.4). For example, the conditioning of an SDS gel with Triton X-100 at low pH (Sect. 2.8.2) results in a reversal of the migration direction of proteins. If uncertainty exists as to the charge situation of the various proteins under the transfer conditions chosen, the gel can be sandwiched between two blotting membranes. Furthermore, several membranes of the same or different filter materials can be placed in succession on the gel in order to trap proteins not bound by the first filter. To control the completeness of electrophoretic transfer, the polyacrylamide gel should be stained after blotting.

Modified glass filters with increased protein affinities as the result of derivatization with ionic, hydrophilic, or hydrophobic groups will play an increasingly important role. At this stage we have only limited experience with such membranes (Sects. 2.1.2 and 2.8.2, see also [7]).

GEL   GF/C FILTER

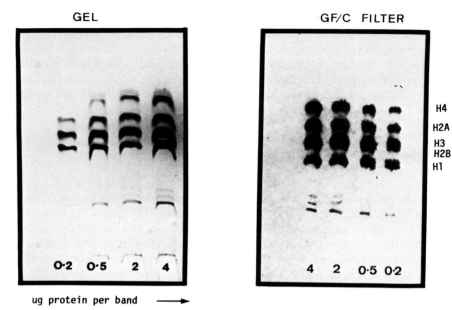

ug protein per band ⟶

Fig. 4. Glass microfiber blot of histones from acidic urea gels. Histones were electrophoresed in 2.5 M urea – 0.9 M acetic acid in 15% polyacrylamide, blotted onto GF/C filter in 0.9 M acetic acid at 6 V cm^{-1} for 60 min and stained (Sect. 2.6)

2.6 Acidic Urea Gels

Acidic urea gels, widely used for the separation of basic proteins with a urea content between 0 and 9 M urea and 0.9 M acetic acid [12] and up to 3 mm thick, are conditioned in 0.7% acetic acid for 1 h, followed by electrophoretic transfer for 1 h at 6 V cm^{-1} toward the cathode (Fig. 4). Proteins in excess of the binding capacity (20 µg cm^{-2} of GF/C filter or approximately 2 µg per band of 0.2×0.5 cm) can be trapped in successively sandwiched filters.

2.7 Non-Ionic Detergent Gels

Polyacrylamide gels developed in 0.37% Triton X-100 or Nonidet-40 at various urea concentrations [13] are conditioned for 1 h in 0.7% acetic acid. Complete transfer toward the cathode is achieved in a field of 6 V cm^{-1} for 1 h (Fig. 5).

2.8 SDS-Gels

2.8.1 Glass Microfiber Filters

Buffers containing SDS are not suitable for the transfer of proteins onto unmodified glass microfiber filters (Sect. 2.2.3). However, histones in SDS-containing

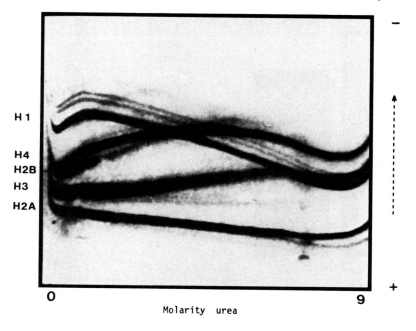

H 1

H4
H2B
H3
H2A

O   9

Molarity urea

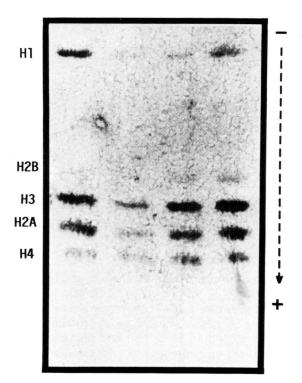

H1

H2B

H3

H2A

H4

Fig. 5. Glass microfiber blot of histones from acidic urea-triton gels. Total histones from 14-h-old sea urchin embryos were separated in 6-mM Triton X-100 – 0.9 M acetic acid in a transverse urea gradient (0 – 9 M) in 15% polyacrylamide. At the end of the run, the gel was conditioned in 0.9 M acetic acid for 1 h and blotted in 0.7% acetic acid onto GF/C membrane for 90 min at 6 V cm^{-1}. The glass fiber membrane was subsequently stained (Sect. 2.7)

Fig. 6. Glass microfiber blot of sea urchin histones from SDS gels. Sea urchin sperm histones were separated on an SDS gel [15] at pH 8.8, blotted toward the anode in 25 mM Tris-192 mM glycine at pH 8.3 at 6 V cm^{-1} for 120 min (Sect. 2.8)

polyacrylamide gel   aminopropyl
                              glass
                              microfibre

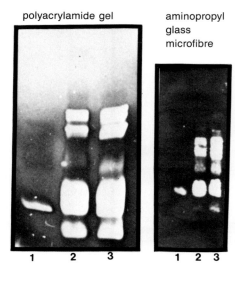

1   2   3   1   2   3

Fig. 7. Aminopropyl glass microfiber blot of fluorescent histones from a SDS gel. Wheat germ histones were labeled with 2-(2-iodoacetamido)-ethylamino-1-naphthalene sulfonic acid (lane 1), fluoresceine isothiocyanate (lane 2) and dansyl chloride (lane 3), separated in 0.1% SDS-phosphate buffer pH 9 and blotted in the Tris-Gly-SDS buffer. The fractions are displayed under UV light on the original polyacrylamide gel and on the glass fiber membrane (Sect. 2.8). *1* (2-(2-iodoacetamido)ethylamino)-1-naphthalene sulfonic acid; *2* fluorescein isothiocyanate; *3* dansyl chloride

gels can be successfully blotted onto glass fiber filters in 15 mM Tris-120 mM glycine (pH 8.3) buffer within 2 h at 6 V cm^{-1} (Fig. 6). During the transfer the free SDS concentration drops rapidly. This can have a pronounced effect on the solubility and mobility of particular proteins present in extracts, e.g., from chromatin, depending on their size and indigenous charge. To avoid such problems SDS can be incorporated into the blotting buffer (15 mM Tris-120 mM Gly-0.1% SDS pH 8.3 with or without methanol) provided a modified glass fiber filter is used, with aminopropyl groups (Sect. 2.1.2 and Fig. 7, see also [7]).

2.8.2 Replacing the SDS with Triton X-100

The SDS in the polyacrylamide gel can be replaced by soaking it in a Triton-containing buffer. In the case of histones the gel is soaked in 0.5% Triton X-100 – 0.7% acetic acid for 2 h. Urea can be incorporated in the buffer if necessary (see Sect. 2.7). The gel is then briefly soaked in the 0.7% acetic acid and blotted in the standard fashion towards the cathode (Sect. 2.6).

2.8.3 Nitrocellulose Membranes

The versatility of nitrocellulose (NC) membranes as an adsorption medium for proteins from SDS gels is well established [2, 3]. However, NC filters are soluble in organic solvents and acids like acetic or formic acid. These properties unfortunately make NC membranes unsuitable for direct sequencing of proteins after blotting. However, NC membranes are resistant to TFA. This acid is an excellent protein solvent (Sect. 2.2.3 and Fig. 2) therefore proteins or peptides blotted onto NC membranes can easily be eluted from such membranes. To achieve this, the

stained polyacrylamide gel   nitrocellulose membrane

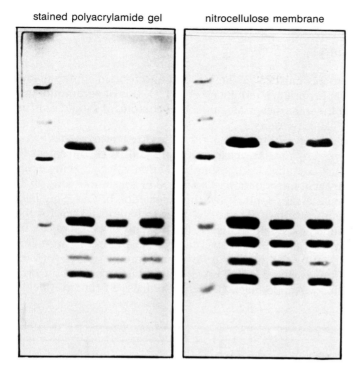

Fig. 8. Blot of stained SDS gel on nitrocellulose membrane. Sea urchin sperm histones were separated on 15% polyacrylamide gel in 0.1% SDS at pH 8.8 (see Fig. 6), stained and subsequently blotted in Tris-glycine-SDS-methanol (see Fig. 3) at 6 V cm^{-1} for 90 min toward the anode (Sect. 2.8.3)

membrane is cut according to the visualized protein fractions. These strips are then eluted descendingly with TFA for further processing. This allows the blotting of stained (Coomassie) polyacrylamide gels irrespective of the buffer used in the preceding separation. If SDS-containing blotting buffers are used, the stain is transferred together with the protein onto the NC membrane, obviating subsequent staining (Fig. 8). Since, however, stain and protein move independently in the electric field in SDS, it still remains necessary to ascertain the completeness of the blotting by restaining the polyacrylamide gel after the transfer.

For electro-blotting, unstained or stained polyacrylamide gels are conditioned for 1 h in 15 mM Tris-120 mM glycine 20% methanol – 0.1% SDS buffer and then blottedd toward the anode at 6 V cm^{-1} for 1 – 3 h (depending on the MW) onto the NC membrane (Fig. 8). Certain proteins and polypeptides, e.g., insulin, are not quantitatively adsorbed to NC membranes on blotting from SDS gels under these conditions.

3 Amino-Acid Analysis

3.1 Glass Microfiber Filters

At this stage of the protocols most of the amino-acid and protein contaminants are introduced due to inappropriate handling (Sect. 2.1.1). It is of paramount importance that handling of membranes, vials, and other equipment should only be done wearing gloves or using tweezers.

Stained protein bands are cut from the filter, inserted into a small vial (4 × 40 mm) and hydrolyzed by the gas-phase procedure [14]. It is convenient to hydrolyze several vials in a container with a suitable Teflon valve allowing evacuation, flushing with nitrogen, introduction of acid (100 µl constant boiling HCl) and sealing (Fig. 9). The necessary controls for the destruction of Cys, Ser, Thr, and Tyr and incomplete liberation of Ile and Val have to be undertaken for the quantification of the amino acids. Alternatively, the proteins or peptides can be eluted from the filter strips (glass fiber as well as NC membranes) chromatographically into a small vial by the application of small amounts of TFA (total 100 – 200 l) until all the stain is eluted. The TFA is removed in vacuo and the protein hydrolyzed as described. Amido black 10 B and Coomassie Brilliant Blue do

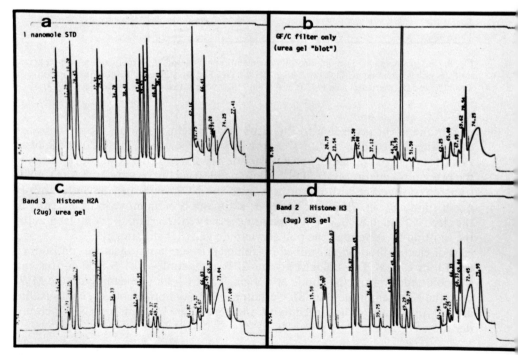

Fig. 9a – d. Amino-acid composition of proteins from glass microfiber blots. **a** Amino-acid standards. **b – d** Correspond to a blank urea, fraction 3 (histone H2A) and fraction 2 (histone 3) respectively of the glass microfiber blot in Fig. 4 (Sect. 3). Amino-acid analysis was done with the OPA-post-column method. Injection volume: 40% gain: ×16

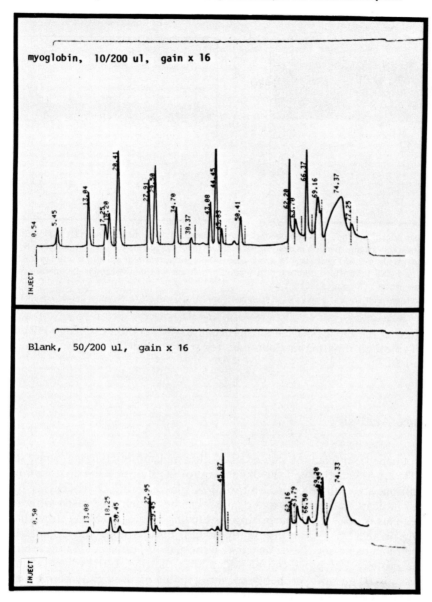

Fig. 10. Amino-acid composition of myoglobin from nitrocellulose membrane blots. Myoglobin was eluted with TFA from a nitrocellulose blot, hydrolyzed, and its amino-acid composition determined (Sect. 3.1)

1st GF/C filter   2nd GF/C filter

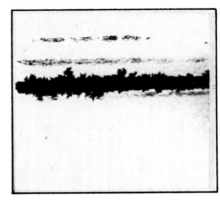

1
3
5

Fig. 11. Glass microfiber blot of *Staphylococcus aureus* protease digest of histone H2B. 100 µg histone H2B from sea urchin sperm was cleaved with *Staphylococcus aurens* protease at glutamic acid
residues, separated on 2.5 M urea – 0.9 M acetic acid – 15% polyacrylamide gel and blotted in 0.9 M
acetic acid onto glass microfiber membrane. A second membrane was sandwiched to the main membrane to trap excess of peptides (Sect. 2.6)

not interfere in the OPA post-column procedure. (Waters ion exchange HPLC
system). Typical chromatograms obtained for protein hydrolysates are given in
Figs. 9 and 10.

4 Sequence Analysis

Discs are punched out from the stained glass filter and subjected to gas-phase sequencing [1, 2]. Alternatively, if the dimensions of the stained band do not allow
this, the cut out area can be placed on the GF/C glass microfiber sequencing filter. Results indicate that proteins and peptides can be sequenced directly without
Polybrene. This is not entirely surprising, as binding studies (Table 1) show that
up to 20 µg protein cm^{-2} filter are tightly bound. Furthermore, the Coomassie
stain which remains during the sequencing on the glass filter may aid in the retention of the protein. Polybrene can be added after blotting to the filter. Of the
usual excess of 100 µg cm^{-2} probably not more than 5 µg are tightly bound (see
Table 1). The excess of Polybrene leads to an increased background of contaminants, some of them unidentified, in the HPLC chromatogram of the PTH
amino acids. Similar observations have been made by Vandeckerckhove et al.
[7]. The addition of Polybrene is, however, necessary for the sequencing of acidic
peptides as these only bind weakly to unmodified glass microfiber filter (Sect.
2.2, Table 1).
  Alternatively, fractions to be sequenced can be eluted from a section of the
glass fiber or nitrocellulose membrane with TFA and subjected to any sequencing
methodology desired.

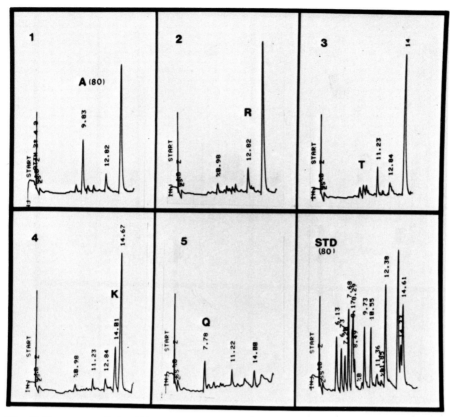

Fig. 12. Sequence analysis of histone H3 from a glass microfiber blot. A glass fiber disc suitably sized for the gas-phase sequencer cartridge was cut from a glass microfiber blot of histone H3 (see Fig. 4). The disc area contained 4 μg histone H3. 50% of the sample from each cycle was injected. The yield of alanine in step 1 was 60%, internal standard: 500 pmol nor-leu. The first five residues were identified as Ala-Arg-Thr-Lys-Gln. No Polybrene had been added to the disc. The protein comprises 135 residues and is very basic (Sect. 4)

Typical sequencing results for a protein, as well as for two peptides generated and blotted, are given in Figs. 12 – 14.

5 Concluding Remarks

A procedure for the isolation of proteins and peptides from polyacrylamide gels via electrophoretic transfer onto glass microfiber filters has been described. The resolving power of this isolation technique exceeds that of most other separation methods. In addition it is quick, very cost-effective as far as equipment is concerned, suitable for small amounts, and allows the handling of large numbers of samples with ease. Proteins and peptides isolated in such a way can be directly

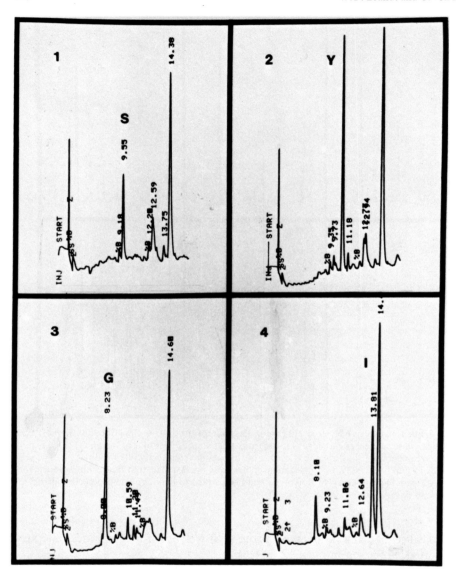

Fig. 13. Sequence analysis of peptide 1 from *Staphylococcus aureus* digest of histone H2B. A disc fitting the cartridge of the sequencer was cut from fraction 1 (Fig. 11, Sect. 2.6) of the glass microfiber blot and subjected to sequence analysis. The disc contained 1-nmol peptide. 50% of the sample from each step was injected, nor-leu standard: 500 pmol, yield of Tyr in step 2: 330 pmol (33%). The first four residues: Ser-Tyr-Gly-Ile. No Polybrene had been added to the disc. The peptide comprises 36 residues and is largely hydrophobic

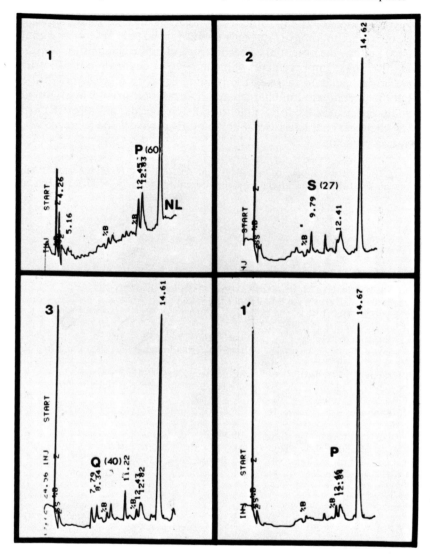

Fig. 14. Sequence analysis of peptide 5 from *Staphylococcus aureus* digest of histone H2B. 200 pmol peptide on a disc prepared as in Fig. 13 from the glass microfiber blot of the digest in Fig. 11 (Sect. 2.6) was subjected to sequence analysis. 50% of the sample was injected at each step, nor-leu standard: 500 pmol. The yield in the first step: 60 pmol (30%). The first three cycles were identified as Pro-Ser-Gln. Frame 1 represents the yield of the second coupling and cleavage. No Polybrene was added to the disc. The peptide comprises 54 residues and is very basic

subjected to structural studies, e.g., amino-acid analysis and gas-phase sequencing. This new micromethodology is not only useful for the partial sequence analysis of proteins, but has the potential to become useful in the elucidation of complete primary structures. Improvements will still have to be made in the blotting methodology, in particular to increase the binding capacity of glass microfiber filters through the introduction of charged and hydrophobic groups or combinations of these, in order to make it possible to transfer peptides and proteins with widely varying chemical characteristics. Such modifications can be achieved with suitable silylation reagents.

Acknowledgments. The work was supported by a grant from the CSIR to C. v. H. We thank J. Barron, M. Chauhan and F. Theiler for their technical assistance and J. Rodrigues for advice and discussion.

References

1. Hewick RM, Hunkapiller MW, Hood LE, Dreyer WJ (1981) J Biol Chem 256:7990 – 7997
2. Brandt WF, Alk H, Chauhan M, von Holt C (1984) FEBS Lett 174:228 – 232
3. Gershoni JM, Palade GE (1983) Anal Biochem 131:1 – 15
4. Gershoni JM (1985) TIBS 10:103 – 106
5. Hamilton PB (1965) Nature 205:284 – 285
6. Wachter E, Machleidt W, Hofner H, Otto J (1973) FEBS Lett 35:97 – 102
7. Vandeckerckhove J, Bauw G, Puype H, van Damme J, van Montagu M (1985) Eur J Biochem 152:9 – 19
8. Bittner M, Kupferer P, Morris CF (1980) Anal Biochem 102:459 – 471
9. Bowen B, Steinberg J, Laemmli UK, Weintraub H (1980) Nucleic Acids Res 8:1 – 20
10. Southern EM (1975) J Mol Biol 98:503 – 517
11. Towbin H, Ramjoue H-P, Kuster H, Liverani D, Gordon J (1982) J Biol Chem 257: 12709 – 12715
12. Panyim S, Chalkley R (1969) Arch Biochem Biophys 130:337 – 346
13. Strickland M, Strickland WN, von Holt C (1981) FEBS Lett 135:86 – 88
14. Bidlingmeyer BA, Cohen SA, Tarvin TL (1984) J Chromatogr 336:93 – 104
15. Laemmli UK (1970) Nature 227:680 – 685

3.5 Protein Blotting from Polyacrylamide Gels on Glass Microfiber Sheets: Acid Hydrolysis and Gas-Phase Sequencing of Glass-Fiber Immobilized Proteins

JOEL VANDEKERCKHOVE[1], GUY BAUW[1], MAGDA PUYPE[1],
JOZEF VAN DAMME[2], and MARC VAN MONTAGU[1]

Contents

1 Introduction

The transfer onto immobilizing membranes of proteins which are separated on polyacrylamide gels (popularly referred to as the Western blot) has become an important tool in protein chemistry (for a comprehensive review see [1]). Originally, transferred proteins were covalently bound onto the membrane [2, 3], but this method showed low coupling yields and was therefore replaced by the introduction of microporous nitrocellulose membranes to which proteins could be bound by noncovalent interactions [4]. This principle was later improved by using a nylon-based, positively charged membrane [5]. Although proteins immobilized in this way could be easily assayed for their antigenicity or their enzymatic or lectin-binding activity [1], the membranes used for these analyses were sensitive to the chemical treatments necessary for acid hydrolysis or for the Edman-degradation-based chemistry. In order to combine the simple, fast, inexpensive, high-resolution purification procedure of polyacrylamide gel electrophoresis with the recently developed technique of protein gas-phase sequencing [6], it was necessary to introduce a support which is both able to bind every kind of protein eluting from a gel, and resistant against chemicals and solvents used in Edman chemistry.

---

1 Laboratory of Genetics, State University Gent, Ledeganckstraat 35, B-9000 Gent, Belgium
2 N.V. Plant Genetic Systems, B-9000 Gent, Belgium

Advanced Methods in Protein Microsequence Analysis
Ed. by B. Wittmann-Liebold et al.
© Springer-Verlag Berlin Heidelberg 1986

The immobilization procedures discussed in this paper are based on the idea that complexes of proteins with ionic detergents will bind onto hydrophobic surfaces that have a charge opposite to that of the detergent used. Thus, sodium dodecyl-sulfate (SDS)-protein complexes will bind to apolar, positively charged surfaces, while cetyltrimethylammonium bromide (cetavlon)-protein complexes will bind to apolar, negatively charged supports.

In this paper we describe different protein immobilization conditions and compare them in terms of protein-binding capacity, ease of operation, and possibilities to be combined with currently used gas-phase sequencing strategies. Immobilization of protein from SDS-containing gels onto glass-fiber sheets coated with a noncovalently bound monolayer of the quaternary ammonium polybase, Polybrene, was found to be the most suitable of all tested procedures. Details for a practical execution of this technique are provided and examples of acid hydrolysis and gas-phase sequencing of proteins immobilized in the 1 – 20 µg range, are shown.

This technique is especially suitable for obtaining sequence information from proteins which are partially enriched by single-step procedures (e.g., by affinity chromatography or immune precipitation) and fractionated or purified by SDS-polyacrylamide gel electrophoresis. It may, therefore, serve as a new tool in obtaining partial sequence information that can either serve as guide for synthesis of specific DNA probes, or confirm the protein sequence predicted from DNA sequencing. In addition, this technique promises to become a major tool in mapping the epitopes of monoclonal antibodies, in allocating the sites of post-translational processing, and in assessing protein homologies on SDS-polyacrylamide gel peptide maps of partially digested proteins.

2 Principles of Protein Immobilization

Proteins which are separated on polyacrylamide gels in the presence of cationic or invert detergents (e.g., cetavlon) are bound to apolar anionic supports. Since unmodified glass already contains negatively charged silanol groups in addition to hydrophobic chemitopes, it has the properties required to bind such cationic protein-detergent complexes. Since proteins are most practically separated in the presence of the anionic detergent SDS, it is necessary to replace SDS by cetavlon prior to the blotting process (this can be achieved simply by washing extensively the SDS-gel in a cetavlon-containing buffer − for details see Table 1). Protein-cetavlon complexes will bind to the glass support by a cumulative effect of the ionic attraction between the positively charged cetavlon heads and negatively charged silanol groups, and by hydrophobic interactions mediated by the cetyl-group and hydrophobic surfaces on the protein and the glass support.

A possible molecular model of the protein-cetavlon-glass immobilization is shown in Fig. 1A.

The step in which SDS is replaced by cetavlon can be avoided when the glass surface is coated with a positively charged, hydrophobic pellicule. This may be obtained by covalently attaching aliphatic chains capped with cationic groups,

Table 1. Procedure of protein blotting from cetavlon-polyacrylamide gels onto unmodified glass fiber

1. Prepare SDS-containing polyacrylamide gels and separate proteins as described in Table 2.

2. When the separation is finished, fix proteins in the gel by washing for 15 min in 50% methanol-water (v/v). Wash the gel for at least 2 h in a solution of 0.5% acetic acid − 0.1% cetyltrimethylammoniumbromide (Janssen Chimica, Belgium). Exchange the solution several times during the 2-h wash.

3. Cut the GF/C (Whatman Ltd., England) glass microfiber paper sheets to the size of the gel. Dip in water and layer onto the gel. The blotting sandwich contains in the following order: a piece of scouring pad, three layers of wet Whatman 3 MM paper, the cetavlon-gel, glass fiber, three layers of Whatman 3 MM paper, and a second piece of scouring pad. The sandwich is pressed between two metal screens and immersed in the blotting apparatus. The protein-cetavlon complex migrates to the cathode. The blotting is carried out for 20 h at 4 V cm^{-1} in 0.5% acetic acid.

4. Detect immobilized proteins by staining for 2 min with Coomassie (0.25% in 45% methanol and 9% acetic acid w/v/v), destaining with water, and finally with 5% methanol, 7.5% acetic acid (v/v).

similar to the preparation of silica-based anion exchangers (see, for instance, Drager and Regnier [7]). More simply, however, one can take advantage of the properties of some polybases which are known to bind tightly, but noncovalently as a very thin layer onto glass surfaces. This can be obtained by immersing the glass in a solution of the polybase in water followed by extensive washings with buffer solution. The type of polybase which will be most efficient in protein binding should contain (1) a density of positive charges high enough to neutralize all silanol groups at the glass surface while still providing an excess of positively charged groups, (2) hydrophobic regions sufficiently large (e.g., long alkyl chains) to impart a significantly hydrophobic nature to the pellicule, (3) bases of sufficiently high pK value so that they remain fully dissociated at pH values around nine, and (4) no functional groups which might yield degradation products during either acid hydrolysis or Edman degradation which can interfere in subsequent amino acid analysis or phenyl thiohydantoin (PTH)-amino acid identification.

Among various commercially available polybases, Polybrene (1,5-dimethyl-1,5-diazaundecamethylene) appears to display most of the proposed requirements: it is a quaternary ammonium polybase and shows a favorable ratio of clustered methylene groups per base, averaging 4.5 methylene bridges per quaternary ammonium ion. Polybrene-coated glass probably shows a structure as depicted in Fig. 1 B and will thus act as a kind of mixed-bed support consisting of a C3/C6 bond phase and a quaternary ammonium anion-exchanger. Protein-SDS complexes migrating out of the gel by electroelution or by diffusion will thus bind to the Polybrene-coated glass, mainly by ionic attraction between opposite charges of the SDS and the Polybrene, cumulating with hydrophobic interactions becoming ordered between the SDS-protein complex and the hexamethylene and trimethylene bridges of Polybrene and hydrophobic chemitopes on the glass surface.

A schematic representation of the way protein-SDS-Polybrene-glass immobilization may be thought to take place is shown in Fig. 1B.

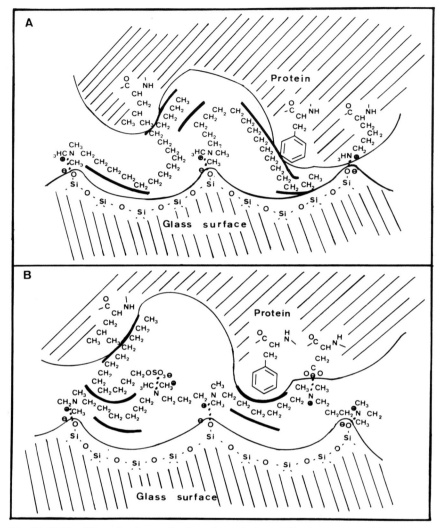

Fig. 1A, B. Possible molecular mechanism of protein immobilization on unmodified glass as protein-cetavlon complex (**A**) or on Polybrene-coated glass as protein dodecyl sulfate complex (**B**). *Solid bars:* hydrophobic interactions; *dotted lines:* electrostatic interactions

3 Detection of Immobilized Proteins

When proteins are immobilized on unmodified glass as protein-cetavlon complexes, detection can be carried out by the classical Coomassie brilliant blue staining (for details see Table 1). An illustration of a glass-fiber blot of the Bio Rad low molecular mass standard proteins is given in Fig. 2.

Proteins bound to Polybrene-coated glass cannot be detected with Coomassie brilliant blue, since this stain binds tenaciously to cationic supports. Here the problem of protein detection can be solved in several ways.

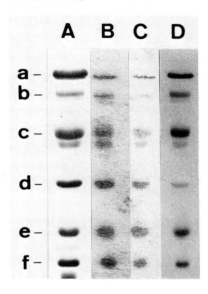

Fig. 2A – D. Blotting of proteins from cetavlon/polyacrylamide gels onto unmodified glass microfiber paper. The Bio Rad low-molecular-mass protein standard mixture was separated on a 12.5% polyacrylamide gel and blotted as described in Table 1. This mixture contained phosphorylase b (*a*), bovine serum albumin (*b*), ovalbumin (*c*), carbonic anhydrase (*d*), soybean trypsin inhibitor (*e*), and lysozyme (*f*). **A** The pattern after staining the gel with Coomassie blue. **B,C** the corresponding glass fiber blots stained with Coomassie blue (first and second layer respectively). **D** The Coomassie-stained pattern of the proteins remaining in the gel after the electroblot

The first possibility is to label fluorescently the proteins prior to separation on gel (e.g., by reaction with fluorescamine, fluoresceine- or rhodamine isothiocyanate). Since this labeling takes place on the NH_2 groups, it is necessary to allow only a small fraction to react so that the Edman degradation can be carried out on the remaining free NH_2 groups. In a one-dimensional SDS-polyacrylamide gel electrophoretic system, labeled proteins migrate similarly to the unlabeled ones. Two-dimensional techniques, however, that separate proteins in one dimension on the basis of their isoelectric point, cannot be used since NH_2-labeling may change the net charge of the protein and thus the position of the tracer molecule relative to that of the unlabeled protein.

A second solution to the problem of visualization is to detect proteins in the gel by classical Coomassie staining prior to transfer. After detection, the gel is washed extensively with transfer buffer (see below) containing 0.1% SDS and transferred and immobilized as described in Table 2. Since Coomassie brilliant blue, like proteins, forms anionic complexes with SDS, it will migrate and be immobilized in the same way, and indeed, in the same place as protein-SDS complexes. Consequently, the position of the immobilized proteins is indicated by the Coomassie blue spots. An illustration of the positions of immobilized standard proteins detected through their Coomassie-SDS-complexes is shown in Fig. 3 (lanes E – G). Like the first method, this technique has the advantage that one is able to judge the quality of protein separation prior to the start of the blotting process. However, it suffers from several disadvantages: it is more time-consuming (gels have to be stained, destained, and finally re-equilibrated), proteins are less efficiently eluted out of the gel, binding capacities are sometimes lower than when proteins are immediately eluted from the gel (results not shown), and finally, subsequent sequence analysis is disturbed by artefactual peaks, due to Coomassie blue-derived contaminants, that appear in the course of PTH-amino acid analysis.

Table 2. Procedure of protein blotting from SDS-polyacrylamide gels onto Polybrene-coated glass fiber

1. Prepare a 1.5-mm-thick SDS-polyacrylamide slab gel with acrylamide concentrations between 7.5 and 20%. Acrylamide concentrations are selected such that most of the proteins to be eluted are present in the lower half of the gel after separation (they will elute more efficiently). Load between 1 and 70 µg of protein in a 9-mm-broad slot and separate proteins as described by Laemmli [9]. Reference proteins (5 µg each) (e.g., Bio Rad low molecular mass protein mixture) are run in parallel and serve as a control for all further manipulations.

2. Cut GF/C (Whatman Ltd., England) glass microfiber paper sheets (46 × 57 cm) with scissors to the size of the gel. Suspend the glass-fiber sheets with paper clamps and allow a 3 mg ml^{-1} solution of Polybrene (Janssen Chimica Belgium) in water to drip from a Pasteur pipet over the glass fiber so that it becomes uniformly coated with the Polybrene solution. Dry the impregnated sheets in the air for at least 1 h. The dried sheets may either be processed immediately or stored for weeks before use.

3. After finishing the gel electrophoresis, equilibrate the gel for at least 2 h in 100 ml transfer-buffer containing 0.1% SDS. The transfer buffer consists of 50 mM sodium borate, pH 8.0, 20% methanol, and 0.02% β-mercaptoethanol (see also below). This equilibration step is necessary in order to avoid extensive swelling of high percentage gels during the blotting, with concomitant distortion of the bands. No serious diffusion of proteins was observed during this washing step.

4. Wash the dried Polybrene glass sheets for 2 min with 100 ml distilled water to remove excess Polybrene and mount the blotting sandwich as for Western blotting [4]: two sheets of Polybrene-coated microfiber glass placed back to back are layered over the gel and pressed on each side between three sheets of wetted Whatman 3 MM paper and a scouring pad (Scotch-Brit) supported by a stiff plastic or metal screen. Immerse the assembly in a blotting apparatus (e.g., a Gel Destainer GD-4, Pharmacia Sweden or a Bio Rad Transblot cell; and carry out the blot for 20 h at 4 V cm^{-1}). The blotting buffer is given above (step 3).

5. At the end of the blotting process, separate the gel from the glass-fiber sheets. Stain the gel with Coomassie (0.25% Coomassie brilliant blue in 45% methanol and 9% acetic acid w/v/v) and destain by diffusion in 5% methanol, 7.5% acetic acid (v/v). Staining of the proteins in the gel allows one to judge blotting efficiency. Wash the Polybrene glass fiber sheets in 20 mM NaCl, 10 mM borate, pH 8.0 by shaking for 5 min in 100 ml buffer. Repeat the washing procedure three times and finally wash the blot with distilled water (1 min). Allow the glass fiber blot to dry in the air by suspending the sheets with paper clamps. Drying can be accelerated by washing the sheets in acetone and drying.

6. Stain the immobilized proteins by dipping the sheets in a freshly made diluted solution of fluorescamine (Fluka AG). Usually concentrations of 1 mg/500 ml are sufficient for visualization of most immobilized proteins. When staining is not sufficient at this highly diluted fluram concentration, a more concentrated solution may be used (up to 1 mg/100 ml). Note that higher fluram concentrations will result in increased blocking of the NH$_2$-terminal groups. Allow the fluorescamine reaction to take place for 10 to 15 min. Visualize immobilized proteins by illumination with a UV mineral light lamp (Ultraviolet Products, San Gabriel, USA) and record results by photography using a Polaroid type 665 P/N film and a Wratten no. 9 filter. Use exposure times varying between 1 min and 2 min. Further characteristics of the procedure are discussed in [13].

So far, the most practical way to detect Polybrene-glass immobilized proteins is by staining with dilute fluorescamine in a way similar to the detection of peptides separated on paper or thin layer [8]. When very diluted solutions are used (1 mg fluram per 200 to 800 ml acetone), only a very small fraction of the NH$_2$-groups is blocked, sufficient for protein detection, but leaving most of these functional groups intact for subsequent Edman degradation. Since most discontinuous SDS-polyacrylamide gels are run using Tris-glycine as the electrode buf-

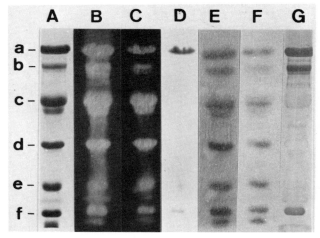

Fig. 3A – G. Blotting of proteins from SDS/polyacrylamide gels onto Polybrene-coated glass fiber. Proteins are as described in Fig. 2. **A** Pattern after staining the gel with Coomassie blue; **B,C** the glass fiber blots of the first (**B**) and second (**C**) layers stained with fluorescamine; **D** Coomassie blue stain of the proteins remaining in the gel after the electroblot. **E,F** the protein patterns on the first **E** and second **F** sheets from the electroblot of the proteins previously stained with Coomassie blue (see **A**), **G** the Coomassie-stained pattern of the proteins remaining in the gel after the electroblot

fer (see for instance Laemmli [9]), the anionic glycinate will also adsorb onto the Polybrene glass, making fluram detection impossible. Thus, bound glycinate must be exchanged by chloride anions (e.g., by washing the blots in NaCl-containing buffers) prior to staining.

The latter procedure has been found to be the fastest, most sensitive and most reproducible method. We will therefore limit further detailed discussions to this technique.

4 Protein Blotting on Polybrene Glass-Fiber as Basis for Acid Hydrolysis and Gas-Phase Sequencing

4.1 Blotting of Various Proteins

Figure 3 shows the results of blotting experiments carried out with the Bio Rad standard protein mixture separated on a 12.5% polyacrylamide gel. One lane of the gel was stained with Coomassie blue and another lane was electroblotted onto two sequentially mounted glass-fiber sheets and stained with fluorescamine. Each protein is sufficient to saturate the first sheet and the to become immobilized on the next sheet. Only subsaturating components (e.g., degradation products of the protein mixture used) are limited to the first layer.

Figure 4 shows the blotting results of another set of proteins including the highly phosphorylated protein, phosvitin, and the basic protein, bovine pancreatic DNase I. Here again, all proteins assayed show quantitative electroelution and adsorption on the glass-fiber sheets.

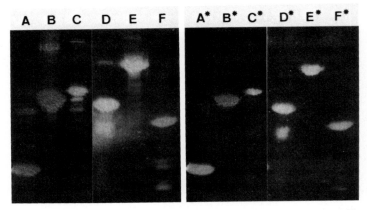

Fig. 4A – F. Polybrene glass blotting patterns of various proteins. **A – F** The fluorescamine stain of the first glass-fiber sheet of immobilized myoglobin, phosvitin, actin, chymotrypsinogen, human serum albumin, and DNase I. **A* – F*** the second layer of the same blot. These blots are used to determine the binding capacity of the Polybrene glass (see Table 4). Approximately $20-40$ µg of each protein is loaded on the gel

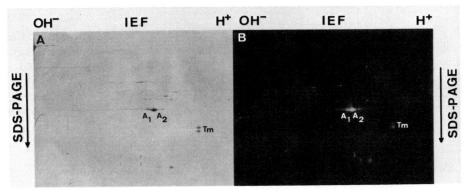

Fig. 5A, B. Two-dimensional Polybrene blot of the proteins from an 18-day chicken embryo skeletal muscle. **A** The Coomassie-stained two-dimensional gel; **B** the corresponding blot of half the amount shown in **A**. A_1, A_2, and Tm refer to β-actin, α-actin and the tropomyosins respectively. H^+ and OH^- indicate the acidic and basic side of the gel. Isoelectric focusing (IEF) is carried out in the horizontal direction. SDS/polyacrylamide gel electrophoresis (SDS-PAGE) in the vertical direction

The best illustration of the general applicability of this technique is shown by blotting a two-dimensional gel separation of proteins present in 18-day embryonic chicken skeletal muscle tissue. In Fig. 5 the pattern observed by Coomassie staining in the gel is compared with that obtained after glass blotting and fluorescamine staining of a parallel gel. The glass blot appears to be a nearly exact copy of the Coomassie-stained pattern. The major spots are α- and β-actin and the tropomyosins. Each spot represents approximately 0.5 µg of protein.

4.2 Acid Hydrolysis Yielding Reliable Compositions from Immobilized Proteins

In order to ascertain whether the amino acid composition obtained by acid hydrolysis of Polybrene glass-immobilized proteins could be used for further protein identification, we have hydrolyzed several immobilized proteins and compared the calculated compositions with those from the known sequences. For these studies we have used most of the proteins shown in Fig. 4. The results of these experiments indicate that Polybrene glass-immobilized proteins can be hydrolyzed (for details on the hydrolysis procedure see Table 3) yielding the correct amino acid composition of the protein. Only methionine values were often found to be too low, while glycine values were often higher than expected. However, the glycine contamination was only a problem for proteins present on the glass sheets at subsaturating levels. Thus even major proteins from two-dimensional gels could be analyzed readily without disturbance from high background levels of glycine. This is illustrated in Fig. 6, where the amino acid analysis of traces of α- and β-actin (the two major spots of the two-dimensional gel shown in Fig. 5) are compared. Although these traces clearly do not allow calculation of the exact amino acid composition (note the high buffer jumps and base-line instabilities due to extreme electronic enhancement), they unambiguously show at a level of less than 7 pmol of protein (only half of the hydrolysate was taken for analysis) that α- and β-actin are true isoforms, differing in only 25 positions out of a total of 375 [10, 11]. It is obvious that amino acid analysis systems, more sensitive than those used in our studies, (e.g. precolumn-derivatization with phenylisothiocyanate followed by HPLC separation of the PTC-derivatives [12] may be used in the future to obtain complete amino acid compositions of most proteins blotted from two-dimensional gels.

4.3 Maximum Binding Capacity of Polybrene Glass Fiber

The fact that proteins, when present in sufficient amounts, pass from the first layer onto the second, indicates that the Polybrene glass has a limited capacity to bind protein. Since it is important to have an idea of this capacity in view of future experiments involving the determination of the amino acid composition and NH_2-terminal sequence, we have determined the saturation value for various proteins. Proteins in saturated areas of glass-fiber sheets were hydrolyzed (see below) assayed quantitatively, and normalized to measure the μg protein cm^{-2} (Table 4). The blotting pattern of most of the proteins used in these studies is

Table 3. Procedure of acid hydrolysis of proteins immobilized on Polybrene-coated glass fiber

Cut out the portion of the glass fiber containing the protein and place in a Pyrex tube 0.9 × 7 cm. Add 400 μl 6 M HCl containing 0.05% 2-mercaptoethanol. Seal tube under vacuum and incubate at 110 °C for 22 h. Open tube and dry quickly in a vacuum desiccator over NaOH. Resuspend residue in 400 μl 0.2 M sodium citrate buffer, pH 2.2. Filter through Millex-HV4 filter (Millipore) and load appropriate amount for amino acid analysis.

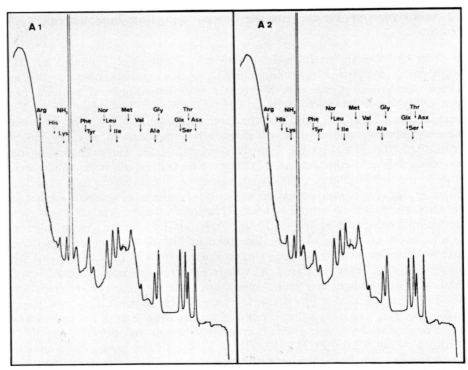

Fig. 6. Amino acid analysis traces for of α-actin (spot A₂ in Fig. 5) and β-actin (spot A₁ in Fig. 5). Amino acid analysis was carried out with a Biotronik amino-acid analyzer equiped with a fluorescence detector. Amino acids were separated by classical ion-exchange chromatography on a Durrum 6A resin 0.6 × 20 cm column and detected after reaction with o-phthaldialdehyde [15]. The chromatograms shown are run on approximately 10 pmol of protein (half the hydrolysate was taken for analysis)

shown in Fig. 4. Values were lowest for phosvitin (8 µg cm^{-2}) and varied for the other proteins between 15 and 28 µg cm^{-2}. Thus at the average, proteins bind to Polybrene glass microfiber with a maximum capacity of ± 20 µg cm^{-2}.

4.4 Gas-Phase Sequencing of Immobilized Proteins

Since gas-phase sequencing can be performed on proteins which are dried on glass filter discs containing a layer of Polybrene [6], proteins transferred from gels and immobilized as described above form suitable samples for this new sequencing strategy.

An illustration of a direct combination of Polybrene-glass blotting with gas-phase sequencing is shown for sperm whale skeletal muscle myoglobin. 30 µg (± 1.8 nmol) of this protein were loaded and run on an SDS-polyacrylamide gel. The blot recovered ± 1 nmol on the first layer of Polybrene glass. The glass-fiber area containing the fluorescamine-detected protein was punched out as a 12-mm

Table 4. Binding capacity of PB-glass for different proteins electroeluted from SDS-polyacrylamide gels (μg cm^{-2})

Protein	Direct blotting followed by fluorescamine detection	Transfer following Coomassie staining in the gel
Myoglobin	21	13
DNase I	21	20
Chymotrypsinogen	15	14
Actin	28	13
Human serum albumin	23	24
Phosvitin	8	5

Table 5. Procedure of gas-phase sequencing of Polybrene glass-immobilized proteins

Use a 12-mm diameter cork-drill to punch out the area of the glass-fiber sheet containing the protein of interest. Contaminating proteins which are not very well separated in the polyacrylamide gel and therefore present at the edges of the disc, are removed by cutting away those portions of the glass fiber. Note that it is not essential that an intact 12-mm diameter disc be mounted in the cartridge of the gas-phase sequenator for proper performance. Mount the glass-fiber disc in the gas-phase sequenator reaction chamber and proceed with the normal program. Identify the stepwise liberated phenyl-thiohydantoin amino acids using published HPLC procedures (e.g., Hunkapiller and Hood [14]).

disc with a cork-bore and mounted in the cartridge of the Applied Biosystem Inc. gas-phase sequenator (Table 5). Eighteen cycles of the Edman degradation were carried out and the PTH-amino acids recovered after each cycle were analyzed and quantitated according to the method of Hunkapiller and Hood [13] (Fig. 7). We found an initial coupling yield of 30% (corresponding with a recovery of 300 pmol of the NH$_2$-terminal residue as PTH-Val). Repetitive yields, calculated from the recoveries of PTH-Leu (on positions 2, 9, and 11) and PTH-Val (on positions 1, 10, and 13), amounted to 91 – 92%. These values are similar to recoveries obtained for other proteins and peptides analyzed with the same sequenator.

The levels of the by-products diphenylthiourea and diphenylurea are extremely low in comparison to values measured in the classical runs. This is probably due to the low amounts of Polybrene present on the electroblotted glass-fiber discs (not more than 100 μg of Polybrene is applied per cm^2; see Table 2).

It is also worth mentioning that the glycine contamination, which as expected from some amino acid analyses might be rather high, has never posed problems of identification of the first NH$_2$-terminal residues. This apparent anomaly may simply indicate that the glycinate which is not exchanged by chloride is covalently bound to the support. During acid hydrolysis these bonds are hydrolyzed giving rise to free glycine, while the milder conditions of the Edman degradation might not break the glycine-support bounds.

A second illustration of the glass-blot-gas-phase sequencing strategy is given for the NH$_2$-terminal sequence determination of the small subunit of the biotin-binding protein of *Streptomyces avidinii*. Ten micrograms of a commercial preparation of this protein (Bethesda Research Laboratories, USA) were located on a

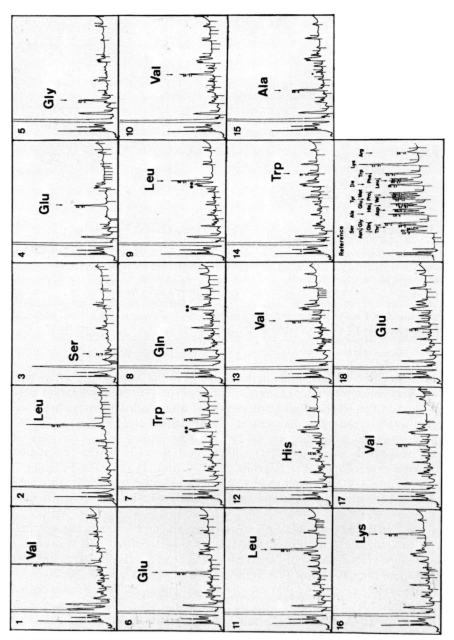

Fig. 7. HPLC traces from an amino-terminal amino acid sequence analysis of Polybrene glass fiber-immobilized sperm whale myoglobin. Amino acid phenylthiohydantoin derivatives were separated by HPLC analysis as described in Table 2. Absorbance was at 254 nm and the detector was set at 0.001 A unit full scale. The positions of the amino acid phenylthiohydantoins, assigned in the traces for cycle 1 through 18, are indicated by *arrows*. Two peaks of uncharacterized byproducts appearing in some chromatograms are indicated by asterisk (see, e.g., steps 7 – 9). Aliquots (50%) from each gel cycle are analyzed. From the calculated yields, initial coupling yields and repetitive yields were calculated. The last HPLC-trace shows the separation pattern of a reference mixture containing 100 pmol of each derivative

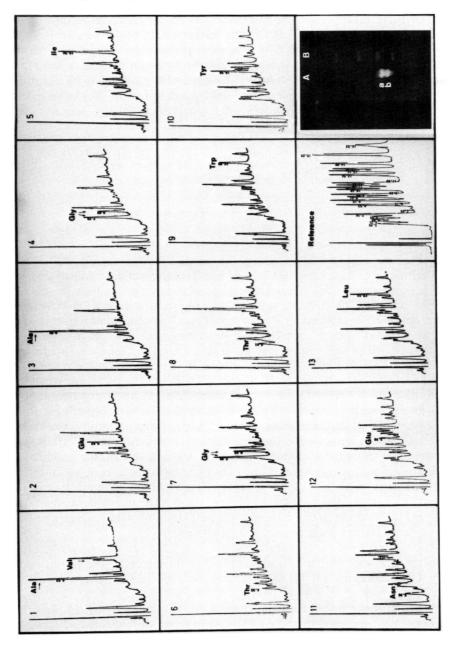

Fig. 8. Separation and sequence analysis of the subunits of streptavidin on a 17.5% SDS-polyacrylamide gel. The corresponding Polybrene glass-fiber blot containing approximately 500 pmol of protein stained with a dilute solution of fluorescamine (1 mg/500 ml of acetone) is shown in the right lower corner (*lane A*). *Lane B* shows the separated Bio Rad reference mixture. The small subunit (*b*) was taken for amino acid sequencing. HPLC traces are arranged and labeled as in Fig. 7

17.5% polyacrylamide gel and the two subunits were separated from each other
and from contaminating material. The fluorescamine-stained blots of this separation pattern is shown in Fig. 8. The glass-fiber surfaces containing the separated subunits were cut out with scissors (pieces of 2×8 mm were obtained) and
mounted on the gas-phase sequenator. The sequentially liberated PTH-amino
acid derivatives of a run on the small subunit (spot b in Fig. 8) are shown in the
traces of Fig. 8. The PTH-derivatives were separated by HPLC analysis (see
above). From these results the following NH_2-terminal sequence of the small subunit of streptavidin could be proposed: Ala or Val-Glu-Ala-Gly-Ile-Thr-Gly-Thr-
Trp-Tyr-Asn-Glu-Leu-Gly-. A similar analysis performed on the large subunit
showed a blocked NH_2-terminus.

5 Conclusion

In the technique of Polybrene glass-fiber blotting, proteins are transferred as
highly charged protein-SDS complexes and bound onto a glass support by media-
tion of the polybase. The binding is thought to be a combination of electrostatic
and hydrophobic interactions. It is very strong, and proteins bound in this way
cannot be removed without breaking the Polybrene-glass support interactions.
When glass-bound proteins are subjected to acid hydrolysis, an amino acid mixture is obtained which accurately reflects the amino acid composition of the protein without interference from the Polybrene glass fiber. The technique of glass-
fiber blotting will certainly find its most successful application in combination
with gas-phase sequence analysis. This is illustrated in the first experiment with
1 nmol of myoglobin and in another experiment with 500 pmol of the small subunit of streptavidin immobilized on glass fiber. The experiments shown in this
paper were not designed to achieve the highest sensitivity of the technique.
Nevertheless, they clearly form the basis for future experiments in which more
sensitive analysis systems will be used (e.g., precolumn derivatization with
phenylisothiocyanate followed by HPLC separation of the phenylthiocarbamoyl
derivatives [12], and the use of more sensitive PTH-amino acid analyzers or the
introduction of fluorescently labeled thiohydantoins).
  Proteins which have been immobilized on glass fiber can also be probed for
antigenicity in a way similar to immunoblotting on nitrocellulose (results not
shown). This possibility may allow the detection of specific antigens among the
immobilized proteins.
  In its actual form, Polybrene glass-fiber blotting is an extremely simple technique, convenient to perform and economical in terms of reagents and time. It
combines the simple, fast, inexpensive, generally applicable high resolution SDS-
polyacrylamide gel electrophoresis with a highly efficient elution system by directly immobilizing the separated proteins on a chemically resistent glass-fiber
support. The glass membrane-bound proteins can now be directly analyzed for
their amino acid composition and partial NH_2-terminal sequence. These new
properties extend the possibilities of previously used blotting techniques.

Acknowledgments. J. V. is Research Associate of the Belgian National Fund for Scientific Research (N.F.W.O.). We acknowledge the help of A. Caplan in the preparation of this manuscript. The work was supported by grants from the N.F.W.O. (in providing the gas-phase sequenator) and a grant of the Deutsche Forschungsgemeinschaft as part of a collaborative project with K. Weber, Göttingen, FRG.

References

1. Gershoni JM, Palade GE (1983) Anal Biochem 131:1 – 15
2. Renart J, Reiser J, Stark GR (1979) Proc Natl Acad Sci USA 76:3116 – 3120
3. Symington J, Green M, Brackmann K (1981) Proc Natl Acad Sci USA 78:177 – 181
4. Towbin H, Staehelin T, Gordon J (1979) Proc Natl Acad Sci USA 76:4350 – 4354
5. Gershoni JM, Palade GE (1982) Anal Biochem 124:396 – 405
6. Hewick RM, Hunkapiller MW, Hood LE, Dreyer WJ (1981) J Biol Chem 256:7990 – 7997
7. Drager RR, Regnier FE (1985) Anal Biochem 145:47 – 56
8. Vandekerckhove J, Van Montagu M (1974) Eur J Biochem 44:279 – 288
9. Laemmli UK (1970) Nature 277:680 – 681
10. Lu RC, Elzinga M (1977) Biochemistry 16:5801 – 5806
11. Vandekerckhove J, Weber K (1978) Eur J Biochem 90:451 – 462
12. Heinrickson RL, Meredith SC (1984) Anal Biochem 136:65 – 74
13. Vandekerckhove J, Bauw G, Puype M, Van Damme J, Van Montagu M (1985) Eur J Biochem 152:9 – 19
14. Hunkapiller M, Hood L (1983) Methods Enzymol 91:486 – 493
15. Benson J, Hare P (1975) Proc Natl Acad Sci USA 72:619 – 622

3.6 Radio-Sequence Analysis; an Ultra-Sensitive Method to Align Protein and Nucleotide Sequences

Nisse Kalkkinen[1]

Contents

1 Introduction

Development of methods and instruments for protein purification, sequential degradation, and amino-acid analysis have made protein sequencing easier, faster, and more sensitive than ever. But even today, when analyses can routinely be performed on the picomolar level, determination of a complete protein sequence by using only protein material is a laborious task.

Genetic engineering has made it possible to investigate proteins also on the nucleotide sequence level and in this way to obtain sequence information on nearly any protein of interest. These studies, however, require combined direct protein analyses to show, for example, where genes start, where signal sequences are cleaved off after membrane penetration, how precursor polyproteins are processed to yield the final mature ones and how proteins are post-translationally modified.

Partial direct protein sequences can be obtained either from unlabeled, chemically pure proteins, or from radiochemically pure proteins metabolically labeled with different radioactive amino acids.

There is no need to use the radio-sequence approach if the protein fragment of interest is available in sufficient amounts and can be purified to chemical

---

1 Recombinant DNA Laboratory, University of Helsinki, Valimotie 7, SF-00380 Helsinki, Finland

Advanced Methods in Protein Microsequence Analysis
Ed. by B. Wittmann-Liebold et al.
© Springer-Verlag Berlin Heidelberg 1986

purity. Many interesting proteins are, however, available only in such small amounts that radio-sequence analysis must be used. In this approach biosynthetically introduced radioactivity is used to elucidate positions of individual amino acids in the primary structure. Many detailed descriptions on radiochemical sequence analysis of biosynthetically labeled proteins have been published (e.g. [1 − 3]). In our laboratory we have used combined partial radio-sequence analysis and nucleic acid sequencing to elucidate polyprotein cleavage sites in Semliki Forest virus-specific structural and nonstructural polyproteins and to investigate where genes start and how signal sequences are cleaved off when bacteria developed by recombinant DNA-technology express and secrete foreign gene products.

2 Labeling of Proteins with Radioactive Amino Acids

2.1 General

Radioactive labeling of a protein involves metabolic incorporation of a radioactive amino acid into the protein of interest. This is usually done in a cell culture or in an in vitro translation system.

In a single-label experiment one type of labeled residue is incorporated into the protein. Release of radioactivity during Edman degradation of the radiochemically pure protein then identifies the positions of the labeled residue in the polypeptide chain. Usually two to three labelings with different amino acids, followed by seperate purifications and Edman degradations, are required to correlate the analyzed protein primary structure to the corresponding nucleotide sequence. The same result can be obtained by one multi-label experiment. Now the protein is labeled simultaneously with up to six to seven different radioactive amino acids. Direct measurement of released radioactivity in Edman degradation of a multi-labeled protein indicates, however, only the position(s) of the labeled amino acids in the primary structure, but the type of labeled amino acid in each positive assignment must be analyzed by radioactivity measurement of PTH-amino acids after separation by HPLC.

No general labeling scheme can be given because of variability in cell types and culture media used for translation of the proteins of interest. The principle, however, is that radioactive amino acids are introduced into the cell culture medium from which they are transported into the cell, mixed with the intracellular pool of unlabeled amino acids and incorporated into the protein during translation. After labeling, the protein of interest is recovered from the cell lysate or culture medium and purified to radiochemical purity.

2.2 Amino-Acid Choice

The choice of amino acids to be incorporated into the protein of interest is important. All 20 amino acids are available both as ^{3}H- and ^{14}C-labeled. For most radio-sequence applications tritiated amino acids are preferred. Their specific activity is high and they are generally less expensive than corresponding amounts of the ^{14}C-labeled ones. ^{35}S-Met and ^{35}S-Cys are, in addition, useful for special applications (e.g., as markers for detection of labeled protein bands in SDS-polyacrylamide gels) due to their high specific activity and high radiation energy. In a single-label experiment, the amino-acid choice is easy. The labeled amino acid should be common in proteins (to obtain as many positive assignments as possible in one sequencer degradation), its incorporation into proteins should be high, it should be metabolically stable and its price should be reasonable. It is also known that ^{3}H-radioactivity is lost from 3,5-^{3}H-Tyr in acid hydrolysis due to H-exchange at these positions [3]. We have with good results used ^{3}H-Val, ^{3}H-Leu, ^{3}H-Ala, ^{3}H-Ile, ^{3}H-Arg, ^{3}H-Phe and ^{3}H-Ser for single-label experiments. If there are special reasons to use long labeling times ($>1-2$ h), metabolic interconversion of the labeled amino acids may occur. At least aspartic acid, asparagine, glutamic acid, glutamine, serine, and glycine are known to undergo significant interconversion during long labeling experiments. The amount of radioactivity per sample needed for a single label radio-sequence analysis depends upon several factors. Generally, $300-500$ cpm per position of the labeled amino acid in the protein is sufficient. In some cases correct positive assignments with less than 100 cpm per position of the labeled amino acid in the protein have been obtained.

The synthesizing system can also be supplemented simultaneously with several radiolabeled amino acids. Advantages and disadvantages using single or multiple labels depend on the circumstances of the experiment at hand. If a single radioactive amino acid is used, usually two to three labelings with different types of amino acid followed by separate purifications and sequencer degradations of the labeled proteins are needed to establish correlation between the protein and its gene. The same result can be obtained by one multi-label experiment. The advantage of a multi-label experiment is that the number of labelings, purifications, and sequencer degradations is minimized. On the other hand, more incorporated radioactivity per residue is usually needed, since now the radioactivity liberated in each degradation cycle is analyzed after separation of the PTH-amino acids by HPLC.

The same factors as in a single-label experiment are important in amino-acid choice for a multi-label experiment. In addition, the radioactive amino acids should be stable as PTH-derivatives and well separatable on HPLC. Thus Ser and Thr are unfavorable, due to their instability as PTH-derivatives. The specific activity of all different labeled amino acids in the final purified multi-labeled protein should be about the same order of magnitude. Thus, a set of labeled amino acids with roughly equal incorporation efficiency into the protein of interest should be chosen or the proportions of the labeled amino acids in the growth medium should be adjusted to such a level that the radioactivity in the different labeled residues in the protein is about equal. Synthetic cell growth media contain large amounts of essential amino acids and usually also others. Labeling of proteins can be accomplished by reducing the amount of the corresponding unlabeled amino acid in the growth medium to less than 5% of the original amount or simply by omitting it from the medium for the time of labeling. Short labelings can be performed in media completely free from unlabeled amino acids.

In a successful labeling experiment, radioactivity should be directed as much as possible toward the protein(s) of interest, while proteins without interest remain unlabeled. This is not always possible, but some viral proteins, for example, can be specifically labeled in mammalian cells (see later) and in vitro translation systems usually label only the desired protein. If SDS-polyacrylamide gel electrophoresis is used to purify the proteins, a separate labeling with ^{35}S-Met or universally ^{14}C-labeled amino acid mixture (usually ^{14}C-labeled protein hydrolyzate) is performed. This preparation is then mixed with the ^{3}H-labeled sample to serve as a marker for detection of the ^{3}H-labeled proteins in the gel (see later).

2.3 Labeling of Semliki Forest Virus-Specific Proteins

In the following an example is given to show how radioactivity can be specifically directed to a limited group of proteins of interest. Because many detailed descriptions have been published (e.g., [4 – 6], only general aspects are discussed here. Semliki Forest virus-specific proteins are usually radiolabeled in virus infected baby hamster kidney (BHK-21) or chicken embryo fibroblast cells, growing on the surface of plastic dishes (for review, see [7]). To get label both in virus-specific nonstructural and structural proteins, high salt treatment is included into the labeling procedure [8, 9]. When virus-infected cells are transferred at the time of maximal synthesis of viral nonstructural proteins to a high salt medium (335 mM NaCl), initiation of all protein synthesis (both virus- and cell-specific) is inhibited but elongation still continues. This procedure removes all ribosomes from the messenger RNA's. When the cells now are transferred back to normal medium, virus-specific protein synthesis recovers faster than the cell-specific one and virus-specific proteins can be selectively labeled. After labeling, cells are dissolved in hot (2%, 60 °C) SDS and the lysate is analyzed by analytical SDS-polyacrylamide gel electrophoresis as shown in Fig. 1. ^{35}S-Met or ^{14}C-labeled proteins can be directly autoradiographed after drying of the gel and ^{3}H-labeled proteins are detected by fluorography [10]. In addition to universal labeling of all Semliki Forest virus proteins, variations in the labeling strategy can be used to direct radioactivity either preferentially to the structural proteins, nonstructural precursors or final proteins or even to some individual viral proteins [4 – 6]. Radiolabeled preparations discussed in the following examples were obtained by a procedure directing radioactivity into all virus-specific proteins [6]. Chicken embryo fibroblast cells were infected with Semliki Forest virus mutant ts-1. The whole procedure was performed in the presence of tunicamycin (to inhibit glycosylation and in this way make purification of some viral proteins easier) and actinomycin D (to inhibit host mRNA synthesis). High salt treatment was 4.5 – 5 h after infection and radiolabel was introduced into the amino acid-free medium (lacking the amino acids used for labeling) at 5 h after infection for 30 min, followed by a 5-min chase (to obtain labeled precursors) or 30-min chase (to obtain radioactive final proteins). Cells from each dish were dissolved after labeling in 500 μl 2% SDS at 60 °C.

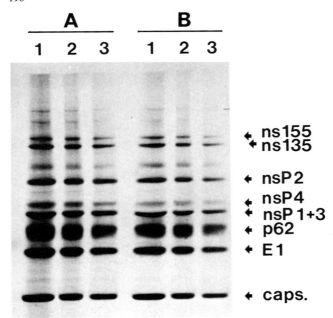

Fig. 1 A, B. Autoradiograms of Semliki Forest virus-induced proteins labeled with ^{35}S-Met (200 µCi) in chicken embryo fibroblast cells for 30 min. Labeling was performed on two dishes **A** and **B** by using a 30-min synchronized time delay in the labeling protocol for **B**. The pulse medium was, after labeling of **A**, transferred to **B**, where labeling was performed for 30 min (note slight dilution of radioactivity in **B** due to use of the same pulse medium as already used for **A**). Cells were disrupted after a 5-min chase in 500 µl 2% SDS at 60 °C. *1* = 20 µl; *2* = 10 µl; *3* = 5 µl aliquots from samples **A** and **B** were analyzed by electrophoresis on a 7.5 – 15% linear polyacrylamide gel in the presence of SDS. After electrophoresis the gel was dried (without treatment with an enhancer) and autoradiographed on a sheet of Kodak X-Omat AR film for 17 h

3 Purification of Radiolabeled Proteins

3.1 SDS-Polyacrylamide Gel Electrophoresis

SDS-polyacrylamide gel electrophoresis is a useful single-step purification procedure to obtain radiochemically pure proteins for subsequent sequence analysis. This method is equally suitable for cell lysates containing SDS as for precipitates obtained from the cell culture medium or in vitro translation mixture. Conventional purification methods, especially high performance liquid chromatography (anion and cation exchange and reverse phase separation) can also be used in most cases. An individual purification procedure must obviously be worked out for each protein of interest. As mentioned before, radio-sequence analysis requires a radiochemically pure protein, which means that the final preparation contains only the desired protein in radiolabeled form. Mostly, however, the preparation contains other unlabeled proteins, which do not affect the radio-sequence analysis but serve as carriers for the labeled one, which is sometimes present only in femtomolar amounts in the sample.

Table 1. Preparative SDS-polyacrylamide slab gel electrophoresis of radiolabeled proteins

1. Starting material: (a) ^{3}H-labeled cell lysate from one 5-cm dish in 500 µl 2% SDS. (b) ^{35}S-Met-labeled cell lysate from one 5-cm dish in 500 µl 2% SDS.

2. Determine the minimum amount of ^{35}S-Met labeled proteins needed to detect the protein bands by autoradiography of the dried gel by overnight exposure. For this, load, e.g., 20 µl, 10 µl and 5 µl of the ^{35}S-Met labeled sample on an analytical Laemmli gel. Dry the gel after electrophoresis and expose overnight on X-ray film (Kodak X-Omat AR, see Fig. 1).

3. Add the determined amount of the ^{35}S-Met labeled sample to the whole ^{3}H-labeled sample from one dish.

4. Load the sample on a preparative 7.5–15% linear gradient SDS-polyacrylamide gel (e.g., 15 cm × 15 cm × 1 mm). Use a preparative slot instead of the ordinary analytical ones).

5. Electrophorese at 60 V until bromophenol blue marker dye reaches the bottom of the gel.

6. Place a sheet of wetted dialysis membrane (slightly smaller in size than the polyacrylamide gel) on the filter paper and place the electrophoresed gel on the filter paper/dialysis membrane [the suggested protein band(s) should be on the dialysis membrane area].

7. Dry the gel in a slab gel dryer. Avoid excess heating. The gel now sticks to the filter paper from the edges, while the protein band area sticks to the dialysis membrane.

8. Mark the surrounding filter paper with radioactive ink in a geometrically unique way to facilitate exact localization of the exposed developed film on the dried gel.

9. Expose the dried gel on a film sheet over night (Fig. 2).

10. Align the developed film on the dried gel/filter paper according to the radioactive ink marker spots.

11. Mark the positions of the protein bands on the dried gel by sticking small holes with a needle (Fig. 2) through the developed film sheet to the dried gel surface.

12. Remove the film and cut out the protein containing gel slices according to the marker holes in the dried gel/dialysis membrane.

13. Store the protein-containing gel slices until elution at −20°C in screw-capped tubes.

Table 1 describes the SDS-polyacrylamide gel electrophoresis procedure for a chicken embryo fibroblast cell extract. The cell lysate from one dish (diameter 5 cm, 4×10^6 cells, 0.5 mg total protein) can be loaded on a preparative SDS-polyacrylamide slab gel (13 cm × 13 cm × 1 mm). Before electrophoresis, ^{35}S-Met or ^{14}C-labeled proteins are added to the corresponding mixture of ^{3}H-labeled proteins to make identification of the desired protein bands easier. We use the Laemmli buffer system [11] for electrophoresis. It contains a high concentration of primary amino groups (glycine) which can serve as a scavenger for reagents in the gel that may react with the amino termini of the proteins of interest. After electrophoresis, the protein bands can be localized with several different methods. We have used autoradiography of a wet gel ("wet exposure") [12], autoradiography of a vacuum dried gel (Fig. 2), slicing of a guide strip from the wet gel followed by solubilization of the slices and counting in a liquid scintillation counter [4]. According to our experience, the most convenient method is autoradiography of a vacuum dried gel. For exposure the gel is dried in an ordinary slab gel dryer without excess heating. The gel is dried on a Whatman 3 MM filter paper but a wetted dialysis membrane is placed between the gel and the filter

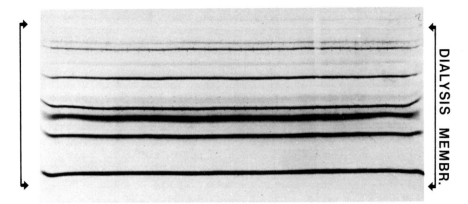

Fig. 2. Preparative SDS-polyacrylamide gel electrophoresis of radiolabeled Semliki Forst virus-specific proteins. The whole ^{3}H-Ala (500 μCi) labeled cell lysate from one dish (in 500 μl, 2% SDS) was mixed with 50 μl of the ^{35}S-Met labeled cell lysate (total volume 500 μl, dish A, Fig. 1). This whole mixture was applied on a preparative (13 cm × 13 cm × 1 mm) 7.5 – 15% linear polyacrylamide gel containing SDS. After electrophoresis the gel was dried on a dialysis membrane/filter paper (dialysis membrane on the *area between the arrows*) and autoradiographed overnight. The developed film was positioned on the dried gel according to radioactive marker ink spots on the surrounding filter paper. Holes were made through the film into the dried gel to mark the protein bands for cutting. The protein pattern in this preparative gel corresponds to that shown in Fig. 1

paper at position(s) of the protein band(s) of interest (see Fig. 2). In this way the protein bands are actually dried on a dialysis membrane, and the rest of the gel sticks on the filter paper. Drying of the gel prior to exposure has some advantages. It is easier and more accurate to cut out narrow, closely locating bands from a dried gel than from a wet one. A dried gel can also be exposed (if needed) for several weeks without affecting the sequencing results, while bands from a wet gel should be cut out as soon as possible (usually within 12 h) to avoid band broadening by diffusion. If the protein is electrophoretically eluted from the gel slice, drying followed by rehydration is a convenient way to change the original electrophoresis buffer to another more suitable for electrophoretic elution.

3.2 Elution of Proteins from SDS-Polyacrylamide Gels

For sequence analysis the radiolabeled proteins are eluted from rehydrated gel slices. This is done either by diffusion or by electrophoretic elution. For elution by diffusion, the gel slice is rehydrated and cut into small pieces (about 1 mm^3). The pieces are then incubated in a volume about five times that of the wet gel slice overnight by slow agitation at room temperature or at 37 °C. We have had good results with 1% formic acid or 0.05% SDS in water for elution. The recovery varies depending, e.g., on the size of the protein but typically (protein size smaller than about 50,000) it should be at least 50%. After elution the protein is in a comparatively large volume of elution buffer and must be concentrated, for example, by precipitation with 15% trichloroacetic acid at 0 °C, followed by acetone wash of the precipitate at −20 °C. Electrophoretic elution is an alternative

Table 2. Electrophoretic elution of radiolabeled proteins from SDS-polyacrylamide gels

1. Rehydrate the protein containing dried polyacrylamide gel slice in 3 ml 10 mM ammonium bicarbonate, 0.01% SDS. The gel slice will swell to its original volume and the dialysis membrane is separated from the gel in about 1 min.

2. Remove the dialysis membrane from the solution and cut the gel piece into small pieces (1 mm^3) with sharp scissors.

3. Transfer the gel pieces and the liquid into the large cup compartment of the ISCO 1750 concentrator and fill the cup to the final volume. Add 100 µg myoglobin (2 mg ml^{-1} elution buffer) to the compartment containing the gel pieces. Use 10 mM ammonium bicarbonate, 0.01% SDS in the cup compartment of the eluator and 50 mM ammonium bicarbonate, 0.01% SDS in the electrode compartments.

4. Elute at 3 W for 3 h. During this time the red-colored myoglobin should reach the final sample compartment.

5. Empty the cup in the order suggested by the manufacturer.

6. Remove the eluted, concentrated radiolabeled protein from the small sample compartment (in 200 µl).

method to that based on diffusion. Different electro-elution methods and equipment have been published (e.g. [13]). Table 2 describes how to elute radiolabeled proteins by using the commercially available ISCO Model 1750 electrophoretic concentrator [14]. The protein is recovered in a 200-µl volume of elution buffer and this whole sample can be applied without further treatment to a spinning cup sequencer. For gas/liquid phase sequencing the protein is further precipitated and redissolved in a smaller volume (e.g., 40 µl 30% formic acid) before application to the glass fiber filter.

3.3 Determination of Recovered Radioactivity

In a single-label experiment, recovery of radioactivity is measured by counting an aliquot of the sample in a liquid scintillation counter. In most cases the sample contains both ^{3}H-radioactivity introduced into the proteins to give partial sequence information and ^{35}S- or ^{14}C-radioactivity to serve for detection of the protein bands in the dried gel. Relative amounts of ^{3}H- and ^{35}S- or ^{14}C-radioactivity can be estimated by counting the sample on two separate channels in the liquid scintillation counter. In a multi-label experiment, the situation is more complicated. Direct measurement of the sample gives an estimation about the total amount of incorporated radioactivity, but does not tell anything about radioactivity incorporated into different amino acids. In this case an aliquot of the multi-labeled protein is hydrolyzed and radioactivity in each type of amino acid used for labeling is determined as described in Table 3. We use mainly precolumn derivatization with Dabsyl-chloride [15] followed by HPLC separation, collection and counting of the amino-acid derivatives. Recently we have also obtained promising results by using precolumn derivatization with phenyl isothiocyanate [16].

Table 3. Determination of incorporated radioactivity in different types of amino acids in a multi-labeled protein

---

1. Transfer an aliquot volume of the electro-eluted protein sample into a hydrolysis tube (4 mm I.D. $\times$ 50 mm). Add amino-acid standard mixture (10 µl, 1 nmol of each individual amino acid). Dry in a vacuum desiccator.

2. Add 50 µl constant boiling hydrochloric acid containing 0.1% phenol. Evacuate the tube and close it on a flame. Hydrolyze 24 h at 116 °C.

3. Open the tube and remove hydrochloric acid in a vacuum desiccator.

4. Add 20 µl 0.1 M sodium bicarbonate, pH 9.1 and 40 µl Dabsyl-chloride (1.3 mg ml^{-1} in acetone). Close the tube with a silicone rubber stopper. Incubate 30 min at 55 °C. Dry in a vacuum desiccator.

5. Dissolve the residue in 30 µl 70% ethanol and separate the Dabsyl-amino acids by HPLC [15]. We use for separation a Spherisorb S5 ODS2 reverse phase column (4.6 $\times$ 250 mm) and a linear 25 min gradient from 20% to 70% acetonitrile in 50 mM sodium acetate, pH 4.110 followed by an isocratic 10 min elution. Flow is 1 ml min^{-1} and detection at 436 nm.

6. Collect the desired amino-acid peaks (about 400 µl each) directly into liquid scintillation counter vials (5 ml). Add the scintillation cocktail and count.

7. Correct the results with a factor obtained when a standard preparation (known amount of radio-activity) is hydrolyzed, derivatized, separated, collected, and counted similarly to the unknown one.

---

4 Sequence Analysis

Usually the radiolabeled protein is subjected to Edman degradation in a sequencer in order to obtain as many positive assignments as possible during one analysis. Manual degradation methods can also be used (e.g., [4]), but now positive assignments should be within 10 – 15 cycles due to the relatively poor repetitive yields (typically around 90%) in this method. Our experience is mainly based on degradation of radiolabeled proteins in spinning cup sequencers (Beckman 890C and 890D). Sensitivity of the sequencer is not the limiting step in a radiosequence analysis because degradation of a radiolabeled sample is always performed in the presence of an added unlabeled carrier protein (about 5 nmol), which also serves as an internal standard to check the sequencer performance. A high degree of initial coupling and good repetitive yields are important in radiosequence analysis in order to yield a good signal-to-noise ratio and to reach positive assignments as far as possible in the sequence. Table 4 shows the strategy of a typical sequencer degradation and following analysis of results. Degradation is mostly performed in the presence of Polybrene [17] but its precycling (two to four cycles in the sequencer) is unnecessary unless it contains impurities which react with primary amino groups and in this way affect the degree of initial coupling. Other impurities, giving extra peaks in the first one to three HPLC chromatograms do not seem to affect the radio-sequence results, while the sequencer performance can be checked also from later HPLC chromatograms. Considerable large volumes of electro-eluted samples containing ammonium bicarbonate and SDS can be loaded into the spinning cup. We routinely load the whole electro-eluted sample (200 µl diluted to a sufficient volume to reach the correct level in

Table 4. Radio-sequence analysis of a biosynthetically labeled radiochemically pure protein

1. Apply 3 mg of Polybrene into the sequencer spinning cup. Dry. Precycle for two to five cycles using a normal degradation cycle if needed.

2. Apply the electro-eluted sample (Table 2) into the sequencer cup in a sufficient volume to reach the correct liquid level in the cup. Dry. Perform one "0-cycle" (normal degradation cycle without addition of phenyl isothiocyanate).

3. Start the degradation (we use Beckman 0.1 M Quadrol programs, Program part No. 345801 and 347336).

4. Convert the extracts (sequencer without autoconverter) in 90 µl 25% trifluoroacetic acid at 55 °C for 25 min and dry. Dissolve the converted sample in 100 µl methanol (single label) or 100 µl methanol containing 10 nmol of each PTH-amino acid corresponding to those used for labeling (multi-label).

5a. (Single-label experiment). Inject 20 µl to the HPLC (PTH-amino acid separation system) and count the rest 80 µl in a liquid scintillation counter after addition of 4 ml scintillation cocktail (e.g., Packard Insta-Gel). Check sequencer performance from the HPLC-chromatograms.

5b. (Multi-label experiment). Count 20 µl directly after addition of the scintillation cocktail. Inject 80 µl into HPLC and collect the desired PTH-amino acids according to the positions of the added PTH-amino acid standards. Count the collected peaks in a liquid scintillation counter. Check the performance of the sequencer by following the carrier protein sequence on the HPLC chromatograms.

6. Plot the obtained radioactivity from each sequencer cycle in a diagram (see Figs. 3 and 4).

the cup) but in this case degradation is started with a "0-cycle" (a normal degradation cycle without addition of PITC) to remove some impurities and nonspecific radioactivity from the cup. Normal degradation is then performed with double coupling in the first cycle. In a single-label experiment an aliquot (usually 20%) of the converted extract from each cycle is analyzed by HPLC to confirm the sequencer performance (to calculate repetitive yields and degree of initial coupling from the HPLC chromatograms). The rest of the sample (80%) is counted in a liquid scintillation counter. Radioactivity then directly identifies the presence of that kind of an amino acid in the given position of the polypeptide chain (see Figs. 3 and 4). Usually two to three single-label experiments are needed to unambiguously correlate the polypeptide sequence with the corresponding gene structure. Partial radio-sequence analyses can also confirm identical amino-termini of proteins suggested to be related to each other as shown for Semliki Forest virus nonstructural proteins nsP3 and its precursor ns135 (Fig. 3A). In a radio-sequence analysis the amount of released radioactivity in positive positions should always correspond to the expected activity per residue and to the repetitive yield of the sequencer (calculated from the carrier HPLC chromatograms). Radioactivity in the positive positions should also roughly correlate to the amount of standard initial coupling when extrapolated to the first cycle. Positive peaks not following the calculated repetitive yield indicate that the analyzed sample is not radiochemically pure and the result consists of labeled amino acids released from a mixture of two or more radiolabeled proteins. In a multi-label experiment a mixture of unlabeled PTH-amino acids (10 nmol of each corresponding to those used for labeling) is added to the converted sample from each degradation cycle to allow collection of the desired peaks. The added amounts of

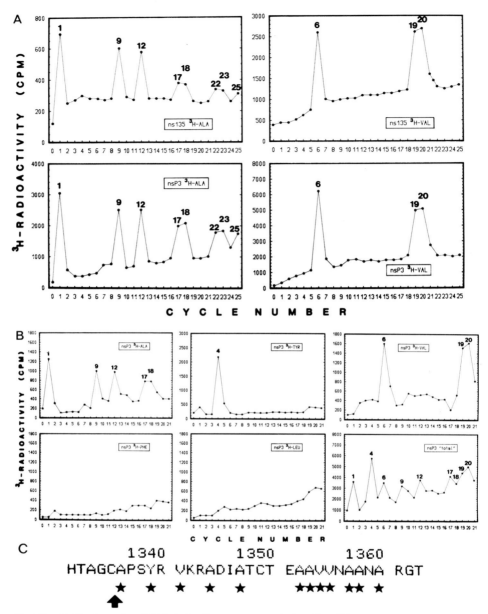

Fig. 3A – C. Single- and multi-labeled radio-sequence analysis of Semliki Forest virus nonstructural protein nsP3 and its precursor ns135. **A** The panels show recovered radioactivity in sequencer degradations of single-labeled nsP3 and ns135. The samples were separately labeled with ³H-Ala and ³H-Val, separately purified and separately degraded in the sequencer. **B** Radioactivity recovered in one sequencer degradation of multi-labeled (³H-Ala, ³H-Tyr, ³H-Val, ³H-Phe, and ³H-Leu) nsP3. Panels show radioactivity in different PTH-amino acid fractions after separation by HPLC and collection. Last panel, "total" shows radioactivity recovered in each sequencer cycle before HPLC separation. **C** Comparison of obtained partial radio-sequence results with the deduced polyprotein amino-acid sequence. *Numbers* indicate amino-acid positions from the amino terminus of Semliki Forest virus nonstructural precursor polyprotein (Takkinen et al., to be published). *Arrow* indicates cleavage site in the nonstructural polyprotein generating the analyzed amino-terminal structures of nsP3 and ns135. Positive assignments in the radio-sequence analyses are indicated by *asterisk*

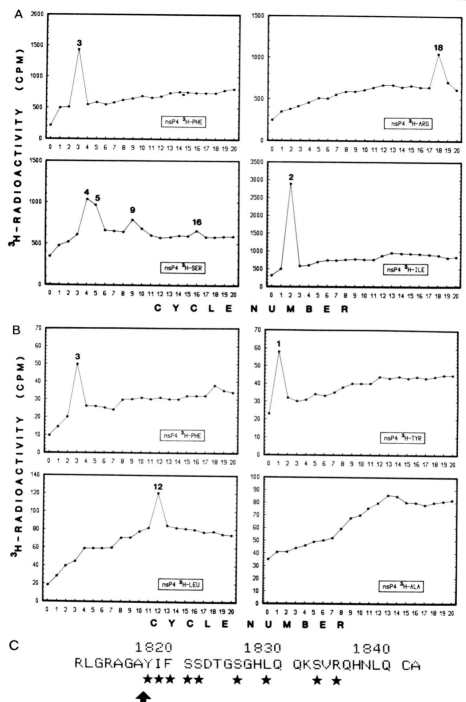

Fig. 4A – C (Legend see p. 206)

PTH-amino acids, however, allow detection of the PTH-amino acids released from the carrier protein for calculation of repetitive yields. An aliquot (20%) of the converted sample is counted directly (Fig. 3B, last panel) to obtain suggestions of positive assignments in the cycles. These suggestions are then confirmed by fractionation of the rest of the sample (80%) by HPLC. Desired PTH-amino acid fractions are collected and counted in a liquid scintillation counter. Results from two multi-label experiments are shown in Figs. 3B and 4B. The latter shows that reliable results can be obtained even with extremely low amounts of radioactivity per position.

For efficient collection of HPLC peaks, the delay from the detector cell to the collection line outlet must be determined. This is easily done, for example, by injecting any kind of PTH-amino acid into the HPLC and starting collection of drops from the outlet at the time of peak maximum in the chromatogram. Each collected fraction (e.g., 2 or 3 drops/fr) is then separately re-injected and the fraction with maximum concentration is determined. The delay can then be calculated from the determined drops/minute value.

Acknowledgments. Thanks are due to Prof. Hans Jörnvall for many kinds of advice, help, and support, and to Dr. Hans Söderlund and Prof. Leevi Kääriäinen for valuable discussions.

References

1. Anderson CW (1982) In: Setlow JK, Hollaender A (eds) Genetic engineering, principles and methods, vol 4. Plenum, New York, p 147
2. Coligan JE, Gates III FT, Kimball ES, Maloy WL (1983) Methods Enzymol 91:413
3. Jörnvall H, Kalkkinen N, Luka J, Kaiser R, Carlquist M, von Bahr-Lindström H (1983) In: Tschesche H (ed) Modern methods in protein chemistry. de Gruyter, Berlin New York, p 1
4. Kalkkinen N, Jörnvall H, Kääriäinen L (1981) FEBS Lett 126:33
5. Kalkkinen N, Laaksonen M, Söderlund H, Jörnvall H (1981) Virology 113:188
6. Keränen S, Ruohonen L (1983) J Virol 47:505
7. Kääriäinen L, Söderlund H (1978) Curr Top Microbiol Immunol 82:15
8. Saborio JL, Pong SS, Kochi G (1974) J Mol Biol 85:195
9. Lachmi B, Kääriäinen L (1976) Proc Natl Acad Sci USA 73:1936
10. Bonner MW, Laskey R (1974) Eur J Biochem 46:83 – 88
11. Laemmli UK (1970) Nature 277:680
12. Kalkkinen N, Jörnvall H, Söderlund H, Kääriäinen L (1980) Eur J Biochem 108:31
13. Hunkapiller MW, Lujan E, Ostrander F, Hood L (1983) Methods Enzymol 91:227
14. Bhown AS, Mole JE, Hunter F, Bennett JC (1980) Anal Biochem 103:184
15. Chang J-Y, Knecht R, Braun DG (1981) Biochem J 199:547
16. Heinrikson RL, Meredith SC (1984) Anal Biochem 136:65
17. Tarr GE, Beecher JF, Bell M, McCean DJ (1978) Anal Biochem 84:622

Fig. 4A – C. Single- and multi-label radio-sequence analysis of Semliki Forest virus nonstructural protein nsP4. A Radioactivity recovered in sequencer degradation of single-labeled nsP4. Four different labelings, purifications, and sequencer degradations were performed. B Radioactivity recovered in sequencer degradation of multi-labeled nsP4. Panels show radioactivity obtained in PTH-amino acid fractions after HPLC separation of the sequencer extracts. In this experiment an extremely low total amount of radioactivity was loaded into the sequencer. C Localization of the amino-terminus of nsP4 in the deduced polyprotein amino acid sequence. Cleavage site is indicated by *arrow* and positive assignments in the radio-sequence analyses are indicated by *asterisk*

Chapter 4
Phenylthiohydantoin Identification, On-Line Detection, Sequences Control, and Data Processing

4.1 Conversion of Anilinothiazolinone to Phenylthiohydantoin Derivatives and Their Separation by High Pressure Liquid Chromatography

AJIT S. BHOWN and J. CLAUDE BENNETT[1]

Contents

1 Introduction

In the isothiocyanate degradation of proteins/peptides, the end product is a very unstable derivative called 2-anilino-5-thiazolinone (ATZ). This thiazolinone is then converted to its isomer and more stable 3-phenyl-2-thiohydantoin (PTH) derivative. The conventional techniques for conversion involve heating (50° – 80°C) the ATZ derivative under nitrogen in acidic conditions [1 – 4]. In the newest commercial sequencers this step has been automated. Two alternative procedures have also been proposed to simplify the conversion step. Guyer and Todd [5] have proposed thermal conversion, while Inman and Apella [6] suggested conversion to phenylthiocarboxyl amino-acid alkylamides.

Following conversion, the final step for each cycle in the amino-acid sequence analysis is the identification of PTH derivatives. The most widely employed method to achieve this is reverse-phase high pressure liquid chromatography (RP-HPLC). A number of reports [7 – 15] have appeared from this and other

---

1 Department of Medicine, Division of Clinical Immunology and Rheumatology, University of Alabama at Birmingham, Birmingham, AL 35294, USA

Advanced Methods in Protein Microsequence Analysis
Ed. by B. Wittmann-Liebold et al.
© Springer-Verlag Berlin Heidelberg 1986

laboratories in recent years, demonstrating complete or near complete separation of PTH derivatives of the 20 generally occurring amino acids.

Existing methods for conversion and identification of the final products suffer from a variety of inherent problems. Conversion with aqueous acid invariably results in deamidation of glutamine and asparagine to their respective acids, while methanolic acid causes esterification of acidic residues. Certain PTH derivatives need adjustment of pH or time for quantitative conversion [16].

Similarly, separation of PTH derivatives by RP-HPLC requires precise pH adjustment for the initial separation of PTH amino acids. Although such separation is based primarily on hydrophobicity, initial separation of acidic amino-acid derivatives and their amides is very much pH-dependent. In fact, it has been our experience that a pH change of 0.01 unit causes acidic amino acids to coelute with amides.

In order to circumvent these problems, Bhown and Bennett [17] have employed a new medium for conversion of thiazolinone to thiohydantoin derivative and a modified method for the separation of PTH derivative of amino acids by reverse-phase high pressure liquid chromatography. The experimental details and results obtained are described in this chapter.

2 Materials

2.1 Chemicals

Chemicals for the sequencer were purchased from Spinco Division of Beckman Instruments (Palo Alto, CA) and solvents for HPLC from MCB Manufacturing Chemists Inc. (Cincinnati, OH). Sequential grade trifluoroacetic acid was purchased from Pierce Chemicals (Rockford, IL). It is important that water to be used for HPLC should be highly purified. Use of the Milli-Q water purification system is recommended.

2.2 Instrumentation

Results of amino-acid sequence analyses shown in Figs. 2 – 7 were obtained on a modified [18] Beckman sequencer 890C and on a Beckman 890M equipped with an autoconverter. Two different HPLC systems were employed for the separation and identification of PTH derivatives; Perkin-Elmer Series 4 microprocessor controlled solvent delivery system with a Rheodyne injector Model 7125, and a Waters HPLC equipped with two 6000A solvent delivery pumps, WISP 710B and a model 440 dual channel absorbance detector. An ultrasphere ODS 5 μm column (0.46 × 15.0 cm) manufactured by Altex Scientific Inc. and marketed by Beckman Instruments (Part #235330) was employed.

3 Amino-Acid Sequence Analysis

The amino-acid sequence analysis may be performed on natural and/or synthetic protein/peptide (1 – 2 nmol) by manual or automated methods. However, if the sequencer is equipped with an autoconverter, make sure that the old conversion medium has been replaced with the one described in this procedure and that all the "0" rings, vent and delivery valve fittings exposed to the vapors are changed accordingly, to make them compatible with the vapors of the new conversion medium.

4 Conversion

This is the last of the series of reactions of isothiocyanate degradation in which the unstable 2-anilino-5-thiazolinone derivative (the cleaved product of the PTC coupled polypeptide) is converted (Fig. 1) by molecular rearrangement to the iso-meric and more stable form called 3-phenyl-2-thiohydantoin (PTH). The conver-sion is carried out under acidic conditions, and the rate increases with the hydro-gen ion concentration [1]. Certain PTH amino acids tend to decompose during the conversion. PTH-serine, which initially undergoes β-elimination and then polymerization, is most rapidly decomposed, followed by PTH-threonine, which exhibits a slower rate of β-elimination. The end product is more stable, exhibiting absorption maxima at 313 nm. The other amino acids difficult to identify at the picomole level are the amides, which tend to hydrolyze to their respective acids. PTH-histidine, PTH-arginine, and PTH-cysteic acid cannot be extracted from the aqueous layer because of their charges.

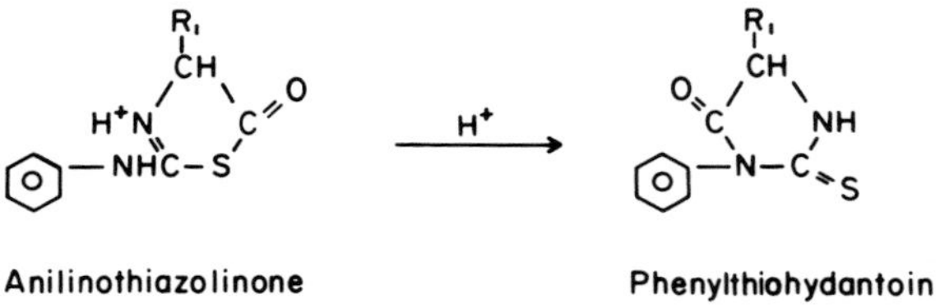

Anilinothiazolinone   Phenylthiohydantoin

Fig. 1. Conversion of anilino-thiazolinone to phenylthiohy-dantoin

4.1 Procedure

The conventional and modified procedure for conversion of thiazolinone to thio-hydantoin derivatives are described below.

4.1.1 Conventional Procedure

In the conventional procedure any of the following conversion mediums may be used:

1. 1.0 N Hydrochloric acid in water.
2. 1.0 N Hydrochloric acid in methanol.
3. 25% Trifluoroacetic acid in water.
4. 25% Trifluoroacetic acid in methanol.

Dry the 2-anilino-5-thiazolinone residue of the cleaved amino acid under the stream of nitrogen. Add 100–200 µl of the conversion fluid to the tube, flush with nitrogen, cover with Teflon stopper and incubate at 80 °C for 10 min. In autoconversion devices, lower the incubation temperature to 50° – 60 °C and increase the time to 15 – 30 min. If aqueous conditions are used for conversion, then add 1.0 ml of ethyl acetate to the tube, vortex, centrifuge, and remove the upper organic layer containing phenylthiohydantoin derivative. Repeat the extraction procedure one more time and pool the organic extracts. Dry both the organic and aqueous layers by freeze drying or under nitrogen. It is advisable to dry the organic layer under nitrogen. However, if acidified methanol is used, then dry the mixture under a stream of nitrogen. The dried residue is then dissolved into the suitable HPLC buffer for the identification of the amino-acid residue as its PTH derivative.

4.1.2 Modified Procedure

As pointed out earlier (Sect. 1), the conventional procedures of conversion suffer from a variety of inherent problems of deamidation and esterification. Obviously, these problems are more troublesome when sequence analysis is being attempted at picomole levels. In the new procedure we have circumvented these problems by using 10% TFA in ethyl acetate. The introduction of ethyl acetate completely and simultaneously suppresses deamidation and esterification (Figs. 2 – 4) without any adverse effects on the chemistry of conversion. The procedure is essentially as described under Section 4.1.1, except use 10% TFA in ethyl acetate and dry the extract under nitrogen.

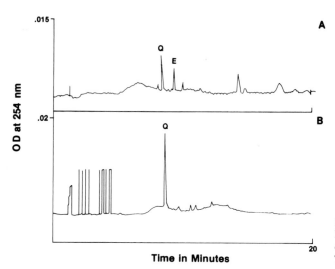

Fig. 2A, B. Identification of PTH glutamine. A 1.0 N HCl conversion medium; B 10% TFA in ethyl acetate [17]

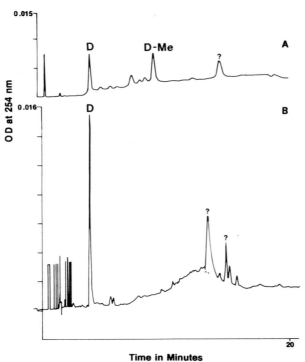

Fig. 3 A, B. Identification of PTH aspartic acid. **A** 25% TFA in Methanol D-Me = Methylester of PTH aspartic acid; **B** 10% TFA in ethyl acetate. ? = Unidentified peaks, which do not correspond to any standard PTH amino-acid peaks [17]

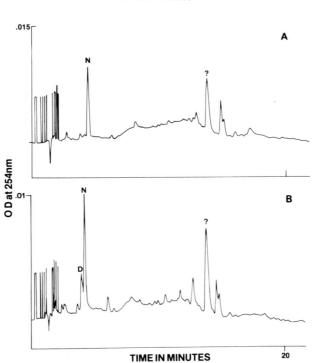

Fig. 4 A, B. Identification of PTH-asparagine. **A** 10% TFA in ethyl acetate; **B** 1.0 N HCl in water [17]

5 High Pressure Liquid Chromatography

Elucidation of the primary structure of proteins involves two fundamental steps: the sequential degradation of proteins by which the NH_2-terminal amino acids are converted into PTH derivatives and the identification of these derived amino acids. Automated sequencers have greatly enhanced the speed of the first step, and high pressure liquid chromatography (HPLC) is a promising addition to the methodology of PTH derivative identification and quantification. Efforts are being constantly directed toward developing new procedures or improving the existing ones for better separation and quantitation of PTH derivatives by reverse-phase high pressure liquid chromatography (RP-HPLC) [7 – 15]. As mentioned earlier in Section 1, although separation of PTH derivatives on reverse-phase high pressure liquid chromatography is primarily based on their hydrophobicity, separation of acidic amino acids and their amides is to a great degree pH-dependent. The existing methods, therefore, warrant a precise pH adjustment to within 0.01 unit, which makes these procedures a little difficult. We have developed [17] a simple system which does not require any pH adjustment. The desired pH is obtained by precise mixing of the solvents A (acidic) and B (basic) by the microprocessor controlled solvent delivery system (pumps) of HPLC, thus creating an accurate and reproducible pH during the course of chromatography. The details are described below.

5.1 Procedure

Prepare the solvents A and B as follows:

1. *Solvent A:* Add 750 – 1000 µl of acetic acid and 350 µl of acetone to 1 l water purified by Milli-Q water purification system or any other equally efficient system and filter through a 0.2 µm unipore polycarbonate membrane (Bio-Rad Laboratories, Richmond, CA).
2. *Solvent B:* To 1 l methanol (HPLC grade) add 900 µl of triethylamine.
   Prepare a standard containing approximately 200 – 250 pmol of PTH derivatives of each of the 20 amino acids in methanol or ethyl acetate. Use an ultrasphere ODS 5 µm column (0.46 × 15.0 cm) manufactured by Altex Scientific Inc. and marketed by Beckman Instruments (Part #235330).

The conditions for column development are:

Initial	– 77% A + 23% B for 2 min
Final	– 54% A + 46% B in 5 min
Hold	– 54% A + 46% for 10 min
Flow	– 1.5 µl min^{-1}
Chart speed	– 1 cm min^{-1}
Detector Sensitivity	– 0.02 full scale deflection.

Equilibrate the column first with initial conditions and then execute a blank run to make certain that nothing elutes from the column by the gradient to be em-

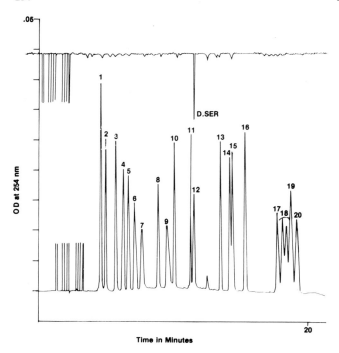

Fig. 5. Separation of 20 standard PTH amino acids (200 – 250 pmol; 0.02 full scale deflection) without any pH adjustment of Solvent A. *1* Aspartic acid; *2* asparagine; *3* serine; *4* threonine; *5* glycine; *6* glutamine; *7* CM-cysteine; *8* glutamic acid; *9* histidine; *10* alanine; *11* tyrosine; *12* arginine; *13* proline; *14* methionine; *15* valine; *16* tryptophane; *17* phenylalanine; *18* iso-leucine and allo-isoleucine; *19* lysine; *20* leucine. Absorbance 254 nm (*bottom*) and 313 nm (*top*) [17]

ployed for the separation of PTH derivatives. Such a blank run should give a flat baseline even at a high sensitivity of UV detection at 254 and 313 nm. Following the blank run, inject a mixture of standard PTH derivatives of amino acids (200 – 250 pmol each). After the separation of standard PTH derivatives has been obtained (Fig. 5), inject unknown sample containing an amount approximately equal to that of standard.

5.2 Quantitation

To quantitate the results, measure the peak heights of standard derivatives and of unknown sample and substitute in the following equation:

$$\frac{\text{Amount in standard}}{\text{Peak height of standard}} \times \text{Peak height of unknown} = \text{unknown sample amount in the volume injected.}$$

6 Precautions

It is strongly recommended that, while working with acids such as TFA and acetic acid, or other reagents, such as triethylamine, laboratory hoods with proper ventilation be used. We have observed that different HPLC systems and different batch columns may require minor adjustments in the initial condition (Sect. 5.1) or in the volume of acetic acid added to the Solvent A. The elution time of PTH-histidine and PTH-arginine can be easily altered by changing the ionic concentration of acetic acid and/or triethylamine, as has been observed by Tarr [11] for a different system. As with any separation by HPLC, proper column care and a constant vigil on the solvent delivery system (pumps) for any kind of leaks, even small ones, must be maintained. The HPLC equipment must be set for proper back-pressure limit to prevent undue pressure on the column, which will otherwise pack the column, creating a void and thus affecting the separation. If possible, the samples should be centrifuged before injection, and finally, use of a guard column is strongly recommended.

7 Conclusions

Sequential release of amino acids by Edman degradation has been refined, in terms of its chemistry, instrumentation, and identification of the end products, to an extent that amino-acid sequence information can now be obtained relatively easily on picomole amounts of proteins and peptides. Unfortunately, conversion of thiazolinone to thiohydantoin derivatives has not received much attention, other than (1) the introduction of an autoconversion device [2] and (2) utilization of aqueous or methanolic TFA. In this standard system, partial deamidation of amides and esterification of acidic residues are anticipated.

Some disadvantages under the existing conversion conditions are that such residues invariably elute as two peaks (Table 1; Figs. 2A, 3A, and 4B) on RP-HPLC analysis. Obviously, this problem is most troublesome when sequence analysis is being attempted at picomole levels. We have successfully avoided this problem by using 10% TFA in ethyl acetate as the conversion medium for both manual and autoconversion systems. The results are shown in Figs. 2B, 3B, and

Table 1. Different conversion mediums and their effects

Residue	Number of peaks in			
	1.0 N HCl in H_2O	25% TFA in H_2O	25% TFA in MeOH	10% TFA in EtAc
Aspartic acid	1	1	2	1
Asparagine	2	2	1	1
Glutamic acid	1	1	2	1
Glutamine	2	2	1	1

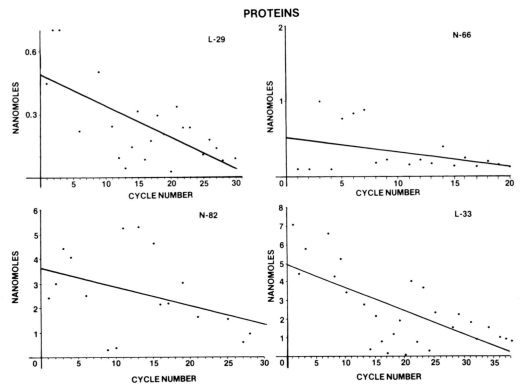

Fig. 6. Quantitation of sequence analysis of proteins employing TFA: ethyl acetate conversion and improved HPLC system [17]

4A and summarized in Table 1. It is clear from Figs. 2B, 3B, and 4A that, when ethyl acetate is used as the conversion medium, then amides and acidic residues elute as individual peaks on RP-HPLC. Concomitantly, one sees a significant increase in the detection limits of these residues. This system is equally efficient for PTH-serine, PTH-threonine, PTH-arginine, and PTH-histidine, which are quantitatively converted.

Figure 5 shows the separation of all 20 PTH derivatives of amino acids (200 – 250 pmol each; 0.02 full scale deflection) on an ultrasphere ODS column using the system described in Section 5.1. Most available methods employ C_{18} or ODS columns with methanol and/or acetonitrile as solvents [8, 10, 17]. The separation is achieved by developing the columns at ambient or elevated temperatures [8, 10, 17]. In practice, however, they suffer from failure to give baseline separation, and/or peaks eluting too closely. In addition the column life is reduced if one uses elevated temperatures. An often frustrating problem with these methods is that a slight change of pH (<0.01 unit) significantly alters the elution pattern. However, the method described in this chapter is reproducible and the baseline separation is achieved within 15 min at ambient temperature

PEPTIDES

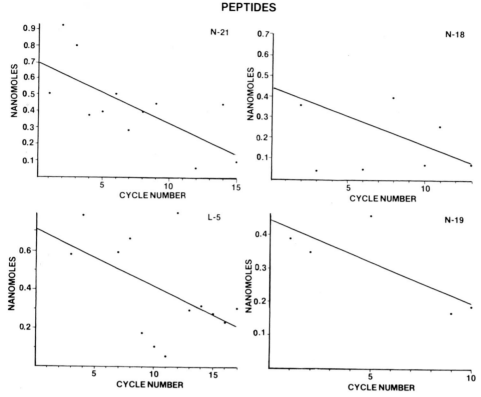

Fig. 7. Quantitation of sequence analysis of peptides employing TFA: ethyl acetate conversion and improved HPLC system [17]

without employing acetonitrile, a potentially toxic chemical. Furthermore, solvent A does not require adjustment for pH. The precise mixing by the microprocessor-controlled solvent delivery systems of HPLC mixes acidic solvent A with methanol containing triethylamine thus creating an accurate and reproducible pH. Figures 6 and 7 show the results of protein and peptide sequence analysis obtained on picomole and nanomole amounts of the sample.

Acknowledgments. The authors wish to thank James Wayland and Lisa Kallman for their expert technical assistance, as well as Sandra Reid for typing this manuscript. This work was supported by Grants AM-03555 and AM-20614. Permission to reproduce figures from [17] by Analytical Biochemistry is gratefully acknowledged.

References

1. Edman P (1970) In: Needleman SB (ed) Protein sequence determination. Springer, Berlin Heidelberg New York, p 215
2. Wittmann-Liebold B, Graffunder H, Kohls H (1976) Anal Biochem 75:621
3. Margolies MN, Brauer A, Omau C, Klapper DG, Horn MJ (1982) In: Elzinga (ed) Methods in protein sequence analysis. Humana Press, Inc., New York, p 191
4. Unpublished observation
5. Guyer RL, Todd CW (1975) Anal Biochem 66:400
6. Inman JK, Apella E (1975) In: Laursen RA (ed) Solid phase methods in protein sequence analysis. Proc 1st Int Conf Pierce Chemical Co, Rockford, IL, p 241
7. Bhown AS, Mole JE, Bennett JC (1981) Anal Biochem 110:355
8. Johnson ND, Hunkapiller MW, Hood LE (1979) Anal Biochem 100:335
9. Fohlman J, Rasak L, Peterson PA (1980) Anal Biochem 106:22
10. Henderson LE, Copeland TD, Oroszlan S (1980) Anal Biochem 102:1
11. Tarr GE (1981) Anal Biochem 111:27
12. Kim SM (1982) J Chromatogr 247:103
13. Black SD, Coon MJ (1982) Anal Biochem 121:281
14. Somack R (1980) Anal Biochem 104:464
15. Cunico RL, Simpson R, Correia L, Wehr CT (1984) J Chromatogr 336:105
16. Beckman Model 890C Sequencer Operating Manual. Sect 5, p 3
17. Bhown AS, Bennett JC (1985) Anal Biochem 150:457
18. Bhown AS, Cornelius TW, Mole JE, Lynn JD, Tidwell WA, Bennett JC (1980) Anal Biochem 102:35

4.2 The Use of On-Line High Performance Liquid Chromatography for Phenylthiohydantoin Amino-Acid Identification

Keith Ashman[1]

Contents

1 Introduction

The chemistry developed by Edman [1] for the sequential degradation of polypeptides leads to the generation of phenylthiohydantoin (PTH) amino-acid derivatives. The entire process has been automated and various machines designed over the last three decades [2 – 7]. The final development has been the incorporation of a reverse-phase high performance liquid chromatography system (HPLC) into the sequencer, thereby automating the identification of the PTH-amino acid derivative generated by the machine [8, 9].

There are two reverse-phase HPLC methods for separating PTH-amino acids, either via a gradient [10] or isocratically [11 – 13]. The latter offers a number of advantages: the baseline is more stable; the resolution and reproducibility are excellent; the cost of equipment and solvents is low. Gradient systems are plagued by the fact that they require a changing eluent composition, which causes constant changes in the refractive index of the solvent mixture, therefore baseline drift, and when mixing is poor, baseline noise. Some gradient systems are also prone to bubble formation, this also causes a noisy baseline, and changes in the retention times. Thus, the gradient formation must be reproducible to obtain constant retention times, a necessity for easy identification. Furthermore,

---

1 Max-Planck Institut für Molekulare Genetik, Abteilung Wittmann, Ihnestraße 63 – 73,
  D-1000 Berlin 33
  Present address: European Molecular Biology Laboratory, Meyerhofstraße 1, D-6900 Heidelberg,
  FRG

Advanced Methods in Protein Microsequence Analysis
Ed. by B. Wittmann-Liebold et al.
© Springer-Verlag Berlin Heidelberg 1986

a continuous supply of the gradient components is necessary, whereas with an isocratic system the solvent can be recycled. The improvements in reverse-phase material, namely the reduction of particle size and the manufacturing of spherical particles, have significantly improved the resolution of the columns such that isocratic separations are as effective as gradient systems.

2 On-Line Analysis of PTH-Amino Acids

There are a number of factors to consider when setting up an on-line HPLC system for PTH separation in sequencers: the support in the column, the column dimensions, the flow rate, the temperature, the eluent, and the equipment. These points are discussed below.

2.1 HPLC Support and Column Dimensions

To achieve a good isocratic separation of PTH-amino acids, a large number of theoretical plates and uniform particles are required. This has led to the use of silica-based reverse-phase supports with small spherical particles. The best separations are obtained with particles of 3 or 4 μ in diameter, and silica-based supports derivatized with an eight-carbon hydrocarbon chain (C8). A particular problem can be peak tailing because the particle size distribution is too large or the particles are irregularly shaped. This results in poor resolution. Column length is also important, generally the longer the column the better the separation; however, too long a column will result in very high back pressures, peak broadening, and longer analysis times. Thus a compromise has to be reached. The columns available for isocratic PTH-amino acid separation are given in Table 1.

Table 1. Chromatography conditions for the isocratic separation of PTH-amino acid derivatives

Temperature:	58 °C
Flow rate:	0.4 ml min^{-1}
Detection:	Fixed wavelength detector at 254 nm and 313 nm or a Variable wavelength detector at 266 nm and 313 nm
Recorder speed:	2 mm min^{-1}
Column dimension:	250×4 mm
Supports:	Spherisorb C8 3 μ from Phase Separation Ltd. (self-packed with a Shandon packing apparatus or purchased from Knauer Berlin) or[a] Lichrocrat HPLC cartridge Supersphere RP8 from Merck
Eluents:	a) Based on 2-Propanol
	2-Propanol : Tetrahydrofuran : 1 M Sodium Acetate pH 5.3 : Water[a]
	195, 220[b] : 10 : 15,5[b] : 805, 780[b]
	b) Based on Acetonitrile
	Acetonitrile : 1,2 Dichloroethane : 1 M Sodium Acetate pH 5.3 : Water
	330 : 7 : 9 : 670

[a] The addition of 10 mg of dithioerythritol to 1 l of the propanol buffer increases the recovery of the basic PTH-amino acid derivatives (e.g., PTH-Lys) on the HPLC-trace; the addition of 0.06 mg sodium azide/l is made to avoid growing of microorganisms.
[b] Indicates the 2-Propanol conditions for the Supersphere column.

The Spherisorb support can be purchased loose or pre-packed into columns (Knauer Berlin). The Lichrocart HPLC cartridge Supersphere RP8 column is only available in a pre-packed HPLC cartridge (Merck).

2.2 Flow Rate and Temperature

The on-line isocratic HPLC is run at a constant low flow rate and the column is heated (see Table 1). The latter has the advantage of reducing the column back pressure, speeding the separation and helps to resolve the PTH-derivatives. However, since HPLC supports can vary from batch to batch, it is important to use the same batch to obtain the same separation pattern. Low flow rates and constant conditions also improve column life.

2.3 Eluents

The aim of an isocratic separation system is to separate all the PTH-derivatives without recourse to a changing eluent composition. For the isocratic HPLC separation of PTH-amino acids, the published systems that resolve most of the derivatives have been based on acetonitrile [11, 12]. Recently an alternative system based on 2-propanol has been described [13]. This is the system used in all three Berlin sequencers. The eluent composition is given in Table 1. It is important to use HPLC grade solvents for the buffer.

The eluent is mixed as follows: first, 195 ml of 2-propanol are made up to 1 l with water (from a Millipore Milli-Q purification system or equivalent) and degased; second, to this mixture are added 10 ml tetrahydrofuran, 10 μl sodium azide (4 mg ml^{-1}) and 15 ml 1 M sodium acetate pH 5.3. This final mixture is then sonicated for 5 min to remove any bubbles. The 1 M sodium acetate pH 5.3 is prepared by titrating 1 M acetic acid with sodium hydroxide.

The separation obtained with the 2-propanol system is shown in Fig. 1a, b and that with acetonitrile in Fig. 1c. Since only picomolar quantities of PTH-derivatives are released in microsequence analysis, the eluent mixture can be recycled. A litre of eluent is sufficient for a week of chromatography. The solvent mixture is purged with a gentle stream of nitrogen and no further degasing is required. Before renewing the eluent, the column should be washed with 90% methanol in water in order to remove degradation products that are not eluted under the isocratic conditions. Thereafter, the column must be re-equilibrated with fresh buffer and a test chromatogram made. A test chromatogram should always be made at the start of each new run with a sequencer.

One further point to note is when installing a new column it is advisable to inject a large amount of the standard PTH-mixture (ca. 20 nmol) in order to saturate the column. Standard PTH-amino acids are available from Pierce and are stable in 100% methanol at −20 °C. They should be diluted to the appropriate concentration with eluent immediately before use. For the on-line injection, the PTH-derivatives are dissolved in the conversion cell in eluent minus acetate (to avoid salt accumulation). It is important to inject in buffer to avoid a large injec-

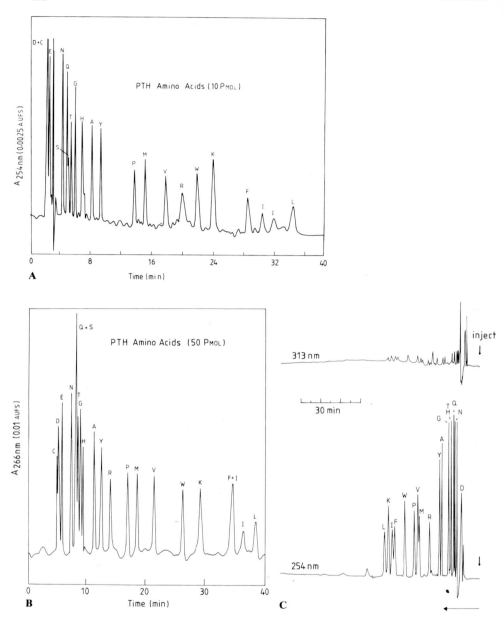

Fig. 1A – C. Isocratic separation of PTH-amino acid derivatives: **A** on Supersphere, **B** on Spherisorb C8 with the 2-propanol system, **C** on a Lichrosphere C8 5 μ column form Merck using the acetonitrile system. The peaks are labeled in single-letter amino acid code (see [9] and Table 1)

tion peak (caused by refractive index effects in the detector cell), which makes the start of the chromatogram difficult to interpret.

As previously described, the elution of PTH-arginine and PTH-histidine is highly dependent on the ionic strength of the buffer (sodium acetate concentration). This fact is illustrated in Fig. 1a, b. The lower ionic strength of the buffer in Fig. 1a causes PTH-histidine and particularly PTH-arginine to elute later. The 2-propanol system also results in the earlier elution of PTH-proline and PTH-lysine. The most difficult derivatives to detect are PTH-serine and PTH-threonine. However, both serine and threonine produce dehydrated derivatives that can be detected at 313 nm, and at this wavelength the other PTH-amino acid derivatives show little or no absorption. The separation is remarkably independent of sample volume, with injection volumes between $10 - 100$ µl, provided the injection is made in 20% 2-propanol and the flow rate is kept low (about 0.4 ml min^{-1}). However, as one would expect, a smaller volume will improve the separation, especially where two peaks are close together.

2.4 Equipment and On-Line Injection

The equipment configuration used for the on-line HPLC system is shown in Fig. 2. Required are: a single HPLC pump (e.g., Gynkotek model 300b or Knauer model 64); a dual-channel UV detector (e.g., Waters model 440 absorbance detector or 2 single-channel detectors connected in series (e.g., 2 Knauer UV photometers); a 2-channel chart recorder (e.g., Kipp and Zonan BD41); an automatic HPLC injection valve (e.g., Rheodyne). A precolumn is included in the circuit (filled with Shandon hypersil 5 µ MOS), to filter the eluent and remove any solid material released from the pump sealings.

The injection is made as follows: the collect line of the conversion cell (C2) is connected via the HPLC valve to the fraction collector. Therefore the solvent

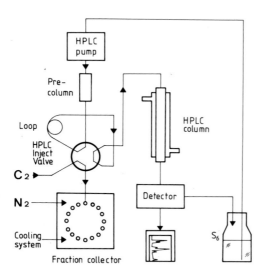

Fig. 2. The configuration of equipment used for the on-line system in the Berlin sequencers. The injection valve is shown in the normal position, i.e., in the pump circuit, and is only connected to C_2 during the sample loop-filling operation, C_2 is the connecting line of the sequencer (from the conversion device) to the HPLC-injection valve, the outlet of which leads towards fraction collector, or the HPLC-column, respectively. Nitrogen, N_2, means the pressurizing of the fraction collector with nitrogen, or pulling of vacuum, respectively (valves not shown)

(eluent minus acetate), under nitrogen pressure transports the PTH-amino acid from the conversion cell toward the fraction collector via the injection valve. During all stages, except the injection into the HPLC column, the HPLC solvent is pumped from the solvent reservoir through the sample loop of the injection valve, the column, the detector, and back to the reservoir. After completion of the conversion and dissolution of the PTH-amino acid the automatic injection valve is operated, thereby connecting C2 via the sample loop to the fraction collecter. Thus the loop can be filled (20 to 100 μl depending on the sequencer type). As soon as the loop is full, the valve is switched back to its "normal" position, thus placing the loop back in the pump circuit and injecting the sample. The remainder of the PTH-amino acid solution is then purged into the fraction collecter with nitrogen.

In order to ensure that the loop is filled with PTH-amino acid solution at the moment of injection and to guarantee an exactly reproducible filling time, all the connecting lines (e.g., C2) must be clear and completely free of liquid droplets before the injection. This is achieved shortly before the dosage of S4 (eluent minus acetate) by blowing nitrogen through the collect line C2 for about 100 s. This guarantees fully reproducible filling times for the loop. Further, during the loop-filling stage it is important that the PTH-amino acid solution is bubble-free. This was achieved by using a 30-s delay (all nitrogen and waste functions of the conversion flask turned off) after S4 is delivered into the conversion flask.

3 Concluding Remarks

In conclusion there are a number of advantages in using an on-line mode of analysis for the PTH amino acids produced by a protein/peptide sequencer:

1. The analysis is easy to perform, and reproducible because no operator intervention is necessary.
2. The released derivatives are identified quickly, thereby reducing the degradation of the samples that occurs when they are kept in a fraction collector.
3. The described isocratic system enables separation of all the common PTH-derivatives. It has the further advantages that the retention times and the recorder baseline are stable. Thus, high sensitivity detection is facilitated. The equipment configuration required is simple and low-cost. Finally, the recycling of the HPLC-buffer system reduces solvent costs.
4. On-line detection allows the performance of the machine to be continuously monitored. As a consequence it is therefore possible to detect problems or the end of a run early. This reduces reagent and solvent consumptions and allows the more efficient use of the sequencer.
5. The introduction of on-line PTH-amino acid analysis into protein sequencers is not only the final stage in automating the Edman degradation, but represents a significant improvement in protein microsequencing technology.

References

1. Edman P (1949) Acta Chem Scand 4:281
2. Edman P, Begg G (1967) Eur J Biochem 1:80 – 91
3. Wittmann-Liebold B (1983) In: Tschesche H (ed) Modern methods in protein chemistry. de Gruyter, Berlin New York, pp 229 – 266
4. Wittmann-Liebold B, Graffunder H, Kohls H (1976) Anal Biochem 75:621 – 633
5. Hewick RM, Hunkapiller MW, Hood LE, Dreyer WJ (1981) J Biol Chem 256:7990 – 8005
6. Hawke DH, Harris DC, Shively JE (1985) (in press)
7. Laursen RA (1971) Eur J Biochem 20:89 – 102
8. Rodriguez H, Kohr WJ, Harkins RN (1984) Anal Biochem 140:538 – 547
9. Wittmann-Liebold B, Ashman K (1985) In: Tschesche H (ed) Modern methods in protein chemistry. de Gruyter, Berlin New York, pp 303 – 327
10. Zimmermann CL, Apella E, Pisano J (1977) Anal Biochem 77:569 – 573
11. Lottspeich F (1980) Biol Chem Hoppe-Seyler 361:1829 – 1834
12. Lottspeich F (1985) J Chromatogr 326:321 – 327
13. Ashman K, Wittmann-Liebold B (1985) FEBS Lett 190:129 – 132

4.3 Device Control, Data Collection, and Processing in Protein Micro-Sequencing and Amino-Acid Analysis

J. Friedrich[1]

Contents

1 Introduction

This paper describes some developments in electronic data processing in the field of protein chemistry. Topics discussed are a control system for analytical devices (sequencer) using a home computer and a program for automatic calculations of amino-acid analyses using a mini or main frame computer.

Before giving details, it seems valuable to note that microcomputers are rather limited for advanced control systems or extensive data processing. Only if the programs used are written in machine language, or in a language which resembles it, are real time processing and successful storage management guaranteed. One possibility is the simultaneous controlling of several devices, displaying and changing of programs.

A microcomputer is not sufficient if, in connection with a data base or a data-screening routine such as the amino-acid analysis calculation, greater tasks are to be performed. However, the microcomputer can work in a useful sense in preparing and transfering data for use in a greater system.

---

1 Max-Planck-Institut für Experimentelle Medizin, Abteilung für Immunchemie, Hermann-Rein-Straße 3, D-3400 Göttingen, FRG

Advanced Methods in Protein Microsequence Analysis
Ed. by B. Wittmann-Liebold et al.
© Springer-Verlag Berlin Heidelberg 1986

Further possibilities are the design of a network of instruments used in a protein chemistry laboratory: identification and integration of chromatographic data [1 – 3], data transfer from local (micro-)computers to main frame devices [2, 3], and programs for final data processing of quantitative sequence analyses [4, 5].

2 Control System for the Sequencer Laboratory

2.1 Device Control

The devices used in the sequencer laboratory in the department of Immunochemistry of the Max-Planck-Institute for Experimental Medicine are a Beckman Liquid Phase Sequencer 890B updated, a Commodore C 64 as controller connected to the sequencer by a self-built interface, Waters Autosampler WISP 710 B and Shimadzu Integrator C-R2AX with INP-R2A for second channel integration. Pump, dual channel photometer, column oven and buffer switching valve (all Waters instruments) are controlled by the same C 64 via a second self-built interface [6] (Fig. 1).

Electronic control of sequencers has been frequently described. Especially the introduction of low-priced microprocessors or microcomputers has encouraged many scientists to update sequencers. For detailed information see [6 – 8].

This sequencer laboratory comprises all parts except the autoconverter and on-line chromatography. The sequencer is controlled via relays. For functioning in a special mode, the control system sets a combination of relays. This setting is held for a specified time. One combination after another directs the two reactions of the Edman degradation, the coupling and cleavage (e.g, by pulling vacuum, pressurizing with nitrogen or allowing for the delivery of reagents and solvents). These two reactions of one degradation cycle are repeated to degrade the protein or peptide successively by one residue. This residue is then analyzed chromatographically at two wavelengths, namely at 254 nm and 313 nm.

Each step of one or more setted combinations can be operated by separate programs or by commands typed in. The control system employed offers 128 relays to be controlled of which 23 are used for the Edman degradation.

The chromatography thereafter has two devices which themselves are programmed. The autosampler works independently giving start signals to the integrator. Then, the integrator handles the dual channel input from the photometer according to a BASIC-programm:

1. Autozero during a run;
2. switching on the second channel (313 nm) for the time needed;
3. chromatogram replotting with changed attenuation (sensitivity) etc.

The chromatographic system is isocratic [9] with a small-bore column (2 mm ★ 250 mm) packed with RP-material (Shandon Hypersil ODS or Spherisorb ODS 2) under a pressure of 900 bar. Column oven, photometer, pump, and the buffer switching valve are controlled in parallel to and independently of the

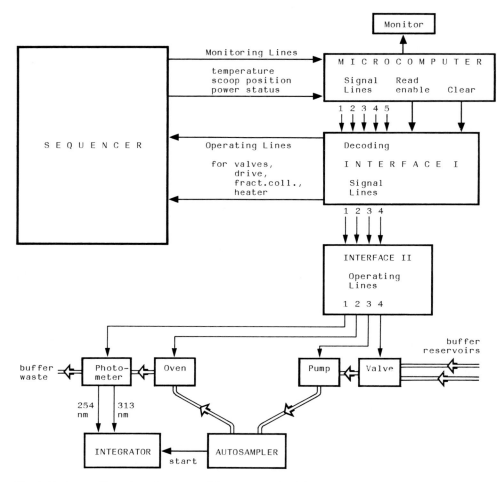

Fig. 1. Device configuration of sequencer laboratory (updated form) controlled by a microcomputer (Commodore C 64)

sequencing procedure. The system allows automatic change of the solvent mixture, e.g., the change of the column rinse by water/methanol (1:1) overnight to the separation buffer. Then, in the morning, it controls the (re)equilibration and keeps the system ready for the next chromatography. It also allows continuation of further PTH-amino-acid analyses and stopping after the last chromatography by changing at first the buffer (for column rinsing) and then switching off the pump, oven and photometer at any desired time (e.g., at night).

This control system is useful for updating the sequencer, because it enables easy changes of the programs. It also facilitates the routines of the identification procedures.

2.2 Problems and Special Precautions

Concerning the reliability of function settings, some important points will be discussed briefly as they have resulted from routine usage. These concern mainly the reliability of function settings. This reliability is especially important for longer sequence runs, e.g., of 40 cycles (or more), which last 40 to 80 h depending on the programs. Therefore, the control system had to be tested thoroughly. Before starting the routine the following precautions were taken (see also Table 1):

1. The decoding chips of the interface are sensitive to very short signals of less than 10 ns: Firstly, the signals sent to the decoders by the five lines are not totally synchronous. Secondly, spikes from the power supply or elsewhere lie in this time range. Therefore, the decoders must only be enabled to read one signal after a lag time. READ ENABLE must go low, while signals must go high. The C 64 interface chip CIA 6225 offers this possibility by a special $\overline{PC}$ flag. One microsecond after each write of the microcomputer the flag goes low for another microsecond. No missettings have ever been observed.
2. The clearing signal for resetting all relays at once (before the next combination) must be filtered by a capacity-resistance combination. This ensures that only such signals are effective which last longer than 300 μs.
3. A power supply line was selected which is free of devices, whose current consumption may be suddenly very high (e.g., starting motors) or which provoke spikes (e.g., refrigerators or heaters). High voltages (220 V) have never been controlled directly, but via relays of at most 24 V.
4. Electrostatic charges can damage the hardware as well as influence software as their values may reach 10,000 V. This would always be deleterious. Therefore, antistatic floors (Semperit) are used which are connected to earth (ground).

Table 1. Problems and their precautions in device control

Problems	Precautions
1. Total shut down of power supply	Provide an additional power supply by means of a battery to retain information about actual conditions. However, operating system and programs can be reloaded if saved on tape.
2. Great instabilities of power supply	Keep the direct line free of devices with sudden very high current consumption.
3. Spikes of power supply provoking a) Missetting relays	Enable decoding of signals only by a special READ ENABLE line.
b) Clearing the relays	Filter the signal by a capacity-resistance combination. Enable only those signals which last longer than 300 μs to clear. Keep the direct line free of devices such as heaters, refrigerators etc. Refresh the combination of set relays with a high frequency without straining the mechanical parts.
4. Electrostatic charges	Use antistatic floors.
5. Failure of devices to be controlled	Provide feedback information for conditional stops (see Sect. 2.3).

Nevertheless, it happened about five times a year that extraction of buffer after coupling was incomplete and accompanied by strong overlapping. This is 'assumed to be caused by uncontrolled resets (they could never be observed directly). Therefore, repetitive refreshing of function settings (20 times per second) was introduced. Refreshing must not strain any relay as its lifetime is "only" about 10 million switchings. Refreshing directs only those relays which have fallen off. Since then, this problem has been solved.

2.3 Future Options

The devices to be controlled may also fail. In this laboratory the following occurred:

1. Improper heating by a misfunctioning fan (result: strong overlapping).
2. Break-down of internal power supply of the sequencer caused by switching on the refrigerator of the fraction collector (result: uncontrolled break of sequencing).
3. Leaking of the high vacuum valve after about 50,000 to 100,000 switchings (no deleterious results because of nitrogen feeding during high vacuum steps).
4. Misfunctioning of the autosampler.

In these cases, no feedback information about the status of the devices was provided. Only the chromatograms of each cycle showed malfunctions. Therefore, it is desirable that the control system monitors the performance and decides about continuation of sequencing. Such possibilities, however, are tested only rudimentarily. They might be installed using the four A/D converters built into the microcomputer. Outside resistances can be measured: one for vacuum, another for temperature control, a third for rough chromatogram recording and a fourth for a series of resistors.

These resistors $-$ up to six $-$ would each have special values and can be bridged by a relay or manual switch (scoop of the reaction cup, power supply, flow meter, drop counter of the waste bottle, etc.) and analyzed. The range of resistance measurement would lie between one and 200 kiloohm. The values of the resistors that would be best are 5, 10, 20, 40, 80 kiloohm and 1 megaohm. By measuring the sum, the operating system can indicated which member of the series is bridged.

Other expansions possible are:

1. Controlling a fraction collector according to peak detection of on-line chromatograms, if samples to be sequenced are radioactively labeled and chromatographed together with a standard.
2. Recording of chromatograms onto discs for storage and data transfer.

3 Computational Evaluation of Amino-Acid Analyses

Following hydrolysis of the (poly)peptide, its derivatization before or after amino-acid separation and its chromatographic identification, exact measure-

Table 2. Formulas used for program execution ("$\star$" means "times")

1. Amount (measured) of an amino acid: mol AA/mol peptide[a].

2. Number (rounded) of residues of an amino acid: rounded "amount".

3. Deviation: positive value of difference between "amount" and "number".

4. Total number: sum of all "numbers".

5. Positions: number of listed amino acids occurring in the analysis.

6. Deviations, divided by number, per position:

$$\sqrt[x]{\sum_{i=1}^{p} (|(amount_i - number_i)/number_i|)^x/positions}$$

for x = 1 (arithmetic deviation), x = 2 (standard deviation), x = 8 and x = 16.

7. Deviations per position:

$$\sqrt[x]{\sum_{i=1}^{p} (|amount_i - number_i|)^x/positions}$$

for x = 1, 2, 8, and 16.

8. Modified mean deviations balancing small and large assumed peptide lengths: First term in formula (10)[b].

9. Correction factor for formula (8): Second term in formula (10)[c].

10. Formula (8) and (9)[d]:

$$\left[(6) + (7) \star \left(1 - \frac{positions}{total\ number} \right) \right]/2$$

$$\star \exp\left[f \star \left(-1 + \frac{total\ amount + total\ amount}{amount\ above\ threshold + total\ amount} \right) \right]$$

11. Lower limit of total number: "positions" minus one

12. Upper limit of total number: either
    a) "positions" $\star$ four   or
    b) molecular weight divided by mean molecular weight per amino acid residue $\star$ (one and a half) or
    c) chosen value.
    The first definition of upper limit is only sufficient for peptides 60 residues of length. For longer (poly)peptides, molecular weights are normally available.

---

[a] Alternative definition: sample value $\star$ correction/standard value.
[b] This formula considers that, at a high total number, the values of deviations, divided by numbers, per position approaches zero. The higher the total number the more mean deviation per position is weighted.
[c] This formula reflects the contribution of amino-acid molar values below a chosen molar limit, namely the assumed impurities. It also reflects the characteristics of most polypeptides in that they have more probably an even distribution of single molar values.
[d] For the factor of the exponent, the value of 4 has been found to be practical and is not changed for purpose of comparison. If, however, this factor equals zero or if the molar limit equals zero then correction factor equals one, i.e., no correction is done.

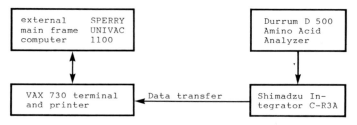

Fig. 2. Instrumentation of amino-acid analyses and calculation of composition

ment of intensities and appropriate corrections enable the calculation of the molar amino-acid content. The evaluation is started by searching for a probable composition of the (poly)peptide, i.e., by searching for a greatest common divisor. Optimally, their quotients should be integers. Within limits the deviations of calculated values from their rounded integers may be used for statistics, for calculation of probable compositions, for optimizations and for listing of best fits.

The limits lie in the probability of measurement imprecisions. With further information such as molecular weights or definite molar (integer) ratios [10], these limits may be expanded. Using an conventional analyzer according to the Stein and Moore method, it was found that analyses can be made down to the 200 pmol level. Combined with an integrator the sensitivity increased to 50 pmol.

A PASCAL program has been developed to aid calculations of peptides as well as of proteins. It provides a statistical value for comparison of suggested possibilities. The results obtained are based on a screening procedure with a formula found to be appropriate for all cases (Table 2).

The instrumentations used for analyses and calculations are (Fig. 2): Durrum Amino Acid Analyzer D 500 and Shimadzu Integrator C-R3A for separation and measurement, DEC VAX 730 for editing, starting runs and printing out and SPERRY UNIVAC 1100 for the calculation runs.

Examples of short to long peptides (proteins) will be discussed.

3.1 Data Preparation

The values are printed out by the analyzer and/or the integrator. If the integrator is used, each chromatogram can be stored. It can be replotted for peak detection points and the areas can be recalculated with changed parameters.

The data for program execution contain the identity of the measured amino acid and a numerical value of its measurement (picomole or combination of sample area/height and reference area/height). The data must be readable for program execution usually as ASCII- or textfile, which may be edited. The data are entered into these files off-line. As a future option, it may be done on-line [2, 3] by means, for example, of a teletype connection. A processing program would have to select the usable data and to start calculation.

The format of such textfile reflects the different possible entries: the comment lines, the lines of molar values and the lines of the areas (Fig. 3).

```
* 70.7 (70'2.21)/3   160'C   30 min          * VNT 15-3.12   150'C   70 min
* OK 6-115237 ; Seq.   SLWNAGT...            * OD 6-119561 ; Seq.   VNTSGF...
*                                            D  <   463    3638
D   >   884.3                                T  <  1775    3734   1.05
T   >  1100.4     1.05                        S  <  1967    4162   1.11
S   >  1489.4     1.11                        E  <  1415    3804
E   >  1320.6                                 P  <   158     139
P   >  2082.5                                 G  <  1037    4248
G   >   498.6                                 A  <  2263    4131
A   >  1984.8                                 V  <  1793    3784
V   >   763.4                                 I  <    44    3980
M   >   424.3                                 L  <   627    4098
L   >  2496.8                                 Y  <   368    3791   1.05
H   >   413.7                                 F  <   659    3650
X   <  3890   11007  1.11                     H  <   295    2402
* Aminoethyl-cysteine, area compared         * accord.to height inst.of area
* with area of 1000 picomole lysine          X  <   275    4352   1.11
* corrected by the factor of 1.11            K  <   169    4352
R   >   759.3                                 R  <   821    3507
```

a   b

Fig. 3. Examples of textfiles ready for program execution with comment lines beginning with " ★ " and data lines beginning with one-letter code and followed by either " > ", picomole value and optionally correction factor or followed by " < ", sample area, standard area and optionally correction factor

An origin-file exists to be filled in by comment statements and by the values. Superfluous lines (of amino acids which did not appear) are deleted.

Such a textfile is ready for calculation.

3.2 Program Execution

The program is called by dialog, by a command file, or by a runstream (Fig. 4). By this, the output of results is directed either to the terminal, to the printer, to another file or to another computer. The name of textfile containing the values to be computed is typed in. The comment lines are read and printed [without leading character " ★ " (Fig. 3)]. Data lines are analyzed for the one-letter code and

```
 1      $ create run.run
 2      §run,z runame,number/user,project,5,50
 3      §sym,c print$.,,mempch
 4      §elt,i aa.25hau2
 5      * HAU   4 microgram    160'C   25 min
 6      D   >   2619.8
 7      T   >   2592       1.05
 8      * ... etc.
 9      K   >   1638.2
10      R   >   1192.7
11      §xqt composition.calculation
12      25hau2
13      l m 21000 a 18000 e 24000 c 7 1.5
14      5
15      /
16      §fin
17      $ netjob run.run   number
```

Fig. 4. Command file of VAX 730 for starting a run on SPERRY UNIVAC 1100 via "$ netjob"

for the type of value presentation. Molar values may be calculated and/or corrected. The results are printed out. Each data line represents a "position".

A threshold value is typed in. A value of zero stops execution. Lower and upper limits representing possible lengths of (poly)peptide are calculated as [number of positions above threshold minus one] and as [number of positions above threshold times four]. The following inputs are optional (formats given in parentheses refer to the program used in this laboratory, running on a SPERRY UNIVAC 1100 machine):

1. Upper limit. ("o" followed by an integer number).
2. Factor of upper limit (" ★ " followed by integer correcting default value of four).
3. Molecular weight. By this, the upper limit is recalculated on the basis of mean molecular weight of residues above threshold ("m" followed by integer number).
4. Exponential correction factor of formula (9) of Table 2 ("f" followed by a number).
5. Calculation limits. These reflect the assumed imprecision of measurement. There is a relative and an absolute limit. Amino acids above the absolute limit in a composition are regarded as not present for statistics. Amino acids above the relative limit are weighted for statistics in relation to this limit ["c" followed by an integer number (relative limit) and by a factor (absolute limit equals factor times relative limit)].

The next options handle screening procedures.

6. Least threshold distance. The routine for finding the next higher threshold examines the relative distance between amino acid to be excluded and lowest amino acid above threshold. The distance is expressed as (lower amino acid divided by higher amino acid). The threshold found may so be discarded ("s" followed by a number).
7. Listing width. Normally, during screening, the calculation of one assumed length is listed only if its least mean deviation is lower than these of assumed length plus one and assumed length minus one. This does not always agree with known compositions. Therefore, neighboring calculations of one assumed length may also be listed if their least mean deviations are worse but not more than times the factor ("r" followed by a number).
8. Percent limit. The screening routine searches for the best fit of compositions to this limit independent of mean deviations ("p" either followed by a number to list screening results or followed by a negative number to print out the compositions according to these fits).
9. Screening quotients ("q" followed by two integers representing a first rough and a second fine screening).

Further options govern the print out of calculated compositions. If molecular weight has been defined and the screening routine has been started, the best fit out of the list of best optima is searched for and printed. All the printings happen once according to the least mean arithmetic deviation and once according to the least mean square deviation. Length of list, number of composition print out, lower and upper limits are the other options.

3.3 Execution Performance

A brief description of the program structure (Fig. 5) is as follows:

1. If textfile name is typed in, program asks for threshold otherwise execution is finished.
2. If threshold (and options) are typed in, composition calculation begins, firstly assuming lowest length of (poly)peptide otherwise next textfile name is asked at level (1).

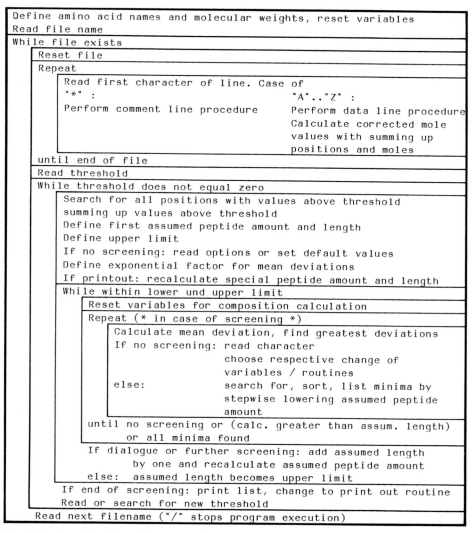

Fig. 5. Rough outline of screening program for calculation of amino-acid composition of (poly)peptides

3. After one result is shown, the program asks for the next calculation (of assumed residue length plus one) up to the upper limit when the program goes back to level (2). Otherwise, program performance is directed either to change to a more detailed printout; to another assumed (poly)peptide length or to another assumed molar amount of (poly)peptide; to change values, to add new values or to add the name and the molecular weight of a new residue; to change lower and upper limit assuming a fixed number of a special amino acid; or to start screening routine.

The screening routine operates similarly. It starts at first assuming a high molar amount of (poly)peptide, dividing the assumed amount step by step by the screening quotient. Two conditions must be fulfilled for a composition of a special (poly)peptide length to enter the list of best minima:

1. The assumed length (after dividing total molar amount of all amino acids by assumed molar amount of peptide) equals the calculated length (sum of rounded values).
2. The best deviations found under condition (1) are better than those of (length plus one) and of (length minus one). In such cases these deviations are regarded as minima. Exceptions may be made by means of listing width [as option (7) in the preceding section].

This (double)list of best minima is limited to at most thirty positions.

All allowed thresholds according to least threshold distance of option (6) in the preceeding section are screened. A new threshold is searched for if the molar amount of (poly)peptide is significantly lower than the actual threshold or if assumed (poly)peptide length reaches upper limit. Then, the program continues the screening routine by assuming a new high molar amount of (poly)peptide. This is repeated until only two amino acids are regarded as components of a possible peptide.

This "extreme" procedure, used for peptides of any length, necessitated the introduction of weighthing the assumed impurity [see formula (9) in Table 2].

The least threshold distance primarily reduces execution time, but it also indicates the purity of the analyzed sample.

3.4 Interpretation of Calculation Results After Screening

The quality of calculations depends directly on the accuracy of the measurement. With an imprecision probability of, say, 10%, it is possible to define contents up to 5 mol of amino acid mol^{-1} (poly)peptide, and of 3% up to 15 mol mol^{-1}.

This is reflected by calculation limits [see option (5) in Sect. 3.2]. If any relative content of a composition exceeds such a limit, interpretation must be done with caution in respect to that amino acid. If there are only a few amino acids having values below this limit, interpretation must be cautious with respect to the calculated composition itself unless there is information about the polypeptide's molecular weight. Limits of this kind of calculation are about 250 residues. Nevertheless, the program can help in suggesting a typical composition even if such limits are far exceeded.

3.5 Screening Results

Seven examples of amino-acid analysis are demonstrated. They were all done on Durrum Analyzers with molar amounts from 50 pmol to 1.6 nmol.

1. Pentapeptide with known sequence. The analysis gives the following values after correction for serine and threonine [pmol]:

> D 399, T 2995, S 961, E 834, P 264, G 701, A 1784,
> V 176, L 328, Y 343, F 1461, H 167, K 1476, R 229.

Upper limit was 56, least threshold distance 0.8.

The four best minima of mean arithmetic deviation are

5 residues,	thresh. 962 pmol,	mod. mean dev. 0.052
39	177	0.083
13	400	0.084
33	177	0.086

The list for mean square deviation

5	962	0.082
39	177	0.100
33	177	0.108
13	400	0.109

The calculations favor strongly the pentapeptide despite the high amount of impurity which does not easily fit into a discrete composition. The composition after fitting to a 10% limit is:

Thr:	amount 1.85	number 2
Ala:	1.10	1
Phe:	0.90	1
Lys:	0.91	1

Amount of peptide: 1.62 nmol.

2. Peptide of known sequence 38 residues in length (Fig. 3a).

The values: D 884, T 1155, S 1653, E 1321, P 2083, G 499, A 1985,
> V 763, M 424, L 2497, H 414, ae-C 392, K 380, R 759.

Upper limit was 60, least threshold distance 0.8.

The four best minima of mean arithmetic deviation are

37 residues,	thresh. 1 pmol,	mod. mean dev. 0.053
32	1	0.112
42	1	0.113
19	499	0.122

The list for mean square deviation

37	1	0.066
43	1	0.127
32	1	0.134
40	1	0.138

The calculations favor strongly a peptide of 37 residues in length. The composition agrees quite well with the known sequence containing one tryptophan which was not measured. All values lie within 8%, with glycine deviating by 20%. Molar amount of peptide was 412 pmol.

3. Protein, not glycosylated, molecular weight of 20 kD.

The values: D 1295, T 257, S 302, E 3410, P 212, G 591, A 752,
             V 325, L 151, K 961, R 348.

Upper limit was 264, least threshold distance 0.7, calculation limits 7 and 10.5.

The four best minima of mean arithmetic deviation are

⩾171 res.,	MW 19,419,	thresh. 1 pmol,	mod. mean dev. 0.072
235	26,684	1	0.075
233	26,457	1	0.083
28	3,193	1	0.088

The list for mean square deviation

235	26,684	1	0.074
232	26,328	1	0.084
⩾167	18,989	1	0.088
242	27,459	1	0.104

Six out of eleven positions were full weighted being lower than any calculation limit. Together with known molecular weight, a polypeptide length of about 170 residues is a reasonable result.

The suggested composition is

Asp:	26.11	26
Thr:	5.18	5
Ser:	6.09	6
Glu:	68.74	69
Pro:	4.28	4
Gly:	11.92	12
Ala:	15.17	15
Val:	6.55	7
Leu:	3.04	3
Lys:	19.38	19
Arg:	7.02	7

Molar amount of polypeptide: 50 pmol.

4. Four different Bence-Jones proteins after sedimentation in ammonium sulfate and gel permeation chromatography (Table 3). The sequences of all four proteins (AU, BAU, HAU, NEI) [16] are known. Compositions are sought according to molecular weight of polypeptide minus content of cysteine and tryptophan. Proteins were not reduced and alkylated. Hydrolyses were performed in TFA/12 N HCl (1:2) 160°C 25 min [11].

Table 3. Molar ratios of amino acids according to known sequences compared with calculated compositions according to best statistical results in the expected range of four Bence-Jones proteins

AA	AU seq'd	pmol 225	BAU seq'd	pmol 169	pmol 328[a]	HAU seq'd	pmol 165	pmol 349[a]	NEI seq'd	pmol 133	pmol 197[a]
Asp	19	20	11	11	11	15	16	14	13	13	
Thr	16	16	22	18		18	16		21	17	
Ser	30	29	29	25		36	33		35	29	
Glu	26	32	23	25		25	28		21	22	
Pro	11	12	16	20		12	12	12	14	16	
Gly	13	16	16	17		12	13	13	17	17	
Ala	13	15	16	16		13	14	13	17	17	
Val	14	14	15	15		15	14	14	17	17	
Met	1	1	1	1		1	–	1	1	1	1
Ile	8	7	6	5	5	8	6	7	5	4	5
Leu	15	16	13	13	13	14	14		11	11	13
Tyr	11	11	11	10	11	10	8	9	10	8	10
Phe	8	9	4	4	4	8	8	8	5	5	6
His	3	3	3	3	3	2	2	2	2	2	2
Lys	13	13	12	11	12	11	10	11	13	11	13
Arg	5	6	3	3	3	7	7	7	5	5	5
Σ	206	220	201	197		207	201		207	195	

List of the best fits according to least mean deviation (number and MW):

1. 123	13362	1. 254	26718	1. 227	24458	1. 46	4787
2. 124	13493	≫2. 197	20721	2. 102	10984	2. 91	9451
3. 285	31070	3. 260	27361	3. 28	2985	3. 90	9314
4. 156	16996	4. 122	12761	4. 101	10846	≫4. 195	20349
≫9. 220	23992	5. 124	12994	≫5. 201	21649	5. 164	17112

[a] Only numbers lower than 15 are listed

In Table 3 the molar ratios according to sequences are compared to the calculated compositions, which have been listed. The best calculation results in the expected molecular ranges are shown. Firstly, it seems that the results are indeed representative of the samples analyzed. Secondly, it may be useful to hydrolyze greater amounts of protein for analysis of less abundant amino acids.

These examples are a very small part of the total results obtained using this program. Since searching for a good fit is always helped by the program, it may be done by dialog. Searching for the best fits, however, would be too time-consuming in dialog sessions. During a screening procedure, 5000 to 20,000 composition calculations are accomplished. On a microcomputer, where this (compiled) program was first used, one calculation required 2 s. Thus, a run would be between 3 and 12 h, instead of between 1 and 4 min on a main frame computer. The author suspects that the program execution time on a microcomputer with BASIC-interpreter is beyond practicality.

The program structure is roughly outlined in Fig. 5. Figure 4 shows a command file of a VAX 730 creating (line 1) a (batch)runstream of lowest priority (line 2) via "netjob" to SPERRY UNIVAC 1100 (line 17). The printout is directed back to VAX 730 (line 3). A textfile is created (line 4) with the content as in Fig. 3a, b (lines 5 – 10). Program execution is started (line 11). The inputs (lines 12 – 15) are:

1. Textfile name.
2. Threshold, molecular weight, printout beginning form molecular weight, printout ending at molecular weight, calculation limit, and factor.
3. "5" represents screening start.
4. "/" represents no further data processing during this run.

Thereafter the run is finished (line 16).

A VAX command file may contain many runs started by "netjob". Overnight, some hundred of runs can be accomplished. (The Gesellschaft für wissenschaftliche Datenverarbeitung Göttingen offers a special slot of batchruns shorter than 5 min CPU time). The data output is yielded with "netget" on VAX 730. The files are shortened down to only the desired information limiting the real print output on paper.

4 Conclusion

Using advanced chromatographic identification and integration [1 – 3, 12, 13] with programs for micro-, mini- or main frame computers and using data transfer equipment [2, 3], it is possible to install a network between analytical procedures (e.g., peptide separation, amino-acid analysis and sequence analysis) and data processing. Especially two-dimensional peptide and protein separation [14, 15] and quantitative amino-acid sequence analysis [4, 5] could have great advantages for further automatization. For example, with computer-aided decisions upon purity and identity in successive chromatograms, separation fractions could be directed to subsequent analyses, to collection and/or to further purifications. Full sequence information may be obtained more easily especially of longer peptides.

Acknowledgment. I am indebted to Mrs. B. Wehle who operated the sequencer and programs. Listings of the computer program for automatic amino acid calculation may be obtained if a cheque of three dollars is provided for mailing.

References

1. Smith RJ (1984) Microcomputers in chromatography. In: Ireland CR, Long SP (eds) Microcomputers in biology. IRL Press, Eynsham, Oxford, England, p 209
2. Pardowitz I, Zimmer H-G, Neuhoff V (1984) Clin Chem 30:1985
3. Kronberg H, Zimmer H-G, Neuhoff V (1984) Clin Chem 30:2059
4. Smithies O, Gibson D, Fanning EM, Goodfliesh RM, Gilman JB, Ballantyne DL (1971) Biochemistry 10:4912

5. Machleidt W, Hofner H (1981) Quantitative amino acid sequencing by high pressure liquid chromatography operating on-line to a solid phase sequencer. In: Lottspeich F, Henschen A, Hupe K-P (eds) High performance liquid chromatography in protein and peptide chemistry. De Gruyter Berlin New York, p 245

6. Friedrich J, Thieme H (1985) Programmable electronic control of sequential procedures of analytical devices. In: Tschesche H (ed) Modern methods in protein chemistry, vol II. De Gruyter, Berlin New York, p 329

7. Begg GS (1980) A microprocessor based controller for the protein sequenator. In: Birr C (ed) Methods in peptide and protein sequence analysis. Elsevier/North-Holland, Amsterdam, p 485

8. Wittmann-Liebold B (1983) Advanced automatic microsequencing of proteins and peptides. In: Tschesche H (ed) Modern methods in protein chemisry. De Gruyter, Berlin New York, p 231

9. Lottspeich F (1980) Hoppe Seyler's Z Physiol Chem 361:1829

10. Creighton TE (1980) Nature 284:487

11. Tsugita A, Scheffler JJ (1982) Eur J Biochem 124:585

12. Fraser PDB, Suzuki E (1969) Anal Chem 41:37

13. Hancock HA Jr, Dahm LA, Muldoon JF (1970) J Chromatogr Sci 8:57

14. Takahashi N, Ishioka N, Takahashi Y, Putnam FW (1985) J Chromatogr 326:407

15. Born J, Hoppe P, Schwarz W, Tiedemann H, Tiedemann H, Wittmann-Liebold B (1985) Biol Chem Hoppe-Seyler 366:729

16. Hilschmann N, Barnikol HU, Kratzin H, Altevogt P, Engelhard M, Barnikol-Watanabe S (1978) Naturwissenschaften 65:616

Chapter 5
Analysis of Cysteine Residues in Proteins

5.1 Analysis of Cyst(e)ine Residues, Disulfide Bridges, and Sulfhydryl Groups in Proteins

AGNES HENSCHEN[1]

Contents

1 Introduction

The presence and the state of cysteine and cystine residues often have considerable influence on the properties, the structure, and the function of a protein. The sulfhydryl groups of the cysteine residues are in most cases the most reactive functional side chains of the protein. They can easily be oxidized or otherwise modified. They are often of importance for the biological acitivity of the protein. The disulfide bonds of the cystine residues contribute in a unique way to the protein's spatial structure and to the stability of this structure. Proteins may contain only cysteine residues, only cystine residues, or a mixture of both. In certain proteins post-translationally derivatized sulfhydryl groups have been shown to exist. Obviously, proteins may also be devoid of cyst(e)ine.

No primary structure analysis of a cyst(e)ine-containing protein can be regarded as complete before the presence and location of sulfhydryl groups and disulfide bridges has been established. The complementary nucleic acid sequence contains no information about the state and connections of the cyst(e)ine residues, only indiscriminate information about their location. The following questions have to be answered during the course of the protein primary structure elucidation:

a) What is the cyst(e)ine content?
b) How many sulfhydryl and disulfide groups are present?

---

1 Max-Planck-Institut für Biochemie, Arbeitsgruppe Proteinchemie, D-8033 Martinsried/München, FRG

Advanced Methods in Protein Microsequence Analysis
Ed. by B. Wittmann-Liebold et al.
© Springer-Verlag Berlin Heidelberg 1986

c) Where are cysteine and cystine residues located in the sequence?
d) Which half-cystine residues are connected by disulfide bonds?

Here will be described and commented on a number of practical approaches to these problems, which can obviously be solved in very many different ways. The methods are intended for the analysis of previously uncharacterized proteins of which a few milligrams or only micrograms of material can be used for this purpose. Excellent, detailed reviews on the chemistry of sulfhydryl and disulfide groups and on their reactivity in proteins are available [1 – 3].

2 Determination of Total Cyst(e)ine Content

The half-cystine content, i.e., the total amount of cysteine and cystine, may conveniently be determined as a part of the analysis for amino acid composition. However, a number of precautions have to be observed as both cysteine and cystine are unstable during acid hydrolysis and analysis. The two forms elute as separate peaks, cysteine between glutamic acid and proline, cystine between alanine and valine in many amino acid analysis systems and as indicated in Fig. 1. Both forms have low color yields, especially cysteine.

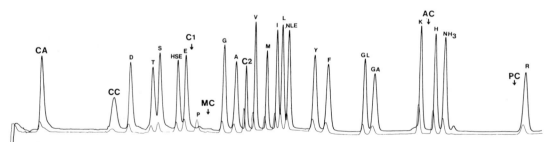

Fig. 1. Chromatogram on amino acid analyzer of standard amino acids, cysteic acid (*CA*), carboxymethyl cysteine (*CC*), homoserine (*Hse*), cystine (*C2*), norleucine (*Nle*), glucosamine (*GL*), galactosamine (*GA*), and with positions of cysteine (*C1*), methyl cysteine (*MC*), aminoethyl cysteine (*AC*) and pyridylethyl cysteine (*PC*) indicated; detection after reaction with ninhydrin at 570 and 440 nm

Cysteine can easily oxidize to cystine during hydrolysis or analysis. Furthermore, it may interfere with proline analysis, as also cysteine has a high 440/570 nm ratio [4]. Cystine, on the other hand, can be reduced to cysteine during hydrolysis, especially when a mercaptan is added to the hydrochloric acid or when mercaptoethane sulfonic acid is used [4]. However, when any cysteine present after hydrolysis is air-oxidized to cystine before analysis, as described in Table 1, the total amount can be determined as cystine. Several cyst(e)ine derivatives can with advantage also be employed for the half-cystine content determination (see below).

Table 1. Determination of total cyst(e)ine by amino acid analysis

1. Hydrolyze protein corresponding to approx. 5 nmol of cyst(e)ine or at least 1 nmol of protein in 200 µl of 6 N hydrochloric acid under vacuum at 110 °C for about 20 h.
2. Remove acid by evaporation.
3. Dissolve hydrolysate in 200 µl of water, add pyridine to pH 6.5 or above.
4. Shake tube continuously or at least frequently for 2 – 4 h at room temperature so that cysteine will be oxidized to cystine by air oxygen.
5. Remove solvent by evaporation.
6. Dissolve sample in buffer for application to amino acid analyzer. Check that pH is sufficiently low, i.e., all pyridine removed.
7. Analyze for amino acid composition.

3 Disulfide Bond Cleavage

The disulfide bridges in proteins are normally cleaved before the separation of peptide chain subunits and also before the isolation of chemically or enzymatically produced fragments. The most commonly used reagents are listed in Table 2. All reactions, except those with performic acid and, possibly, sodium borohydride are highly specific for disulfide bonds. It is often necessary to denature the protein in order to achieve a complete disulfide cleavage. The reaction with sulfite, mercaptan, sodium borohydride, and sometimes also phosphine is therefore performed in the presence of 6 M guanidine hydrochloride, 8 – 9 M urea, or 1% sodium dodecylsulfate. The preferred denaturing agent is guanidine hydrochloride, as it gives no side reactions and can be quantitatively removed. All reductive disulfide cleavages are carried out under a protective layer of nitrogen or freon, as otherwise the sulfhydryl groups formed may re-oxidize. After the reductive cleavages the sulfhydryl groups are normally modified (see below) to prevent them from being oxidized.

Table 2. Reactions for disulfide bond cleavage

Reagent	Derivative formed
Performic acid	Cysteic acid
Sulfite	S-sulfo cysteine + cysteine
Sulfite + oxidant	S-sulfo cysteine
Mercaptan	Cysteine
Sodium borohydride	Cysteine
Phosphine	Cysteine

Performic acid treatment of a protein will result in a cleavage of the disulfide bonds and a stable modification of the cyst(e)ine residues [5]. A suitable procedure for the reaction is given in Table 3. The cleavage is quantitative. The reagent is volatile and can be removed by evaporation. It may be used as a vapor, e.g., to oxidize peptides on a paper electropherogram [6]. A serious disadvantage is, however, the low degree of specificity of the reaction. Thus, methionine residues are oxidized to the methionine sulfone form, and tryptophan as well as tyrosine residues are at least partly destroyed.

Table 3. Performic acid oxidation of cyst(e)ine residues to cysteic acid

---

1. Prepare performic acid reagent by mixing 100 µl of 30% hydrogen peroxide with 900 µl concentrated formic acid, incubating the mixture at room temperature for 1 h.
2. Treat 1 – 2 mg of protein with 1 ml of performic acid at 0 °C for 4 h.
3. Remove solvent by evaporation, e.g., after dilution with cold water. (Note: Performic acid may be recognized by the chlorine-like smell).

Sulfite alone will cleave disulfide asymmetrically, whereby S-sulfo cysteine and cysteine residues are formed. In the presence of oxidizing agent, such as copper ions [7] or tetrathionate [8], a quantitative conversion to S-sulfo cysteine residues is achieved also in proteins, provided that guanidine hydrochloride or other denaturing agents have been included in the reaction mixture. The sulfitolysis conditions are mild and the reaction is highly specific. However, the reaction product is only moderately stable (see below).

Mercaptans of several different types are for most purposes the disulfide cleavage reagents of choice, as they convert protein cystine residues quantitatively and completely selectively to cysteine residues in the presence of denaturing agents. The most commonly used mercaptans are 2-mercaptoethanol and 1,4-dithiothreitol, i.e., DTT, or 1,4-dithioerythritol, i.e., DTE. DTT and DTE are needed only in lower concentrations as compared to mercaptoethanol, since their redox potentials are more favorable. However, mercaptoethanol is convenient to use as it a water-miscible liquid. The mercaptolysis should be performed under the protection of an inert gas to prevent re-oxidation. A procedure for the quantitative cleavage, even of unusually stable disulfide bridges, is presented in Table 4. The molar excess of reagent over protein disulfide is 100- to 1000-fold.

Table 4. Disulfide bond cleavage by mercaptoethanol

---

1. Dissolve 1 – 10 mg of protein in 200 µl of 6 M guanidine hydrochloride – 0.1 M Tris pH 8.5 with hydrochloric acid.
2. Flush sample with nitrogen or freon.
3. Add 10 µl of 2-mercaptoethanol. Close vessel tightly. Incubate at 40 °C over night or at 50° for 4 h.
4. Block the sulfhydryl groups formed with one of the reagents in Tables 6 – 9.

Sodium borohydride quantitatively reduces protein disulfides to sulfhydryls in presence of denaturing agents. Disadvantages connected with the use of this reagent for preparative purposes are the low stability in solution, the high pH resulting from the decomposition and, possibly, peptide bond cleavage. It may, however, be used with advantage for analytical purposes (see below, Table 10).

Tri-n-butylphosphine is a powerful and specific reagent for the cleavage of disulfide bridges, only an approximately 1.2-fold molar excess being needed for a complete reaction, provided that a pure, oxide-free reagent is used [9].

4 Sulfhydryl Group Modification and Identification of Derivative

The sulfhydryl groups in proteins, both those present in the native protein and those produced by reductive cleavage of disulfide bonds, are normally converted into a more stable form before the separation of fragments and sequence determination. During amino acid sequence analysis neither cysteine nor cystine will give rise to a phenylthiohydantoin derivate, as their products are decomposed by β-elimination in a manner similar to, but faster than that of serine. Accordingly, unmodified cyst(e)ine residues will only show up as gaps. However, several cysteine derivatives can be well identified by thin-layer or high-performance liquid chromatography (HPLC) during sequence analysis, as shown in Fig. 2.

A very large number of different reagents for sulfhydryl group modification have been described, the most commonly used being listed in Table 5. With most reagents it is difficult to find or keep reaction conditions which lead to a quantitative and selective derivatization of sulfhydryl groups. Many reagents will also react to some extent with methionine, histidine, lysine, tyrosine, or tryptophan residues, although the reactions with these residues are considerably slower.

An ideal reagent for preparative purposes should obviously give a quantitative and highly selective modification of sulfhydryl groups, and the corresponding cysteine derivative should be easily identifiable in high yield by amino acid

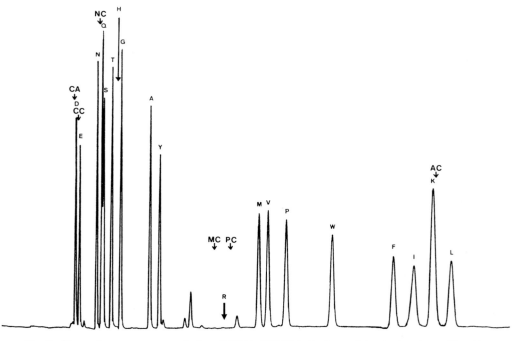

Fig. 2. Chromatogram on reversed phase HPLC of PTH derivatives of standard amino acids with positions of cysteic acid (*CA*), carboxymethyl cysteine (*CC*), carbamoylmethyl cysteine (*NC*), methyl cysteine (*MC*), pyridylethyl cysteine (*PC*) and PTC-aminoethyl cysteine (*AC*) indicated; column: Spherisorb ODSII; elution: isocratic; detection: 254 nm

Table 5. Reactions for sulfhydryl group modification

Reagent	Derivative formed
Performic acid	Cysteic acid
Sulfite + oxidant	S-sulfo cysteine
Iodoacetate	S-carboxymethyl cysteine
Iodoacetamide	S-carbamoylmethyl cysteine
Methyl iodide	S-methyl cysteine
Methyl nitrobenzene sulfonate	S-methyl cysteine
Ethylenimine	S-aminoethyl cysteine
Bromoethylamine	S-aminoethyl cysteine
Vinylpyridine	S-pyridylethyl cysteine
N-ethylmaleimide	S-ethylsuccinimido cysteine
Dithiobis(nitrobenzoic acid)	Mixed disulfide
Nitrothiocyanobenzoic acid	S-cyano cysteine

analysis in the composition, as well as by high-performance liquid chromatography of the phenylthiohydantoin during sequence analysis. However, when a reagent is to be used for purely analytical purposes, e.g., for estimating number or reactivity of sulfhydryl groups, the requirements are different.

Performic acid will both cleave cystines and oxidize all cyst(e)ines to cysteic acid residues. The procedure is described in Table 3. The product formed can be identified by amino acid analysis, although cysteic acid elutes unretained (Fig. 1), and therefore together with any other unretained material. The phenylthiohydantoin derivative can also be identified, but elutes from high-performance liquid chromatography column too close to other components (Fig. 2). An additional disadvantage of the performic acid is that it destroys several other types of amino acid (see above), so that a heterogeneous protein preparation is obtained, which is also resistant to certain types of cleavage, e.g., by cyanogen bromide due to the oxidation of methionine residues.

Sulfite, in the presence of an oxidizing agent, gives a highly specific and selective conversion of all cyst(e)ines to S-sulfo cysteines. However, the product decomposes completely during acid hydrolysis and during sequencing. A special property of the S-sulfo group, which may occasionally be useful, is the ease with which it can be removed, e.g., by mercaptoethanol treatment, whereby the free sulfhydryl group is obtained back. The S-sulfo group can therefore be used as a temporary blocking group.

Iodoacetate, together with iodoacetamide, has probably been employed more often than any other reagent for protein sulfhydryl group modification. A suitable procedure is outlined in Table 6. The resulting S-carboxymethyl cysteine is readily quantified by amino acid analysis (Fig. 1), when care is taken to exclude any oxygen during the hydrolysis. The derivative obtained during sequencing is sufficiently stable, but elutes in many systems too close to other components (Fig. 2). In order to overcome quantification and identification problems, radio-

Table 6. Sulfhydryl group modification by iodoacetate

1. Cleave disulfide bonds in 200 μl of 6 M guanidine-5% mercaptoethanol according to Table 4.
2. Add a solution of 24 mg iodoacetic acid and 30 mg of Tris base in 200 μl of 6 M guanidine hydro-chloride pH 8.7 – 9.0. (Note: The iodoacetic acid should be white). Leave the sample at room temperature in the dark for 15 min. Check that the pH stays above 8.0.
3. Add 10 μl mercaptoethanol to assist in removing unreacted iodoacetate.
4. Isolate protein immediately from reagents, e.g., by reversed phase HPLC after acidification, by gel filtration, by dialysis or by precipitation.

active iodoacetate is often used. The commercially available iodoacetic acid may sometimes contain iodine, which gives a yellowish color to the material. It is then necessary to recrystallize the reagent from an organic solvent before use.

The carboxymethylation of protein sulfhydryl groups quickly reaches completion, but the reaction may be accompanied by several serious side reactions [10, 11]. Thus, S-carboxymethyl methionine may easily form, which introduces heterogeneity by adding a positive charge to the protein, and which prevents cyanogen bromide cleavage. The methionine derivative is decomposed during acid hydrolysis, giving rise to homoserine(lactone), S-carboxymethyl homo-cysteine, and unmodified methionine. The yield of methionine in amino acid analysis may therefore be misleading, but the presence of homoserine (see Fig. 1) and S-carboxymethyl homocysteine (appearing between proline and glycine) serve as a warning. Histidine and lysine residues may also be affected [11]. It seems that using a somewhat lower amount of iodoacetate than mercaptan on a molar basis, i.e., 0.90 or 0.95, keeping the pH of the reaction above 8, and removing the reagent after the reaction (all as in Table 6) will handle the side reaction problems.

Iodoacetamide will react with sulfhydryl groups to form the S-carbamoylmethyl (often also called S-carboxamidomethyl) derivative. Much of what has been mentioned for S-carboxymethylation (see above) also applies to this modification reaction. S-carbamoylmetyl cysteine is converted into S-carboxymethyl cysteine during acid hydrolysis before amino acid analysis. In a manner analogous to asparagine and glutamine, two different phenylthiohydantoin derivatives are obtained during sequence analysis, i.e., that of S-carbamoylmethyl and that of S-carboxymethyl cysteine, which is a disadvantage, especially since both elute in crowded parts of the chromatography [12], as shown in Fig. 2. (A special, colored iodoacetamide-related reagent, 4-dimethylaminoazobenzene-4'-iodo-acetamide, is described in Section 3 of this chapter).

Methyl iodide is a most useful reagent for derivatizing sulfhydryl groups [13]. A protocol for this reaction is shown in Table 7. The derivative, S-methyl cysteine, elutes between proline and glycine on the amino acid analyzer (Fig. 1), and the yield can easily be determined. Also the corresponding phenylthiohydantoin (Fig. 2) is well separated from other components [12, 14]. S-methyl cysteine is stable during both acid hydrolysis and sequencing. Methyl iodide reacts specifi-

Table 7. Sulfhydryl group modification by methyl iodide

---

1. Cleave disulfide bonds in 200 μl of 6 M guanidine-5% mercaptoethanol according to Table 4.
2. Add 10 μl of colorless methyl iodide. Shake the sample repeatedly at room temperature for 15 min to disperse the reagent. Control that the pH remains above 8.0 by addition of Tris base.
3. Remove protein solution from undissolved methyl iodide at the bottom of the vessel.
4. Add 10 μl mercaptoethanol to assist in removing unreacted methyl iodide.
5. Isolate protein as in Table 6, last part.

---

cally with cysteine residues [13], methionine being the only residue causing problems, as it is quickly destroyed by S-methylation especially at a pH below 7. It is therefore advisable to modify the methionine residues, e.g., by cyanogen bromide cleavage, before the cysteine residues. Radioactive methyl iodide is readily available.

Methyl-p-nitrobenzene sulfonate converts cysteine residues quantitatively and selectively to S-methyl cysteines [15]. As already mentioned, this cysteine derivative shows excellent properties both in amino acid and sequence analysis. However, the reagent is for most purposes less convenient to use than methyl iodide, as it is solid, bulky, and insufficiently soluble, unless an organic solvent, like acetonitrile, is included in the reaction mixture [15].

Ethylenimine, or aziridine, is another highly useful reagent for sulfhydryl groups [16]. The modification procedure is described in Table 8. The product, i. e., S-β-aminoethyl cysteine, emerges between histidine and lysine during amino acid analysis (Fig. 1) and close to the corresponding phenylthiocarbamyl-phenylthiohydantoin of lysine [12, 14] during sequence analysis (Fig. 2). The yields in both types of identification method are high, but the resolution of S-aminoethyl cysteine from lysine may be problematic in both systems. The reaction with ethylenimine seems to be limited to cysteine residues under the alkaline conditions recommended, even though the molar excess over mercaptan is eightfold [16]. A special property of S-aminoethyl cysteine is that it functions as a lysine analog with respect to proteolytic enzymes, i.e., trypsin [16, 17], *Armillaria* protease and endoproteinase Lys-C will cleave peptide chains after both of these residues.

Table 8. Sulfhydryl group modification by ethylenimine

---

1. Cleave disulfide bonds in 200 μl of 6 M guanidine-5% mercaptoethanol according to Table 4.
2. Add 20 μl of ethylenimine three times at 10 min intervals.
3. Isolate protein as in Table 6, last part.

β-Bromoethylamine will give rise to the same derivative as ethylenimine [16, 17]. However, the reaction is considerably slower, and possibly less specific for cysteine residues.

4-Vinylpyridine at present appears to be the optimal reagent for sulfhydryl modification [18], although it has not yet been extensively used. The protocol, as indicated in Table 9, is simple. The product, S-β-4-pyridylethyl cysteine, is easily identified in high yield both during amino acid analysis (Fig. 1) and sequence analysis (Fig. 2). In the first type of chromatography it elutes just before arginine [18], and in the second it may elute in a similar position as S-methyl cysteine [12]. The reaction seems to be specific for cysteine residues [18]. An equal molar amount or a twofold excess of vinylpyridine compared to total mercaptan is recommended. An additional advantage of S-pyridylethylation is that the reaction introduces a group absorbing specifically at 254 nm, which may be utilized for identification of S-pyridylethyl cysteine-containing peptides during high-performance liquid chromatography [19].

Table 9. Sulfhydryl group modification by vinylpyridine

1. Cleave disulfide bonds in 200 μl of 6 M guanidine-5% mercaptoethanol according to Table 4.
2. Add 15 μl of vinylpyridine. Leave the sample at room temperature in the dark for 90 min.
3. Isolate protein as in Table 6, last part.

N-Ethylmaleimide, or NEM, was formerly often used for quantification by absorption measurement at 305 nm and for derivatization of sulfhydryl groups [20]. However, the reaction may be only moderately specific and the product, S-ethylsuccinimido cysteine, insufficiently stable. On acid hydrolysis S-succinyl cysteine is formed [20].

5,5'-Dithiobis(2-nitrobenzoic acid), or DTNB, i.e., Ellman's reagent, has long been extensively employed for the specific and sensitive estimation of sulfhydryl groups by absorption measurement at 412 nm [21], e.g., as outlined in Table 10. The sulfhydryl group itself is converted into a mixed disulfide with one half of the symmetrical reagent, the other half of the reagent being released as mercaptonitrobenzoate, giving rise to the yellow color. The reaction is not suited for preparative purposes.

2-Nitro-5-thiocyanobenzoic acid, or NTCB, reacts with sulfhydryl groups to form S-cyano cysteine residues and release mercaptonitrobenzoate [22], i.e., the same yellow compound as released from DTNB (see above). The cyanylated peptide chain may be cleaved by β-elimination and cyclization on the N-terminal side of the S-cyano cysteine residue [22]. However, conditions to obtain high yields in this series of reactions have not yet been firmly established.

Reagents introducing fluorescent or strongly light-absorbing groups by reaction with sulfhydryl groups are available in large numbers [23, 24]. Several of these have been used for protein modification, but mainly for analytical purposes, e.g., labeling of reactive groups or detection in chromatographies. Many of the reagents are derivatives of iodoacetate, ethylenimine, or N-ethylmaleimide.

Radioactive reagents for sulfhydryl modification have been used extensively, mainly in the form of iodoacetate and iodoacetamide. It should, however, be noted that most reagents are not completely selective for cysteine residues, and that occasionally certain histidine, methionine, or other residues show enhanced reactivity. Radioactive label alone may therefore not provide sufficient evidence for the presence of a cysteine residue.

5 Determination of Cysteine and Cystine Separately

In proteins which may contain both sulfhydryl groups and disulfide bonds, it is often important to estimate the number of each. Furthermore, it may be of interest to establish which cyst(e)ine residues in the sequence contain a free sulfhydryl group and which are disulfide linked.

For the first purpose it is usually sufficient to determine the number of sulfhydryls and the total number of cyst(e)ines, and then calculate the number of disulfides by difference. This can be achieved by measuring the reaction with Ellman's reagent before and after treatment with sodium borohydride, as in Table 10. Alternatively, the incorporation of carboxymethyl or other sulfhydryl-modifying groups before and after treatment with mercaptoethanol, and evaluation by amino acid analysis or other techniques may be employed (see Tables 4 and 6 – 9). However, when the form of each individual cyst(e)ine residue along the sequence has to be elucidated, it is advantageous to introduce different modifying groups on original and disulfide-derived sulfhydryls, e.g., as suggested in Table 11. In order to ensure selective labeling it seems necessary to remove the reagent reacting with the original sulfhydryls before the disulfides are cleaved.

Table 10. Determination of total sulfhydryl + disulfide content with Ellman's reagent

1. Evaporate peptide sample, e.g., chromatography fraction corresponding to 1 – 10 nmol of sulfhydryl or disulfide groups, to dryness.
2. Reduce disulfide groups by incubation with 100 μl of freshly prepared 2.5% solution of sodium borohydride in water at 50°C for 1 h.
3. Destroy remaining borohydride by addition of 50 μl of 1 N hydrochloric acid, leaving sample at room temperature for 30 min.
4. Add 1 ml of freshly prepared 0.13% solution of 5,5'-dithiobis (2-nitrobenzoic acid), i.e., Ellman's reagent, in 0.04 M sodium hydrogenphosphate pH 8.1 with sodium hydroxide.
5. After 5 min measure the absorption of the yellow color at 412 nm against a peptide-free blank. Analyze a calibration standard in parallel. (Note: 10 nmol of disulfide should give rise to an absorption of approx. 0.3 per cm). Sulfhydryl groups alone may be determined by starting with 100 – 200 μl of peptide solution and omitting steps 2 and 3.

Table 11. Separate modification of original and disulfide-derived sulfhydryl groups

1. Dissolve 1 – 10 mg of protein in 200 μl 6 M guanidine hydrochloride-0.1 M Tris pH 8.5 containing one of the reagents from Tables 6 – 9, but using only approx. 1/10 of the amount in order to modify free sulfhydryl groups.
2. After the reaction time, isolate protein as in Table 6, last part.
3. Redissolve protein in 200 μl guanidine hydrochloride-0.1 M tris pH 8.5, and cleave disulfide bonds with mercaptoethanol as in Table 4.
4. Block the disulfide-derived sulfhydryl groups with a different reagent from Tables 6 – 9 as described.
5. Isolate protein as in Table 6, last part.

6 Establishment of Disulfide Linkage

For the identification of a protein's disulfide linkage pattern it is necessary to cleave the protein and isolate fragments containing the single disulfide bonds [25]. Before the cleavage all sulfhydryl groups and traces of contaminating mercaptans must be blocked off, as they will otherwise easily catalyze disulfide interchange, especially at alkaline pH, and the half-cystine-containing peptides will be recovered in more than one combination.

Suitable fragments for disulfide bond identification are often obtained by cyanogen bromide or trypsin cleavage. The fragments may preferably be isolated by high-performance liquid chromatography, each peak being analyzed for disulfide content, e.g., as described in Table 10. It may often be informative to compare the patterns obtained before and after disulfide cleavage, as cystine-free fragments will keep their positions in the chromatogram, but cystine-containing fragments are likely to shift.

The cystine-containing fragments are characterized by N-terminal sequence and composition. The fragments will contain two sequences when two different segments of the peptide chain(s) are disulfide-linked. They contain single sequences when the disulfide bridge connects two different residues along the sequence in a loop-like structure but also when the two corresponding residues in a dimeric protein participate in a symmetrical disulfide bridge. Additional evidence for the nature of the disulfide bridge can be obtained by disulfide cleavage of the isolated fragment and rechromatography.

It should be noted in this context that certain proteins have been shown to occur as mixed disulfides with glutathione [26] or cysteine.

References

1. Liu T-Y (1977) In: Neurath H, Hill RL (eds) The proteins, 3rd end, III. Academic Press, London New York, pp 239 – 402
2. Torchinsky YM (1981) Sulfur in proteins. Pergamon, Oxford, pp 1 – 294
3. Lundblad RL, Noyes CM (1984) Chemical reagents for protein modification. I. CRC Boca Raton, pp 55 – 98
4. Gardner MLG (1984) Anal Biochem 141:429 – 431
5. Moore S (1963) J Biol Chem 238:235 – 237
6. Brown JR, Hartley BS (1966) Biochem J 101:214 – 228
7. Pechère J-F, Dixon GH, Maybury RH, Neurath H (1958) J Biol Chem 233:1364 – 1372
8. Bailey JL, Cole RD (1959) J Biol Chem 234:1733 – 1739
9. Rüegg VT, Rudinger J (1977) Methods Enzymol 47:111 – 116
10. Henschen A, Edman P (1972) Biochim Biophys Acta 263:351 – 367
11. Gurd FRN (1972) Methods Enzymol 25:424 – 438
12. Han KK, Belaiche D, Moreau O, Briand G (1985) Int J Biochem 17:429 – 445
13. Rochat C, Rochat H, Edman P (1970) Anal Biochem 37:259 – 267
14. Lottspeich F (1980) Hoppe-Seyler's Z Physiol Chem 361:1829 – 1834
15. Heinrikson RL (1971) J Biol Chem 246:4090 – 4096
16. Raftery MA, Cole RD (1966) J Biol Chem 241:3457 – 3461
17. Lindley H (1956) Nature 178:647 – 648

18. Friedman M, Krull LH, Cavins JF (1970) J Biol Chem 245:3868 – 3871
19. Fullmer CS (1984) Anal Biochem 142:336 – 339
20. Riordan JF, Vallee BL (1972) Methods Enzymol 25:449 – 456
21. Riddles PW, Blakeley RL, Zerner B (1983) Methods Enzymol 91:49 – 60
22. Degani Y, Patchornik A (1974) Biochemistry 13:1 – 11
23. Imai K, Toyo'oka T, Miyano H (1984) Analyst 109:1365 – 1373
24. Perrett D, Rudge SR (1985) J Pharm Biomed Anal 3:3 – 27
25. Henschen A (1978) Hoppe-Seyler's Z Physiol Chem 359:1757 – 1770
26. Spector A, Wang G-M, Huang R-RC (1986) Curr Eye Res 5:47 – 51

5.2 Identification of the Heme-Binding Cysteines in Cytochromes c Without Radioactive Labeling

JOZEF VAN BEEUMEN[1]

Contents

1 Introduction

In spite of the recent improvements in manual and automatic Edman degradation technology, the identification of cysteine residues remains difficult, mainly because of the instability of its phenylthiohydantoin (PTH) derivative. A way to circumvent the problem is to modify the cysteine into a more stable derivative. Many products are known to react with sulfhydryl groups such as ethylenimine, 2-bromoethyltrimethylammonium bromide, methyl-p-nitrobenzenesulfonate, sodium sulfite, acrylonitrile and 2-nitro-5-cyanobenzoate (for references, see [1]). The most commonly used reagent, however, appears to be iodoacetic acid or its amide iodoacetamide, giving rise to carboxymethylcysteine (CMC) or carboxyamidomethylcysteine. The reagents are commercially available as C^{14} radiolabeled products, and those who have the facilities to deal with isotopes prefer to identify PTH-CMC by radioactive counting. It is also possible to identify nonlabeled PTH-CMC, either by thin-layer chromatography or by high pressure liquid chromatography (HPLC), the method mostly used in conjunction with automatic sequence analysis. Depending on the HPLC system used, PTH-CMC may not always be identifiable as a separately isolated peak. In the present contribution we present a simple method to recognize PTH-CMC even though it coelutes with another PTH residue. The method is illustrated for the two heme-binding sites of the diheme cytochrome c-552 from the denitrifying bacterium *Pseudomonas perfectomarinus*.

---

1 Laboratory of Microbiology and Microbial Genetics, State University Gent, Ledeganckstraat 35, B-9000 Gent, Belgium

Advanced Methods in Protein Microsequence Analysis
Ed. by B. Wittmann-Liebold et al.
© Springer-Verlag Berlin Heidelberg 1986

2 Preparation of Cytochrome c Prior to Sequence Analysis

The structurally most characteristic feature of a cytochrome of the c-type is the fact that the prosthetic heme group is covalently bound to the polypeptide chain by means of two thioether linkages. These linkages originate from two cysteines which are themselves separated by two other amino acids. The second cysteine is always followed by a histidine residue so that the evolutionary invariant heme-binding site can be written as -CYS-x-y-CYS-HIS-. For the so-called class I cyto-chromes c, to which all mitochondrial cytochromes c belong, the heme-binding cysteines occur near the N-terminus of the polypeptide chain starting around position 13 [2 – 4].

It has been experienced that the efficiency of the Edman degradation strongly decreases as soon as the heme-binding cysteines are reached. The heme should therefore be removed prior to N-terminal sequence analysis. Also if the complete primary structure has to be determined, deheming is strongly advised because heme-binding peptides tend to streak upon purification by (e.g., paper) electro-phoresis [5] or to give broadended peaks during reverse-phase HPLC.

There are two main procedures for heme removal. The first is based on treat-ment of the native protein with acid in the presence of H_gCl_2 and urea [5], the second uses BNPS-skatol [6]. The latter reagent is also kown to cleave peptide bonds after tryptophan residues. The author prefers the first treatment because it leaves the polypeptide chain intact and avoids the problem of separating the heme from a peptide mixture. The detailed procedure is given in Table 1. The yield of apoprotein after the gel filtration step is usually 90 to 95%. Determina-tion of the cysteines after performic acid oxidation and subsequent amino acid analysis of the apoprotein yields on the average 1.5 instead of the expected number of 2.0 cysteic acid residues. Performic acid oxidation of the native cytochrome mostly allows to quantify only 40 – 60% of the number of cysteines.

It should be added here that the acidic acetone treatment which is commonly used for heme removal from hemoglobines is not efficient for cytochromes c [7].

Table 1. Procedure for heme removal from cytochromes c

1. Lyophilize the protein solution (0.5 µmol or more) in a 5 or 10 ml tube with screw cap. The cap should preferably have a Teflon inlay.
2. Prepare a fresh solution of 8 M urea; per 50 ml of this solution, add 0.42 ml of concentrated HCl (0.1 M final concentration).
3. Add 2 ml of the acidic urea to the lyophilized protein.
4. Add an amount of $HgCl_2$ equal to the weight of the amount of cytochrome.
5. Close the tube and incubate overnight at 37 °C under mild shaking.
6. Separate the heme from the apoprotein by gel filtration over a column of (e.g.) Sephadex G-25, fine (1.5 × 20 cm), washed and eluted with 5% formic acid. Follow the elution of the apoprotein by detection at 280 nm, (e.g., with a Uvicord II). The salt, which also absorbs at the given wavelength, should normally be baseline separated from the protein. The heme tends to stick to the column but can quickly be removed with 50% pyridine.
7. Lyophilize the apoprotein. The powder obtained is ready for carboxymethylation.

Amounts smaller than 500 pmol can be treated in the same way but sizes of glass ware and quantities of reagents should be appropriate.

3 Identification of PTH-CMC by HPLC

The HPLC analysis system used by the author since the arrival of the gas-phase sequenator is similar to the one described by Hunkapiller and Hood [8]. A typical separation and details of the equipment and of the gradient as we use them in our laboratory is given in Fig. 1 and its legend. We never succeeded in obtaining the nearly baseline separation for all the PTH-amino acids as reported in [8], even not with a new column. The chromatogram of Fig. 1, however, was obtained after more than 4000 analyses on the same column and was found to be still sufficiently good for identification purposes. Note that the separation between the first three PTH's and between PTH-Gln and PTH-Gly is slightly better when the sample is dissolved in aqueous acetonitrile (ACN/H_2O, 1/2, as suggested by Applied Biosystems in the Sequencer Installation Instructions) instead of 100% ACN alone, at least with injection volumes of 10 µl or more. Since we use methanolic HCl instead of aqueous trifluoroacetic acid as the thiazolinon conversion reagent, the PTH-derivative of glutamic and of aspartic acid is obtained as the methyl ester. The same is true for PTH-carboxymethylcysteine.

In the separation system given, PTH-carboxymethylcysteine methyl ester (PTH-CMC-OMe, or "CMC-1") has the same elution time as PTH-Pro. It is not possible to separate these two compounds, e.g., by changing the gradient program or the relative concentration of the components of solvent B, without badly influencing the separation between some of the other PTH-amino acids. However, the major characteristic feature of the chromatogram is seen at 313 nm, the wavelength which is usually set to identify the dehydration products of PTH-Ser and PTH-Thr. PTH-CMC-OMe shows an asymmetrical pattern of unseparated peaks of which the last, but also the smallest one, corresponds to the major peak CMC-1 at 254 nm; the ratio OD_{254}/OD_{313} is 7.3. On the contrary, the major peak at 313 nm ("CMC-2") corresponds to a smaller peak at 254 nm which is detected 86 s earlier than CMC-1; the ratio OD_{254}/OD_{313} for CMC-2 is 0.32.

→

Fig. 1 A – F. HPLC chromatogram of reference PTH-amino acids (**A**, **B**) and of methylated PTH-carboxymethylcysteine (**C** to **F**). **A** 60 pmol of each PTH-amino acid (Pierce standard) dissolved in CH_3CN; injection volume: 10 µl. **B** Idem, but sample dissolved in CH_3CN/H_2O (1/2). **C** 350 picomol of PTH-CMC-OMe (methylated Pierce standard) dissolved in CH_3CN. **D – F** as for **C** but sample dissolved in 90% solvent B; the times indicated as ΔT represent the period from the moment of sample dissolving to the moment of CMC-1 elution. Experimental conditions: Waters Interlink System with two 6000 A pumps, a 710 WISP automatic injector, a 720 System Controller, a 730 Data module and a dual channel 440 fixed wavelength detector set at 254 and 313 nm; a 5 µ cyanopropyl column (4.6 × 250 mm, IBM, Connecticut) equipped with a Permaphase ETH precolumn (4.6 × 50 mm), DuPont, Delaware) both incubated in a forced-air column oven set at 37 ± 1 °C. Solvent A: 100% CH_3CN (Carlo Erba, Milan), solvent B: 20 mM sodium acetate (Merck, Darmstadt) pH 5.30 containing 3.75% tetrahydrofuran (Carlo Erba); both solvents are continuously degassed with helium. Gradient elution program: from 36% to 52% A in 5 min (linear), from 10% to 36% A in 12 min (convex gradient no 5), from 36% to 52% A in 5 min (linear) and from 52% to 10% A in 3 min (linear); reequilibration at 10% A for 10 min; flow rate: 1.0 ml min^{-1} at a working pressure of ca 1100 psi. The chromatograms of PTH-CMC-OMe are shown aligned with those of the other references; PTH-CMC-OMe itself (CMC-1) has the same elution time as PTH-Pro; the registration at 313 nm (*top*) is displaced versus the one at 254 nm (*bottom*) as shown

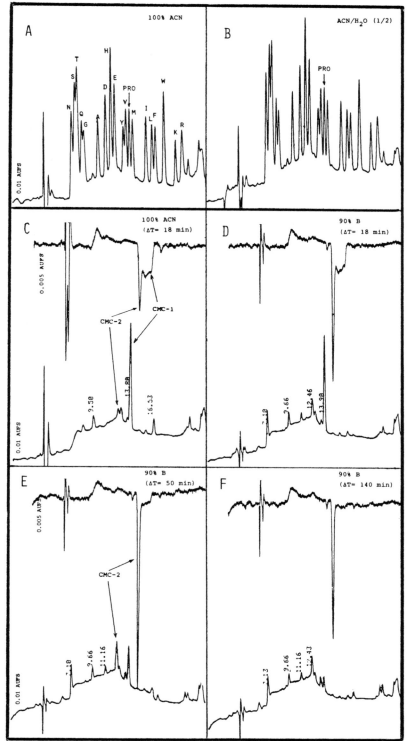

Fig. 1A – F

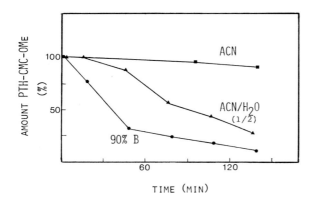

Fig. 2. Decomposition of PTH-carboxymethylcysteine (methylated form) in different solvents. *B* is the buffer solvent used in the HPLC-analysis

TIME (MIN)

Table 2. Detection of PTH-CMC-OMe by HPLC

1. Dry the PTH sample as early as possible after it has been delivered to the fraction collector of the sequencer. Use a Speed Vac Concentrator if at hand. Drying from MeOH with a water jet pump takes about 25 min at room temperature. If 25% TFA instead of methanolic HCl has been used for the conversion reaction in the sequencer, methylate the sample for 25 min at 55° with methanolic HCl. This solution is made as suggested in [8] by mixing CH_3COCl and MeOH. Keep this solution at 4° when not in use; prevent its trapping moisture by putting it under a blanket of nitrogen.
2. Dissolve the sample in an appropriate volume of 100% CH_3CN, mix carefully and carry out the HPLC analysis with the system of your choice. If you use a cyano column (IBM) for the separation, the experimental conditions given in the legend of Fig. 1 may be useful. PTH-CMC-OMe can be recognized by its asymmetrical peak pattern at 313 nm.
3. If PTH-CMC-OMe coelutes with another PTH-amino acid, dry the sample again in the Speed Vac, dissolve it in a wanted volume of 90% of the buffer solution used in the HPLC-analysis and reanalyze it within a period of 2 h. A decomposition of PTH-CMC-OMe is expected to occur as given in Fig. 2.

The second major feature of the CMC-chromatogram is apparent when the compound is dissolved in 10% solvent A/90% solvent B, the composition of the eluent at the start of the HPLC-analysis. Figure 1 D shows that already 18 min after the sample has been dissolved, CMC-2 at 313 nm has increased by a factor of 1.9, whereas the height of the asymmetrical peak region has decreased. After 50 min of incubation, CMC-2 has increased by a factor of nearly 3 compared to the height in 100% ACN, whereas CMC-1 has come down by approximately the same factor. After 140 min, CMC-1 has lost 88% of its original height, but also CMC-2 has started to decrease (for some 30%) compared to the maximal value observed after 50 min.

The presence of the sodium acetate salt is apparently not the prime agent for causing the decomposition of CMC-1 in favor of CMC-2. Similar effects are seen when PTH-CMC-OMe is dissolved in ACN/H_2O (1/2), although the effect appears to be less drastic. For comparative purposes, we have summarized the data for CMC-1 (254 nm) in Fig. 2. There is also some decomposition in pure acetonitrile; the rate at which this occurs (ca. 5%/h) is small but still substantially higher than for any of the other reference PTH-amino acids. As the result of our findings, we propose a practical procedure for measuring PTH-OMC-OMe as given in Table 2.

4 Detection of a "Classical" and an "Unusual" Heme-Binding Site in a Diheme Cytochrome C-552

Pseudomonas perfectomarinus is a marine denitrifying bacterium which contains several soluble c-type cytochromes. Amongst them there is a small molecular weight cytochrome c-551 ($M_r = 7660$) and a larger diheme cytochrome c-552 ($M_r = 25000$). In the latter protein, the redox potential of the two heme groups is different, $+0.17$ and -0.18 V respectively [9, 10].

In Figs. 3 and 4 we illustrate our method for the recognition of the two heme-binding sites of the cytochrome c-552 by showing the HPLC-analyses for the two different heme peptides. The carboxymethylation of the dehemed protein was carried out according to Crestfield [11] with an initial addition of a tenfold molar excess of dithiothreitol (DTT) over the expected number of cysteines. The peptides investigated here were obtained after cleavage of the carboxymethylated apoprotein with *Staphylococcus aureus* protease followed by purification of the peptide mixture by HPLC on a Vydac C_4 column. The sequence of the first peptide S.a. 1 can be interpreted from Fig. 3 as Thr-CMC-Ala-Gly-CMC-His, which is typical for a "classical" heme-binding site. The chromatographic pattern for residue 2 shows the asymmetrical pattern at 313 nm typical for PTH-CMC-OMe, but in this particular case, it also shows the superimposed peaks of PTH-dehydro-threonine (ΔT_1 and ΔT_2). The latter are to be explained as the carry-over of PTH-Thr from the N-terminal residue of the peptide.

The chromatograms for the residues 3 to 9 of the second hemepeptide S.a. 2 are given in Fig. 4. The typical chromatographic features of PTH-CMC-OMe are found at the positions 4 and 8. The bottom figures also illustrate the rapid decomposition of CMC-1 in favor of CMC-2, as expected, when the sample is analyzed after dissolving in 90% solvent B. The possibility that the new peaks at the positions 4 and 8 are derived from PTH-Pro has therefore to be completely excluded. The sequence of the second heme region can thus be read as -Gly-CMC-Trp-Gly-Ser-CMC-His- (chromatogram of the histidine not shown). At position 7 one should also notice, along with PTH-Ser at 7.57 min, the DTT-addition product of PTH-dehydro-Ser after 13.22 min. This elution time is very close, but not identical, to that of PTH-dehydro-Ser itself which elutes 38 s earlier. Reference analyses have shown that the latter compound has the same elution time as CMC-2.

The occurrence of three residues between the two heme-binding cysteines has not yet been found in any of the more than 200 class I cytochromes c sequenced so far [2 – 4]. Also the fact that one of the residues between the cysteines is a tryptophan is unique. One may wonder in how far the presence of this strongly hydrophobic residue is responsible for the large difference in redox potential between the two heme groups of this cytochrome c-552. Awaiting the complete

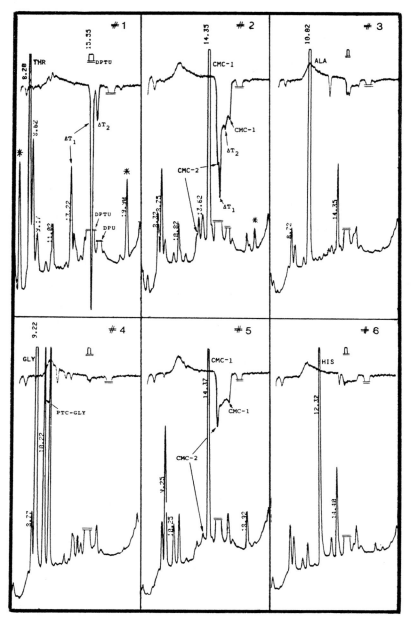

Fig. 3. HPLC-analyses of the first six Edman cycles of the heme-peptide S.a. 1 from *P. perfecto-marinus* cytochrome c-552. The sequencer run was carried out on 6.9 nmol of peptide (10 residues) without prewash of the polybrene (3 mg). Peaks indicated with an asterisk are impurities originating from the polybrene. The peaks of diphenylthiourea (DPTU) and in some cases of diphenylurea (DPU) were removed from the picture in order not to blurr the absorbance trace at 313 nm. The ana-lyses were run on a different column (but of the same type) as the one used for the reference chro-matograms of Fig. 1; retention times are therefore slightly different. Other experimental conditions: as in Fig. 1

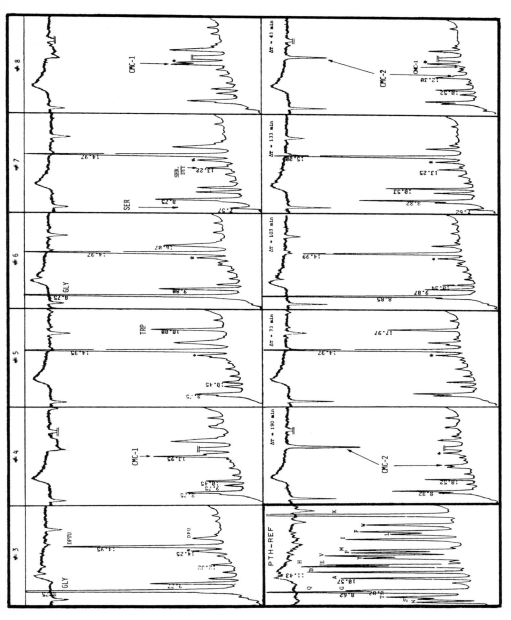

Fig. 4. HPLC-analyses of the Edman cycles 3 to 8 of the "unusual" heme-peptide S. a. 2 from *P. perfectomarinus* cytochrome c-552. Sequencer analysis was carried out on 3.7 nmol of peptide (11 residues). The top figures and the reference chromatogram were obtained after dissolving the samples in 100% acetonitrile; for the peptide, aliquots of 33% were injected on the HPLC. The remaining amount of each sample was dried under vacuum and redissolved in 10% ACN/90% solvent B; aliquots of 40% were injected at the moments indicated as ΔT in the bottom figures. For residues 4 and 8, the peak of DPTU was removed from the picture in order to show the typical asymmetrical pattern of PTH-CMC-OMe and the peak of CMC-2 more clearly. Peaks indicated with an asterisk are impurities from the Edman chemistry. The solution of the reference PTH-amino acids contained an excess of PTH-Lys and no PTH-Arg. All recordings were done at detector settings of 0.005 AUFS

amino acid sequence, it may at least be concluded that *Pseudomonas perfecto-marinus* cytochrome c-552 is a diheme cytochrome with a "classical" and an "unusual" heme-binding site.

Although the method has been shown here to be useful in the case of cytochromes c, it is evident that it can be applied to any protein or peptide that has been initially carboxymethylated. An HPLC system with a separation on a cyanopropyl column as the author has used is absolutely no prerequisite. Each investigator should try to work out first a method by which he can separate PTH-CMC or PTH-CMC-OMe from the other PTH's without losing resolution between the latter. If this turns out to be impossible, as the author has experienced with his system, PTH-CMC-OMe can nevertheless be recognized by its typical asymmetrical pattern at 313 nm. Even if a second UV-detector is not available, PTH-CMC-OMe can still be identified through its decomposition into CMC-2 (which is most likely identical to PTH-dehydro-Ser) on condition that the sample is dissolved in an aqueous buffer. Those who use automatic injectors should be aware that they might miss PTH-CMC-OMe if they dissolve their samples in ACN/H_2O (as in Fig. 2B) and the cysteine-containing Edman cycle has to wait too long before being analyzed. The problem of decomposition of CMC-1 is of course avoided if the PTH-derivatives can be analyzed "on-line" with the sequenator. An on-line analyzer has recently become available from Applied Biosystems (USA), but several research groups have developed their own system (see, e.g., Ashman's contribution).

References

1. Glazer AN, Delange RJ, Sigman DS (1975) In: Work TS, Work E (eds) Chemical modification of proteins. Elsevier/North Holland, Amsterdam, p 101
2. Ambler RP (1982) In: Robinson AB, Kaplan NO (eds) From cyclotrons to cytochromes. Academic Press, London New York, p 263
3. Dickerson RE (1980) Sci Am 242:237
4. Van Beeumen J (1980) In: Peeters H (ed) Protides of the biological fluids, vol 28. Pergamon Press, Oxford, p 61
5. Ambler RP, Wynn M (1973) Biochem J 131:485
6. Fontana A, Vita C, Toniolo C (1973) FEBS Lett 32:139
7. Rossi Fanelli A, Antonioni E (1958) Biochim Biophys Acta 30:608
8. Hunkapiller MW, Hood LE (1983) In: Hirs CHW, Timasheff SN (eds) Methods in enzymology, vol. 91. Academic Press, London New York, p 486
9. Liu MC, Payne WJ, Peck HD, LeGall J (1983) J Bacteriol 154:278
10. Liu MC, Peck Jr HD, Payne WJ, Anderson JL, Dervartanian DV, LeGall J (1981) FEBS Lett 129:155
11. Crestfield AM, Moore S, Stein WH (1963) J Biol Chem 238:622

5.3 Micro-Isolation of Polypeptides Precolumn Labeled with Hydrophobic Chromophore

JUI-YOA CHANG[1]

Contents

Abbreviations: DABITC = dimethylaminoazobenzene isothiocyanate; DABTC = dimethylamino-azobenzene thiocarbamoyl; DABTH = dimethylaminoazobenzene thiohydantoin; DABIA = dimethylaminoazobenzene iodoacetamide; DABCAM = dimethylaminoazobenzene carboxyamido-methyl; PITC = phenylisothiocyanate; PTH = phenylthiohydantoin; TFA = trifluoroacetic acid; HPLC = high performance liquid chromatography.

1 Introduction

Detection of polypeptides is a crucial technique associated with every chromato-graphic methods of peptide isolation. Conventionally, peptides are detected, after chromatographic separation, by their reaction with reagents like ninhydrin [1, 2], fluorescamine [3, 4] or o-phthaldialdehyde [5, 6]. Alternatively, peptides can be detected through their intrinsic absorption of amide bonds at low UV region (200 nm – 230 nm) [7, 8]. This low UV detection system in combination with the use of reversed phase HPLC is at present the most versatile technique for the

---

1 Pharmaceutical Research Laboratories, Ciba-Geigy Limited, CH-4002 Basel, Switzerland

Advanced Methods in Protein Microsequence Analysis
Ed. by B. Wittmann-Liebold et al.
© Springer-Verlag Berlin Heidelberg 1986

isolation of peptides. To enhance the sensitivity of detection, peptides can also be labeled with chromophore preceding chromatographic separation. For instance, dansylated peptide mappings [9, 10]. We have developed two precolumn labeling techniques for micro-isolation of peptides. (1) A method using DABITC labeling for isolation of peptides with free N-termini [11]. (2) A method using DABIA labeling for selective isolation of cysteine containing peptides [12].

Both DABITC and DABIA (Fluka catalog No. 39061) are commercially available from Fluka (Switzerland). Double-recrystallized DABITC can also be obtained from Pierce (USA).

2 DABITC Labeling Method

2.1 Chemistry

The chemistry of the DABITC technique is shown in Fig. 1. The DABTC-peptides can be detected in the visible region and sequenced after recovery from HPLC. The method is most suitable for separating mixtures of small peptides, e.g., tryptic or chymotryptic peptides.

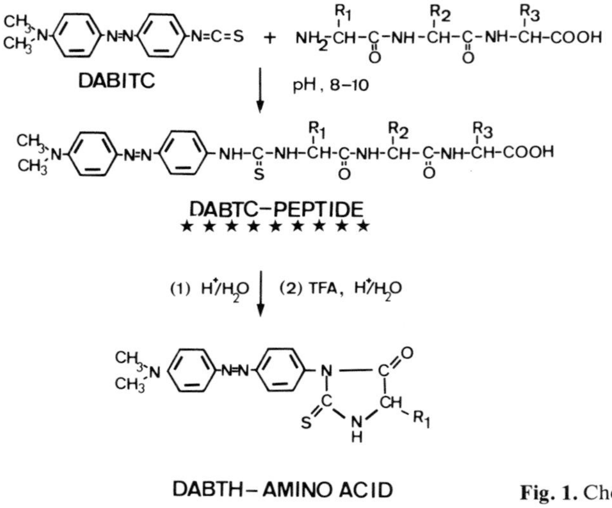

Fig. 1. Chemistry of the DABITC technique

2.2 DABITC Labeling

- Approximately 0.1 – 5 nmol of peptide mixture, freeze-dried and preferably salt-free.
- Dissolve in 50 μl of water.
- Add 100 μl of DABITC solution (10 nmol μl^{-1} in pyridine).
- Heat at 70 °C for 1 h.
- Extracted with 4×500 μl of heptane/ethylacetate (2:1, v/v).

— The remaining aqueous phase (and precipitate, if any) which contains DABTC-peptides is evaporated with nitrogen or freeze-dried.
— Redissolve in 40 – 250 µl of 20% aqueous pyridine, stored at – 20 °C and used as the stock sample.

2.3 HPLC Isolation of DABTC Peptides

The detection limit of DABTC-peptide is around 1 pmol. Thus, if one has more than 1 nmol of stock DABTC peptides, it is advantageous first to perform a few analytical runs at 5 – 10 pmol level. Analytical runs are used to search for an optimal chromatographic condition before the total sample is loaded for the preparative isolation. Optimal condition is usually the one which gives maximum separated peaks. When the analytical runs are to be performed, aliquot of the stock sample is diluted to the concentration of approximately 1 pmol µl^{-1} with 50% aqueous acetonitrile.

Since DABTC peptides are detected in the visible region (420 nm – 436 nm), a variety of organic solvents and buffers can be chosen to achieve the separation. Adjustment of the ionic strength and the pH of the aqueous buffer is of particu-

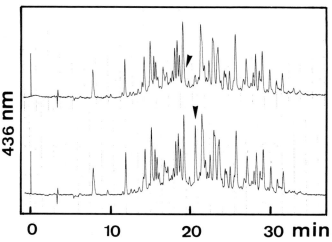

Fig. 2. Analytical DABTC-peptide mappings (6 pmol, trypsin-digested) of a normal human antithrombin III (*top panel*) and a genetically abnormal antithrombin III (*bottom panel*). These two antithrombin III were isolated from the same patient. They have indistinguishable MW (58,000) and progressive antithrombin activities. However, abnormal antithrombin III has reduced heparin cofactor activity. DABTC-peptides mappings indicates that there is one peptide with different retention time (indicated by *arrows*). Approximately eight analytical runs were performed and this presented chromatogram was selected for the preparative isolation because the abnormal peptide was eluted at the most desirable position. Eventually, approximately 800 pmol of this abnormal peptide was isolated and sequenced by the DABITC/PITC method [18]. It gave the sequence of Ile-Leu-Glu-Ala-Thr-Asn-Arg which corresponds to the positions of 40 – 46 in the normal antithrombin III, but with one substitution of Pro by Leu at position 41 [16]. Solvent A was 10 mM acetate, pH 5.5. Solvent B was acetonitrile. The gradient was 20%B to 70%B in 40 min. The column was Vydac C-18 (5 µm). Column temperature was 37 °C. Detector was set at 436 nm, 0.01 AUFS [16].

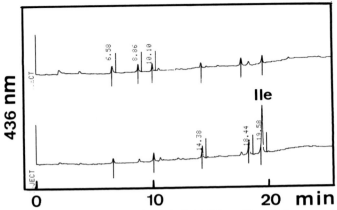

Fig. 3. Amino-terminal analysis of the DABTC peptides which represent the structural difference be-tween normal and abnormal antithrombins (from fractions indicated by arrows in Fig. 2). The DABTC peptides which were isolated from 200 pmol of the mixtures, were dried with nitrogen evap-oration. Their amino-termini were released as DABTH amino acids (see Fig. 1) by direct incubation with aqueous acid (acetic acid/5N HCl, 2:1, v/v; at 54°C for 45 min). The released DABTH amino acids were analyzed by the HPLC conditions described in [18]. When sequence analysis of DABTC peptides is required, the amino-termini should be released first by TFA instead of aqueous acid [16]

larly effective in resolving overlapping DABTC peptides. Acetonitrile and sodium acetate buffer (5 mM – 35 mM, pH 5 – 7) are the most frequently used solvents in our laboratory [13 – 16]. Pyridine-acetate buffer and ammonium acetate buffer also gave satisfactory results. In our experience, phosphate salt in-terferes with the subsequent sequence analysis of the DABTC peptides and is not recommended. Buffers with pH lower than 4.5 should also be avoided. The re-covered DABTC peptides are dried with nitrogen and stored at −20°C. Sequence determination of DABTC peptides by either automatic sequenator [17] or manual method [18, 19] begins with TFA cleavage. A successful application of the DABITC labeling method is shown in Figs. 2 and 3.

2.4 Comments

2.4.1 Advantages

a) High sensitivity of detecting DABTC peptide allows isolation of low pmol of pure peptide.
b) Detection of DABTC peptide at the visible region allows selection of a wide range of solvents and buffers for HPLC separation.

2.4.2 Limitations

a) The precolumn derivatization method is not suitable for isolation of biologi-cally active peptides.

b) Peptides with N-termini blocked or mono-alkylated [20] can not be isolated with this method.

c) Peptides with more than 2 to 3 lysines usually (also depending on the sequences) become rather insoluble after DABITC derivatization. However, for tryptic or chymotryptic peptides, this case rarely occurs.

3 DABIA Labeling Method

3.1 Strategy

The strategy for the selective isolation of cysteine-containing peptides is shown in Fig. 4. (a) Sulfhydryl groups of a protein are specifically labeled with DABIA. (b) The labeled protein is then digested with trypsin or chymotrypsin. (c) When the peptide mixtures are separated by the reversed-phase HPLC, only cysteine-containing peptides are detected in the visible region. Furthermore, these cysteine-containing peptides are generally more hydrophobic than noncysteine-containing peptides and can easily be separated from each other.

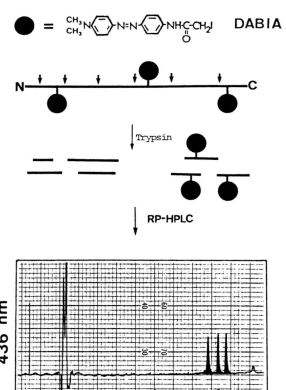

Fig. 4. Selective isolation of cysteine-containing peptides

3.2 DABIA Labeling

Free sulfhydryl groups or reduced disulfide linkages are both labeled with the same protocol [12]. When reduction of disulfide linkages is performed, reduced protein should be separated from excess dithiothreitol or mercaptoethanol preceding DABIA labeling. This can be done by desalting the reduced sample through gel-filtration (Sephadex G-25 or G-10, column size 1.2 cm i.d. ×25 cm) using 1 M acetic acid as eluent.

- Two to 10 nmol of protein (with free −SH groups or reduced −S−S− bonds) in freeze-dried form.
- Dissolve in 400 μl of 5 M-guanidinium chloride/0.5 M-Tris/HCl buffer, pH 8.4.
- Add 400 μl of DABIA solution (0.5 mg ml^{-1} in dimethylformamide).
- Labeling is performed at room temperature in the dark for 1 h with magnetic stirring.
- The sample is acidified with 20 μl of TFA.
- Desalt through a G-25 column (1.2 cm i.d. ×25 cm) using 50% acetic acid as eluent. DABIA-labeled protein is eluted as a colored band in about 5 − 7 ml.
- Freeze-dry.
- Redissolve in 200 μl of 50% formic acid, then 400 − 600 μl of water. Transferred to a small tube suitable for enzyme digestion and freeze-dried again.

The DABIA-labeled protein can be directly used for sequencing (Fig. 5) or digested by enzyme in order to isolate cysteine-containing peptides.

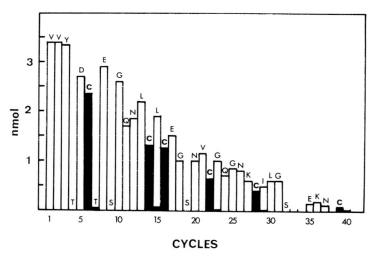

CYCLES

Fig. 5. Yields of PTH-amino acids from a sequenator run performed with 3.5 nmol of hirudin which was reduced-carboxymethylated with DABIA. The amino-terminal sequence was determined by a Beckmann 890C sequenator. At steps 6, 14, 16, 22, 28, and 39, red-colored PTH-derivatives of amino acids were observed in the fraction collector tubes of the sequenator. These red-colored derivatives represent six cysteine residues (S-DABCAM-PTH-Cys). Quantitative analysis of S-DABCAM-PTH-Cys was performed with HPLC (see Fig. 7) [21]

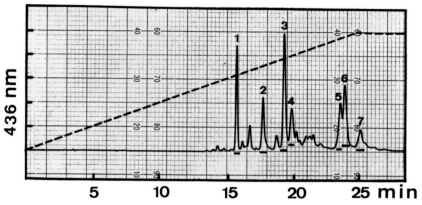

Fig. 6. Preparative isolation of cysteine-containing peptides (2 nmol) derived from the tryptic cleavage of reduced, DABIA-labeled immunoglobulin 7S34.1 light chain [anti-(streptococcal group A polysaccharide) light chain]. This light chain has an extra cysteine at position 43 (contained in peptide No. 7). A total of 7 major DABIA-labeled peptides were recovered from the light chain 7S34.1. Solvent A was 10 mM phosphate buffer, pH 7.0, containing 2% of dimethylformamide. Solvent B was acetonitrile. The gradient was 10%B to 60%B in 25 min. The column was μ-Bondapak C-18 (Waters). Column temperature was 22 °C. The detector was set at 436 nm and 1.0 AUFS [12]

3.3 Enzyme Digestion of DABIA-Labeled Proteins

- One to 10 nmol of DABIA-labeled protein.
- Dissolve in 200 – 300 μl of 0.1 M ammonium bicarbonate solution and digested with 10 – 20 μg of trypsin or chymotrypsin.
- Digestion is carried out at 37 °C overnight with magnetic stirring (using a 6-mm magnetic stirring bar). The DABIA-labeled proteins are not always soluble in ammonium bicarbonate buffer. However, they usually become soluble after enzyme digestion has taken place.
- After digestion, the sample solution can be injected directly into HPLC or freeze-dried and then redissolved in a suitable volume of 20% pyridine for HPLC.

3.4 HPLC Separation of DABIA-Labeled Cysteine-Containing Peptides

The strategy for isolation of DABIA-labeled peptides by HPLC is similar to that for DABTC-peptide isolation. It is preferable to perform a few analytical runs at 5 – 10 pmol level with varying chromatographic conditions, particularly with varied pH and salt concentration of the aqueous buffer. After the optimal condition has been selected, the total sample was then injected for the preparative isolation.

The solvents and buffer systems available for the isolation of DABIA-labeled peptides are even less restrictive than those for the DABTC-peptide isolation because DABCAM derivative is stable in both acid and alkaline solutions. How-

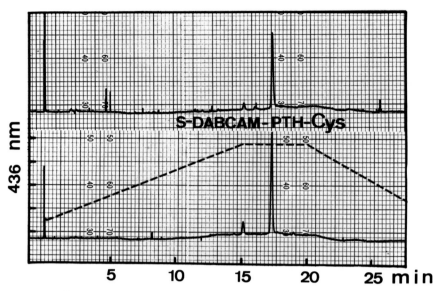

Fig. 7. HPLC analysis of the standard S-DABCAM-PTH-Cys (*bottom panel*, 15 pmol) and the S-DABCAM-PTH-Cys released from the sequence analysis of a cysteine containing peptide (derived from the light chain 7S34.1, see Fig. 6) (*top panel*). Solvent A was 35 mM acetate, pH 5.0. Solvent B was acetonitrile. The gradient was 45%B to 78%B in 15 min, stayed at 78%B till 20 min. The column was Zorbax ODS (5 µm). Column temperature was 22 °C. The detector was set at 436 nm and 0.01 AUFS. The standard S-DABCAM-PTH-Cys was prepared by the following method. Cysteine (500 nmol) was dissolved in 100 µl of water, and to it was added 50 nmol of DABIA in 100 µl of dimethylformamide. The mixture was incubated for 1 h at room temperature and dried in vacuum. The dried sample was dissolved in 300 µl of aq. 50% pyridine, and to it was added 25 µl (approx. 100 µmol) of PITC. After reaction for 30 min at 50 °C, the mixture was again dried and the solid was redissolved in 150 µl of aq. 50% TFA. The acid solution was heated for 40 min at 50 °C and then dried in vacuum, and the solid was redissolved in 1 ml of aq. 50% acetonitrile for HPLC analysis

ever, it should be observed that the wavelength of the maximum molar extinction coefficient of the DAB chromophore is dependent upon the pH. Between pH 5 – 7 the wavelength is around 425 nm. A successful application of the DABIA-labeling method is described in Fig. 6. When the isolated DABIA peptides were sequenced by automatic Edman degradation, cysteine residues were released as colored derivatives of S-DABCAM-PTH-Cys (Figs. 5 and 7).

3.5 Comments

3.5.1 Advantages

The DABIA-labeling method represents a new approach for the selective isolation of cysteine-containing peptides. The method is efficient and sensitive. Furthermore, during the sequence determination of DABIA-labeled peptides, the cysteine residues are released as colored PTH-derivatives which can be visually detected at the picomole level.

3.5.2 Limitations

The hydrophobic character of DABIA also represents disadvantages. Because of the limited solubility of DABIA in the labeling solution, reducing agents like dithiothreitol and mercaptoethanol have to be removed from the reduced protein before DABIA labeling. This requires an extra desalting step as compared to the standard carboxymethylation [22].

References

1. Heinz N, Hultin T (1983) Methods Enzymol 91:359–366
2. Hermann AC, Vanaman TC (1977) Methods Enzymol 47:220–236
3. Lai CY (1977) Methods Enzymol 37:236–243
4. Samejima K, Dairman W, Udenfriend S (1971) Anal Biochem 42:222–229
5. Roth M (1971) Anal Chem 43:880–884
6. Bensen JR, Hare PE (1975) Proc Natl Acad Sci USA 72:619–622
7. Machleidt W, Otto J, Wachter E (1977) Methods Enzymol 47:210–220
8. Hermodson M, Mahoney WC (1983) Methods Enzymol 91:352–359
9. Tichy H (1975) Anal Biochem 69:552–557
10. Zanetta JP (1970) J Chromatogr 51:441–458
11. Chang J-Y (1981) Biochem J 199:537–545
12. Chang J-Y, Knecht R, Braun DG (1983) Biochem J 211:163–171
13. Chang J-Y, Herbst H, Aebersold R, Braun DG (1983) Biochem J 211:173–180
14. Chang J-Y, Knecht R, Ball R, Alkan SS, Braun DG (1982) Eur J Biochem 127:625–629
15. Chang J-Y, Knecht R, Maschler R, Seemuller U (1985) Biol Chem Hoppe-Seyler 366:281–286
16. Chang J-Y, Tran TH (1986) J Biol Chem 261:1174–1176
17. Edman P, Begg G (1967) Eur J Biochem 1:80–91
18. Chang J-Y (1983) Methods Enzymol 91:455–466
19. Chang J-Y, Brauer D, Wittmann-Liebold B (1978) FEBS Lett 93:205–214
20. Chang J-Y (1978) FEBS Lett 91:63–68
21. Dodt J, Muller H-P, Seemuller U, Chang J-Y (1984) FEBS Lett 165:180–184
22. Gurd FRN (1972) Methods Enzymol 25:424–438

Chapter 6
Methods of Analyzing Protein Conformation

6.1 Synthetic Immunogens for Secondary Structure Assignment: Conformational Sequencing of Proteins with Antipeptide Antibodies

KONRAD BEYREUTHER, HEINRICH PRINZ, and URSULA SCHULZE-GAHMEN[1]

Contents

1 Introduction

Antibodies to proteins can be induced by immunizations with short synthetic peptide immunogens. They have been used for the identification of protein products from open reading frames of nucleic acid sequences of cloned protein genes, structure-function studies, localization of proteins at their functional compartment, topological studies of membrane proteins, virus neutralization, and for experiments on the specificity of the immune system (reviewed in [1, 2]). The high frequency with which protein-reactive antipeptide antibodies can be generated allows the isolation of antibodies of predetermined specificity to most

---

1 Institut für Genetik, Universität zu Köln, Weyertal 121, D-5000 Köln 41, FRG

Advanced Methods in Protein Microsequence Analysis
Ed. by B. Wittmann-Liebold et al.
© Springer-Verlag Berlin Heidelberg 1986

regions of proteins. Not only regions comprising surface-exposed residues, but also sequence stretches from the interior react with their cognate antipeptide antibodies [3, 4]. Therefore, the resulting sum of the immunogenicity of the pieces of a protein represented by the synthetic peptide immunogens by far exceeds the immunogenicity of an intact protein, whereby immunogenicity refers to the ability of an antigen, here either a peptide or an intact protein, to induce antibodies. Recognition by an antibody is referred to as antigenicity. Since only a limited part of an intact protein is immunogenic and since antigenic determinants of proteins are often [5] but not always [6] composed of amino-acid residues far apart from each other in the sequence, these determinants are called discontinuous, conformational, and more recently, assembled topographic determinants [5].

On the average there is about one immunogenic site for each 40 – 50 amino-acid residues (MW about 5000) of a native protein, whereas sufficient structural information is contained in peptides as small as 6 – 8 amino-acid residues plus two or three spacer residues for coupling to carrier proteins to produce protein reactive antibodies at high frequency [1, 2]. The same number of residues is also sufficient for the binding of antibodies to the immobilized peptide [7]. For such antigenic determinants the antibody binding sites were calculated to have dimensions not exceeding $30 \times 25 \times 15$ Å [1, 4], which fall into the size ranges measured by X-ray analysis for antibody-combining sites [8].

The results on antipeptide antibodies suggested that antibody binding occurs with conformations in proteins which are different from that of the native. This would provide an explanation for the already mentioned finding on the discrepancy of protein immunogenicity and antigenicity. There are regions of proteins which are antigenic but not immunogenic. This could reflect both an atomic mobility component of protein antigenicity [9 – 11] and the already mentioned conformation dependency of protein immunogenicity [5].

We have tried to make use of the success with which short synthetic peptides can generate antibodies that react with the cognate part in folded proteins and the findings on the nature of immunogenic sites of proteins for the design and synthesis of small peptide immunogens that elicit antibodies with specificity for a given conformation in a sequence-specific manner. Such antipeptide antibodies would be of potential use to test secondary structure predictions for proteins of known sequence which are not readily available in crystallized form, for neutralizing and inhibitory antibodies, and for other structural and biological aspects of protein-protein interaction.

We have tested the ability of antibodies raised against short peptides of predetermined β-turn or loop conformation to interact with their cognate sequences in model peptides [12] and intact proteins [17] in order to address the question of the specificity of the immune response for determinants (epitopes) in β-turn or loop conformation and of the possibility of assignment of secondary structures by antipeptide antibodies.

We have also designed and analyzed model peptides to raise antibodies with specificity for protein regions in helical or β-pleated sheet conformation. The corresponding peptide immunogens included only those protein residues which are brought into proximity by the secondary structure folding of the protein chain. We linked these residues by spacer amino-acid residues in order to adjust the translation of the selected protein residues in the model peptides used for immunizations to that in a helical or β-sheet folding [13, 14]. Antibodies to these synthetic peptides resembling assembled topographical determinants of intact proteins were shown to react with the folded intact proteins and to represent a new type of immuno reagent with specificity for segments of proteins in helical or β-sheet conformation. The strategies outlined here should provide us with suitable tools for "conformational sequencing" of protein regions of biological and medical interest.

2 Design of Model Peptides

2.1 β-Turn Model Peptides [12]

Antibodies with specificity for a sequence in β-turn conformation will possibly recognize all four residues constituting the turn, since its side chains are all expected to be exposed to the surface. Accordingly, the synthetic immunogen with an exposed β-turn region presented to the immune system is expected to elicit antibodies reactive with the contiguous residues of the β-turn sequence in a conformation-dependent manner. We therefore designed a pair of peptides with homology restricted to a tetrapeptide sequence of high β-turn propensity. The β-turn model peptide should present the homologous region folded as β-turn (Fig. 1a), whereas the control Gly-peptide should adopt a completely different conformation (Fig. 1b). The β-turn of the model peptide was inserted between two segments of high propensity to adopt an antiparallel β-sheet structure to stabilize the β-turn. For the choice of the amino-acid sequence of the three structural building blocks β-sheet-β-turn-β-sheet we took into consideration the prediction rules of Chou and Fassman [15] and the observations of Brack and Orgel on the tendency of polypeptides with alternating hydrophilic and hydrophobic residues

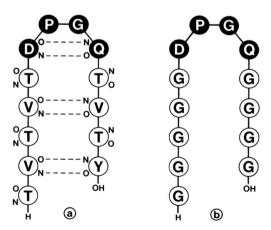

Fig. 1a, b. β-Turn model peptide and control Gly peptide. Sequence and putative secondary structure of β-turn model peptide (**a**) and control Gly peptide (**b**). Hydrogen bonds between the NH and CO groups of the backbone are indicated by *dotted lines*. The Gly peptide is thought to adopt none of the known secondary structures

to adopt β-conformations [16]. The selected β-turn tetrapeptide model sequence (D-P-G-Q) has a turn potential of 1.38 and the β-sheet cartridges with alternating threonine and valine residues have a sheet potential of 1.39 [15]. The side chains of the threonine and valine residues (Fig. 1a) should point to opposite sides of the sheet plane and are expected to result in dimerization by forming bilayer structures [16]. The Gly control peptide (Fig. 1b) contains homology to the β-turn model peptide only in the turn sequence. The turn flanking regions, consisting of Gly residues only, should endow the peptide with high flexibility, minimize unspecific interactions between the peptide and antibodies, and thus allow us to study the contribution of sequence and secondary structure to antibody specificity.

2.2 Loop-Structured Model Peptides [17]

We intended to approximate the loop-structured region 140 – 146 of the HA₁ (hemagglutinin) molecule of the influenza virus [6] by inserting this sequence into an unrelated artificial β-sheet structure similar to that described in the previous section [12, 14]. Providing that a specific amino-acid sequence largely determines its own three-dimensional structure, we thought the amino-acid residues 140 – 146 could adopt a similar structure in the HA₁-molecule and in a synthetic peptide, if the peptide design considers forces promoting loop formation. Based upon the experience with the β-turn model peptide, the hemagglutinin sequences (140 – 146) of two influenza variants were introduced into a peptide skeleton consisting of alternating hydrophobic and hydrophilic residues of β-sheet potential of 1.21 (N-terminal arm) and of 1.26 (C-terminal arm) (Fig. 2a). The ends of the synthetic peptides were equipped with cysteine residues to assure loop formation by a disulfide bridge. Again a control peptide was also designed, carrying the same framework amino-acid residues as the loop peptide but, instead of the HA₁-sequences, a stretch of glycine residues (Fig. 2b).

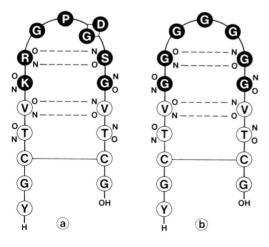

Fig. 2a, b. Loop-structured model peptide and control peptide. Sequence and potential conformation of the immunogenic loop-structured model peptides (a) and the control-peptide (b). *Solid line* between the two cysteine residues indicates a disulfide bridge; *dotted line* hydrogen bonding. The negatively printed sequence stretch corresponds to the hemagglutinin sequence 140 – 146 of strain A/Aichi/2/68 with Gly at 144 [6] and of strain A/Victoria/3/75 with Asp at 144 [6] (a) or to a similar presented control-sequence (b) of Gly residues

2.3 Helix Model Peptides [13, 14]

In contrast to the fully surface-exposed side chains of residues constituting turn or loop structures in intact proteins, helices have only a fraction of the residues exposed to the protein surface. If such residues form an antigenic site it will be a conformation-dependent determinant not involving all contiguous residues of the helical segment. What would be an appropriate peptide model for such an assembled topographical determinant? We chose a design for the helix model peptide which took into account only superimposed residues of a putative helix in the correct distance. Every fourth helical residue fulfils this requirement and their distance corresponds to the helical pitch of 5.4 Å [18]. In the model, approximately the same residue translation can be achieved if a spacer amino acid between each pair of selected helical residues is inserted assuming an extended conformation for the model peptide and a residue translation of about 3 Å. As spacer residue we chose alanine throughout or the neighboring protein amino acid. These considerations allowed the selection of two helix model peptides of different sequence for the same putative helical segment formed by residues 324 – 344 of the integral membrane protein lactose permease of *Escherichia coli* [13, 14]. From the sequence *Phe*-Glu-Val-Pro-*Phe*-*Leu*-Leu-Val-*Gly*-*Cys*-Phe-Lys-*Tyr*-*Ile*-Thr-*Ser*-*Gln*-Phe-Glu-*Val*-*Arg* (permease residues 324 – 344) we deduced the helix model peptides *Phe*-Ala-*Phe*-Ala-*Gly*-Ala-*Tyr*-Ala-*Gln*-Ala and Leu-Gly-Ser-Tyr-Ile-Ser-Gln-Val-Arg. The selected permease residues are shown in italics. They correspond to residues 324, 328, 332, 336, and 340 for the model peptide with alanine spacers and to residues 329, 332, 333, 336, 337, 339, 340, 343, and 344 for the model peptide using permease residues (residues 332, 336, 339, and 343) as spacers. In the latter peptide, cysteine 333 of the permease was replaced by serine to facilitate peptide synthesis. Assuming an alternating orientation of the side chains, the extended form of the model peptide with the alanyl spacers possibly forms a single antigenic determinant for permease, whereas the other model peptide may form two. Both sequence designs share no significant homology to known protein sequences.

2.4 β-Sheet Model Peptides

In a β-pleated sheet the side chains point alternatively to either side of the plane provided by the hydrogen-bonded peptide backbone of the polypeptide chains [19]. The residue translation is 3.2 to 3.4 Å for parallel and antiparallel sheets, respectively. A putative assembled topographic determinant formed by a single β-strand would consist of those residues having side chains pointing to the same side. Accordingly, a β-sheet model immunogen would have to include every second residue of a sheet sequence in the appropriate distance of about 6.4 – 6.8 Å. As already discussed for the helix model peptides, the latter requirement should be fulfilled by the insertion of spacer residues as linker between the protein residues in an alternating fashion. Provided the model peptide adopts an extended conformation, the protein residues are expected to point to one side and the spacer residues to the opposite. For choosing the spacer residue we again took into account the prediction rules of Chou and Fasman [15] and selected alanine throughout.

Two β-strands of concanavalin A [20] corresponding to residues 33 – 42 (*Arg*-Ser-*Lys*-Lys-*Thr*-Ala-*Lys*-Trp-*Asn*) and 108 – 120 (*Ser*-Trp-*Ser*-Phe-*Thr*-Ser-*Lys*-Leu-*Lys*-Ser-*Asn*-Ser-*Thr*) were sources for the protein sequences to be tested by the β-sheet model peptide approach. The selected concanavalin A residues are shown in italics. The corresponding model peptides have the sequences *Arg*-Ala-*Lys*-Ala-*Thr*-Ala-*Lys*-Ala-*Asn* (peptide Con Aβ2 with residues 33, 35, 37, 39 and 41 of Con A) and *Ser*-Ala-*Ser*-Ala-*Thr*-Ala-*Lys*-Ala-*Lys*-Ala-*Asn*-Ala-*Thr* (peptide Con Aβ1 with residues 108, 110, 112, 114, 116, 118 and 120 of Con A).

3 Synthetic Peptide Immunogens

3.1 Synthesis of Peptides

The protected peptides were assembled by solid-phase synthesis [21] on cross-linked polystyrene supports starting with loading of 0.4 mmol (manual procedure) to 1 mmol (automated procedure) of the C-terminal amino acid per g of resin. The chemical syntheses were performed manually using chloromethylated styrene-1%-divinyl-benzene beads as described previously [12, 14] or with a fully automated peptide synthesizer (Applied Biosystems model 430 A) using chloromethylated-phenylacetoamidomethyl-polystyrene resins [22]. N-t-butyloxycarbonyl (tBoc) protected amino acids with appropriate side chain-protecting groups were coupled as the highly reactive symmetric anhydrides [23] formed prior to use. For the protection of the side chains of aspartic acid, glutamic acid, serine, threonine, and tyrosine we used the benzyl group, for cysteine the p-methoxy-benzyl-group, for lysine the benzyloxy-carbonyl groups and for the guanidino function of arginine the nitro group. After assembly of the protected peptide chain, the anchoring bond between peptide and resin was cleaved and the side chain-protecting groups were removed by anhydrous hydrogen fluoride in the presence of a thioether scavanger [24] and a scavanger for radicals (10% anisol in HF) [25]. The cleaved deprotected peptide was precipitated and washed with ether, extracted from the resin with 50% aqueous acetic acid, filtered and lyophilized.

Disulfide bridges were formed by stirring the reduced peptide at concentrations below 0.05 mg at pH 8.0 (0.1 M acetic acid adjusted with 0.1 M NaOH) for 3 – 5 days at ambient temperature until the oxidation was complete and no free thiols were detected with the Ellman assay [26]. We were also able to separate the reduced and oxidized forms by HPLC analysis on C18-reverse-phase (Fig. 3) and thus to follow the oxidation process. Figure 3a – c demonstrates as a representative example the proceeding oxidation of the loop control peptide. Figure 3a shows the fully reduced form of the crude loop control peptide and Fig. 3d the completely oxidized form after purification by gel permeation chromatography.

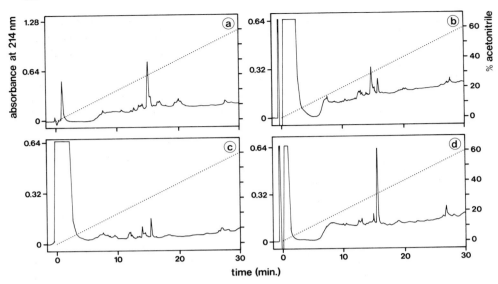

Fig. 3a–d. Reverse-phase HPLC of the loop-structured control peptide. Different stages of the oxidation process are shown: **a** Completely reduced, unpurified; **b** after 16 h or air oxidation; **c** after 48 h oxidation; **d** completely oxidized and purified by gel permeation chromatography on Biogel P4. The highly diluted peptide solution used for oxidation required a large injection volume for **b** and **c**. Separation was on C-18 (0.46 × 25 cm, 300 A, Baker). Solvent A is 0.1% TFA in water and B 0.1% TFA in 60% acetonitrile at a flow rate of 1 ml min^{-1} and a gradient of 2% B min^{-1}. Absorbance was recorded at 214 nm

The fully reduced and fully oxidized peptides were collected and subjected to a qualitative Ellman assay [26] to confirm that the peptide transformation documented by a shift in retention times of about 1 – 2 min is due to oxidation.

3.2 Peptide Purification

Lyophilized peptides were purified by gel permeation chromatography on Biogel P4 (200 × 2 cm) in 0.1 M acetic acid which usually resulted a 90% pure peptide compound as determined by HPLC analysis, amino-acid analysis and sequencing. The glycine-rich control peptides, however, did not elute in a single major peak from Biogel P4 and an additional chromatographic step on DEAE-cellulose or reverse-phase (Fig. 3) was required for the β-turn and loop control peptides, respectively [12, 17].

3.3 Chemical and Biophysical Characterization of the Synthetic Peptides

Since the design of the β-turn model peptide and the two loop-structured peptides included alternating hydrophobic and hydrophilic residues to form β-sheet structures, and these were expected to dimerize by building bilayer structures [16], their approximate molecular size was determined by exclusion chromatography on Biogel P4 in 1 M acetic acid (Fig. 4a) and phosphate buffered saline (PBS) (Fig. 4b). All three peptides behaved identically and, as shown for the β-turn model peptide in Fig. 4b, an aggregated form of apparent molecular weight of 3000 – 35,000 is observed in PBS. This aggregate could be composed of two or three peptide monomers of MW 1379, but the most reasonable model for aggregation by hydrophobic interactions favors the existence of a dimer. For the helix model peptides, β-sheet model peptides and the β-turn control peptide no discrete aggregates were found in PBS.

The circular dichroism spectrum of the β-turn model peptide came closest to the 31% β-turn, 69% β-sheet predicted according to the rules of Chou and Fasman [15] and resulted in 2% helix, 33% β-sheet, 20% β-turn and 45% unordered structure [12, 14]. We cannot distinguish whether the prediction is fully realized or in only one half or two thirds of the molecules without further evidence by laser raman or nuclear magnetic resonance spectroscopy. The circular dichroism spectra of the β-turn control peptide [12, 14], the helix model peptides and the β-sheet model peptides differ quite strongly from that of the β-turn model peptide, indicating a different conformation. The structures of these molecules can be estimated to be mainly unordered and flexible. This is in line with our findings that the spectra of these peptides remained unchanged upon addition of 8 M urea l^{-1} as expected for flexible molecules [12 – 14].

Reverse-phase HPLC of purified peptides was applied as one of several criteria to assess the homogeneity of the material used for further studies. Peptides eluting as a single, symmetric peak corresponding to 90% of the sample loading were considered to be of sufficient purity for immunizations. For these HPLC separations C-8 or C-18 columns (0.46 × 25 cm) (Du Pont, 70 Å pore size or Baker, 300 Å pore size) were routinely applied and developed at 1 ml min^{-1} with a gradient system of 0.1% TFA and 0.1% TFA/60% acetonitril (Fig. 3).

The purified peptides were chemically characterized by amino-acid analysis (Beckman 121M or Biotronik LC 5001) and sequencing. Automated Edman degradation was done in an updated Beckman 890B sequencer with 1 – 20 nmol

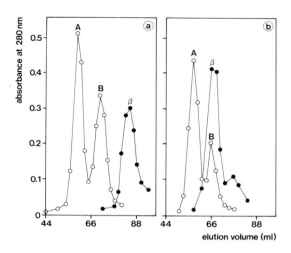

Fig. 4a, b. Gel permeation chromatography of β-turn model peptide (*filled circles*) and MW standards (*open circles*) on Biogel P4 (180 × 1 cm) in 1 M acetic acid (a) and PBS (b). *A* aprotinin, MW 6500; *B* insulin B, oxidized, MW 3500; *β* β-turn model peptide (monomer MW 1379)

peptide using a 0.2 M Quadrol program [27] and a gas-phase sequencer [28] (Applied Biosystems 470 A) with 0.2 – 2 nmol peptide. Both systems employed 1.5 – 3 mg Polybrene. The phenylthiohydantoin derivatives were identified by isocratic elution from cyanopropyl columns (Du Pont, 0.46 × 26 cm) using 0.1 – 0.2 M sodium acetate pH 4.7, containing 23.56% tetrahydrofurane, 6.60% acetonitrile and 2.84% methanol as eluent or by gradient elution from C-18 columns as described in [29].

4 Immune Response Against the Synthetic Peptides

4.1 Coupling of Synthetic Peptides to Carrier Protein

The β-turn model peptide was coupled to keyhole limpet hemocyanin (KLH) or bovine serum albumin (BSA) with glutaraldehyde (GDA) [2]. The two helix model peptides were also coupled to BSA using GDA. The other peptides were coupled to KLH or BSA with 1-ethyl-3-(3'-dimethyl-aminopropyl) carbodimide HCl (ECDI) according to [30]. The coupling efficiency (Table 1) was determined by comparing amino-acid analyses of the carrier proteins before the conjugation procedure and the theoretical amino-acid compositions of BSA and KLH with amino-acid analyses of the respective peptide-protein conjugates.

4.2 Preparation of Antibodies

Antisera were obtained by immunizing rabbits (or female Balb/c mice) with 0.1 to 1 mg (mice: 50 – 100 μg) of peptide-protein conjugate, emulsified 1 : 1 (v/v) in complete Freund's adjuvant. The s.c. injection (mice: i.p.) were repeated after day 28, 49, and 70 with the same immunogen, mixed with incomplete Freund's adjuvant. Blood samples were taken on day 0 and after 2, 5, 8, and 11 weeks. For isolation of monospecific antipeptide antibodies the antisera were subjected to

Table 1. Coupling efficiency of synthetic peptides to carrier proteins

Peptide	Coupling procedure	Carrier	Peptides/carrier (mol mol^{-1})
β-Turn model peptide	GDA	BSA or KLH	4 – 9
Loop model peptide	ECDI	KLH	3
Helix model peptides	GDA	BSA	4 – 6
β-Sheet model peptides			
Con Aβ1	ECDI	BSA or KLH	3
Con Aβ2	ECDI	BSA	10

GDA = glutaraldehyde; ECDI = 1-ethyl-3-(3'-dimethylaminopropyl) carbodiimide-HCl; KLH = keyhole limpet hemocyanin; BSA = bovine serum albumin.

affinity-chromatography on the corresponding peptides, coupled to CNBr-activated Sepharose 4B. For this, the antibody fraction of 10 ml serum was incubated with 2 ml peptide- Sepharose overnight at 4 °C and transferred onto a 0.7 × 5 cm column. The flow-through volume was saved for controls. The bound antibody fraction was eluted with 0.1 M glycine/HCl pH 2.1 in 1-ml fractions which were neutralized immediately by addition of Tris base. The pooled peptide specific antibodies were dialyzed against PBS and stored at − 20 °C.

4.3 Solid Phase Radioimmunoassays

Microtiterplates were coated with free peptides (20 μg ml^{-1}) or peptide-protein conjugates (2 μg ml^{-1}) in 15 mM Na_2CO_3, 30 mM $NaHCO_3$ pH 9.6 overnight at 4 °C, washed with 0.05% Triton or Tween 20 in PBS, saturated with 1% ovalbumin/PBS for 1 h at ambient temperature, washed again and incubated with 50 μl of serial dilutions of antibodies in 1% ovalbumin for 3 h [12]. After an additional wash, the specifically bound antibodies were detected by ^{125}I-labeled goat anti-rabbit or anti-mouse antibodies. The preimmune sera served as controls.

The specificity of the antibodies was analyzed as inhibition of antibody binding by various free peptides. To reduce the favored cooperative antibody binding to the fixed conjugates, the peptide-protein-conjugate coating was performed with dilute solutions of 1 μg mol^{-1}. The antibodies were diluted to give half-maximal binding without inhibition and were preincubated with serial dilutions of free peptide at 4 °C overnight before addition to the sensitized microtiterplates.

4.4 Analysis of Epitope Specificity by Western Blotting

A Western blot refers to the electrophoretic separation of the antigen on polyacrylamide gels followed by its subsequent transfer to nitrocellulose paper and incubation with monospecific antibody and then with labeled protein A or second antibody [31].

For the analysis of the specificity of the anti-helix model peptide antibodies total membrane proteins (3.6 mg) of *Escherichia coli* strain T206 containing a total of 160 μg of permease were electrophoresed on a 12.5% SDS polyacrylamide slab gel (PAGE). After washing of gel and nitrocellulose paper (NC) in 10 mM Na-acetate, pH 5.2 for 10 min, the proteins are transferred to the NC electrophoretically using the same buffer at 3 A (30 V) for 1.5 to 2.5 h with the NC placed at the anode side of the gel. The nitrocellulose sheet is incubated for 2 h in 100 mM Tris/HCl, 1% gelatine pH 7.4 at 37 °C to block off all the remaining sites and then in 50 mM Tris/HCl, pH 7.4 containing affinity purified anti-peptide antibody (105 − 185 μg/blot) for 16 h at 4 °C. The sheet is washed several times in 50 mM Tris/HCl, pH 7.4 followed by incubation with ^{125}I-protein A for 2 h, washed again and dried for subsequent autoradiography.

4.5 Characterization of β-Turn Model-Peptide Specific Antibodies

The analysis of the antibody specificities from the primary and hyperimmune response was performed by inhibition experiments in the form of solid phase radioimmunoassays (Fig. 5a – d). The possibility of raising antibodies against the specific sequence adopting a β-turn structure is clearly demonstrated by the results summarized in Fig. 5d. The antibodies of the hyperimmune response (Fig. 5c, d) and those of the primary immune response after affinity purification (Fig. 5b) reveal a more restricted binding pattern than the total primary response (Fig. 5a). The difference in antibody affinity to the immunogenic β-turn model peptide or possibly β-pleated sheet, β-turn structure and to the Gly-control-peptide T14 (see legend to Fig. 5) of possibly unordered structure [12, 14] amounts to about three orders of magnitude. This encourages us to proceed toward β-turn examination of native proteins using this antipeptide-antibody approach.

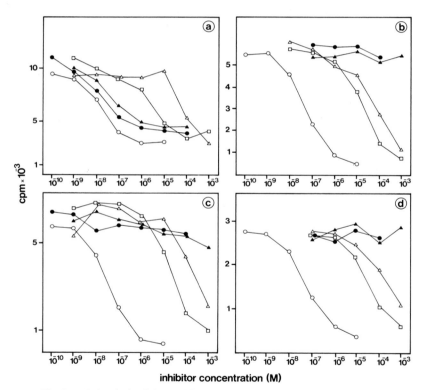

Fig. 5a – d. Analysis of the specificity of the immune response to the β-turn model peptide. Inhibition assay in the form of a solid-phase radioimmunoassay on antibodies of the primary immune response before (**a**) and after (**b**) and of the hyperimmune response before (**c**) and after (**d**) affinity chromatography. The antibody concentration were: **a** 85 μg ml^{-1}; **b** 130 ng ml^{-1}; **c** 6 ng ml^{-1}; **d** 12 ng ml^{-1}. The sequences of the five analyzed peptides are ○———○ β-turn mode peptide (TVTVT*DPGQ*TVTY); △———△ Gly-control peptide (GGGGG*DPGQ*GGG); □———□ T14 peptide (T*T*SSGT*T*SSTTSSG); ▲———▲ V10 peptide (VFGDEKA(T/S/R-NO$_2$)FY); ●———● A13 peptide (AKYDYYGSSYFDY)

4.6 Immunogenicity of Loop-Structured Model Peptides

Short synthetic peptides including residues of the loop-structured region 140−146 of the influenza hemagglutinin (HA_1) failed to induce an anti-protein immune response [1, 3, 6]. In Fig. 6 we show that a 16 amino-acid residue peptide, containing only 7 residues of the HA_1-molecule (residues 140−146) in an appropriately designed loop structure (Fig. 2), elicits an immune response which is focused on the loop sequence in the synthetic peptides and which contains antibodies that react with the influenza virus [17]. The loop structure of the synthetic peptides clearly forms the immunodominant epitope, which was demonstrated by the unreactivity of the antisera with the control peptide.

Since the antibodies show the same clearly positive but undifferentiating reactivity with the two virus strains (Fig. 6a, b), we conclude that the antibodies recognize a somewhat different determinant, when the 140−146 sequence is presented in context of the native hemagglutinin molecule. For sterical reasons the highly specific antibodies might simply not fit with their binding crevice in the cognate sequence in the protein molecule.

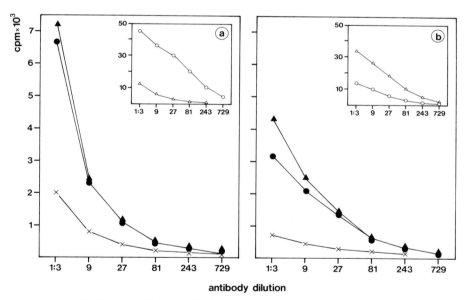

Fig. 6a, b. Binding of affinity purified anti-loop-structured model peptide antibodies with their homologous peptides and with influenza virus strains. The antibodies elicited against the conjugate of the peptide with Gly at position 10 (HAG) (**a**) and with Asp at position 10 (HAD) (**b**) (see legend to Fig. 2) were affinity purified over the respective immunogenic peptides. Direct binding to the homologous peptides HAG ○———○ and HAD △———△ is shown in the insets. The body of the figures depict binding to virus A/Aichi/2/68 (144-Gly) ●———● and A/Victoria/3/75 (144-Asp) ▲———▲ and to 1% BSA ×———×. Antibody concentration at 1:3 is 120 μg ml^{-1} (**a**) and 50 μg ml^{-1} (**b**), respectively

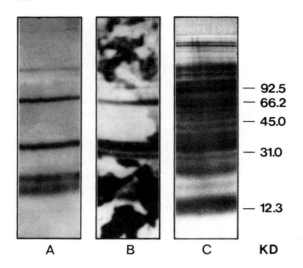

Fig. 7A – C. Western blot analysis of the anti-protein immune response elicited by the helix-model peptides. Total membrane protein of *Escherichia coli* of the permease-overproducing strain T206 separated on a 12.5% SDS polyacrylamide slab gel was stained with Coomassie brilliant blue (**C**); transferred to nitrocellulose paper (**A, B**) and incubated with affinity purified antibody elicited by the peptide with the alanyl spacer residues (**A**) and antibody elicited by the peptide completely composed of permease residues (**B**). The autoradiography after incubation with ^{125}I-protein A is shown (**A, B**)

4.7 Anti-Protein Immune Response Induced by Helix Model Peptides

The immune response elicited by the two helix model peptides was analyzed by Western blotting (Fig. 7). The two affinity-purified antibody preparations of the hyperimmune response react with the major permease band migrating with an apparent MW of 33 kDa. In addition, two proteins of higher MW (65 kDa and 95 kDa) and two proteins of lower MW (20 kDa and 18 kDa) also bind the antibodies (Fig. 7). Pretreatment of the antibody used in the experiment shown in Fig. 7a with the corresponding peptide completely abolished the reaction with all five bands, suggesting that the antibody possibly binds to permease (33 kDa), permease aggregates (65 kDa and 95 kDa) and permease breakdown products (20 kDa and 18 kDa). The preimmune sera showed no reactivity.

These experiments suggest that residues 324 – 340 of lactose permease constitute an epitope which is shared with the synthetic helix model peptides. Since the synthetic immunogens were assembled from residues which are not contiguous in the amino-acid sequence of the permease but could be brought together by folding of the corresponding protein region into a helix, we conclude that the determinant recognized on Western blots is helically folded and that helix examination should be possible using this antipeptide approach.

4.8 Characterization of the Anti-Protein Immune Response Elicited by the β-Sheet Model Peptides

The two β-sheet model peptides included alternating residues of concanavalin A from different β-sheet regions of the protein. Model peptide Con Aβ1 was assembled from residues 108 – 120, some of which participate in aggregation of subunit I with subunits II and III [20]. In contrast, the Con Aβ2 model peptide includes mainly surface-exposed residues from the β-sheet region 33 – 42. As

Table 2. Reaction of anti-β-sheet model peptide antibodies with concanavalin A

| Antibody | Tetramers | Relative reactivity[a] | | |
		Dimers[b]	Monomers[c]	Denatured[d]
Anti-Con Aβ1	0.12	–	1.00	0.08
Anti-Con Aβ2	0.35	0.52	1.00	0.05

a Antibody-binding was measured by solid-phase radioimmunassays. Both affinity purified antibodies (2 mg ml^{-1}) were tested at dilutions of 1 : 10^4; a concanavalin A solution of 2.5 mg ml^{-1} was used for coating.
b Assay performed under condition of dimer formation (pH 5.0).
c Protein loaded to wells in 1% SDS.
d Protein loaded to wells in 8 M urea.

expected, the Con Aβ1 peptide specific antibody recognizes readily monomeric concanavalin A. Antibody binding to tetrameric or denatured concanavalin A was comparatively weak (Table 2). The antibody elicited by the β-sheet model peptide Con Aβ2 recognized tetrameric, dimeric, and monomeric concanavalin A but again, only weakly the denatured protein (Table 2). The binding to the tetrameric and dimeric form was about one third and half of that for the monomer, respectively, suggesting steric hindrance of neighboring subunits also for antibody binding to this region.

The binding of both antibody preparations to the concavalin A monomer was completely abolished by preincubation of the antibodies with the corresponding free model peptides.

It follows that the antigenic determinants, shared between monomeric concanavalin A and the corresponding β-sheet model peptides, have a similar constellation of interacting side chains. This complementarity to the antibody combining site requires folding of the corresponding concanavalin A region into a β-sheet and recognition of the surface-exposed side chains of alternating residues.

5 Conclusions

We think that the important feature of the three major protein secondary structures could be roughly approximated by artificially designed synthetic peptide-immunogens, which indeed raised protein reactive antibodies with specificity for topographic determinants. Improvements like those for the dimerizing Thr-Val skeleton as presented for the experiments on the loop-structured model peptides, enlargement of the homologous turn region from four to six residues for the β-turn model system, and the use of monoclonal antibodies for affinity measurements should provide us with suitable tools for probing secondary structure predictions of proteins of unknown three-dimensional structures.

Acknowledgments. This work was supported by the Deutsche Forschungsgemeinschaft through SFB 74, the BMFT, the Minister für Wissenschaft und Forschung of Nordrhein-Westfalen and the Fonds der Chemischen Industrie.

References

1. Lerner RA (1984) Adv Immunol 36:1
2. Walter G, Doolittle RF (1983) In: Setlow JK, Hollaender A (eds) Genetic engineering, vol 5. Plenum, New York, p 142
3. Green N, Alexander H, Wilson A, Alexander S, Shinnick TM, Sutcliff JG, Lerner RA (1982) Cell 28:477
4. Wilson IA, Niman HL, Houghten RA, Cherenson A, Conolly ML, Lerner RA (1984) Cell 37:767
5. Benjamin DC, Berzofsky JA, East IJ, Gurd FRN, Hannum C (1984) Annu Rev Immunol 2:67
6. Wilson IA, Skehel JJ, Wiley DC (1981) Nature 289:373
7. Smith JA, Hurrell JGR, Leach SJ (1977) Immunochemistry 14:565
8. Amzel LM, Poljak RJ (1979) Annu Rev Biochem 48:975
9. Westhof E, Altschuh D, Moras D, Bloomer AC, Mondragon A, Klug A, Van Regenmortel MHV (1984) Nature 312:123
10. Tainer JA, Getzoff ED, Alexander H, Houghton RA, Olson AJ, Lerner RA, Hendrickson WA (1984) Nature 312:127
11. Dyson HJ, Cross KJ, Houghton RA, Wilson IA, Wright PE, Lerner RA (1985) Nature 318:480
12. Schulze-Gahmen U, Prinz H, Glatter U, Beyreuther K (1985) EMBO J 4:1731
13. Bieseler B, Prinz H, Beyreuther K (1985) Ann NY Acad Sci 465:309
14. Beyreuther K, Schulze-Gahmen U, Bieseler B, Prinz H (1986) In: Blöcker H, Franke R, Fritz H (eds) Chemical synthesis in molecular biology, GBF monographs, vol 8. Verlag Chemie, Weinheim (in press)
15. Chou PY, Fasman GD (1978) Adv Enzymol 47:46
16. Brack A, Orgel LE (1975) Nature 256:383
17. Schulze-Gahmen U, Klenk H-D, Beyreuther K (1986) Eur J Biochem (in press)
18. Pauling L, Corey RB, Branson HR (1951) Proc Natl Acad Sci USA 37:205
    Perutz M (1951) Nature 167:1053
19. Pauling L, Corey RB (1951) Proc Natl Acad Sci USA 37:729
20. Reeke GN, Becker JM, Edelman GM (1975) J Biol Chem 250:1525
21. Merrifield RB (1963) J Am Chem Soc 85:2149
22. Mitchell AR, Kent SBH, Engelhard M, Merrifield RB (1978) J Org Chem 43:2845
23. Hagenmaier H, Frank H (1972) Hoppe-Seyler's Z Physiol Chem 253:1973
24. Tam JP, Heath WF, Merrifield RB (1983) J Am Chem Soc 105:6442
25. Lenard J, Robinson AB (1967) J Am Chem Soc 89:181
26. Ellman GL (1959) Arch Biochem Biophys 82:70
27. Beyreuther K, Raufuss H, Schrecker O, Hengstenberg W (1977) Eur J Biochem 75:275
28. Hewick RM, Hunkapiller MW, Hood LE, Dryer WJ (1981) J Biol Chem 256:7990
29. Users Manual, PTH-Analyzer Model 120A, Applied Biosystems (Version 1.0) June 1981
30. Tamura T, Bauer H, Birr C, Pipkorn R (1983) Cell 34:587
31. Towbin H, Staehelin T, Gordon J (1979) Proc Natl Acad Sci USA 76:4350

6.2 Aspartyl-tRNA Synthetase-Induced Aspartylation of Proteins: a Fingerprint Approach to Map Accessible Domains in Protein

H. Mejdoub, D. Kern, R. Giege, Y. Boulanger, and J. Reinbolt[1]

Contents

1 Introduction

Aspartic acid can be covalently linked to yeast aspartyl-tRNA synthetase and to other proteins in the absence of tRNA, under conditions where the synthetase activates the amino acid into aspartyl-adenylate, i.e., in the presence of ATP and $MgCl_2$ [1, 2]. The aspartyl adenylate is poorly bound to the enzyme.

$$E + Asp + ATP \xrightleftharpoons{MgCl_2} E \cdot Asp \sim AMP + PPi$$
$$E \cdot Asp \sim AMP \rightleftharpoons E + Asp \sim AMP .$$

After dissociation from the synthetase, the free aspartyl-adenylate is able to react with various nucleophilic acceptors present in the incubation mixture:

---

1 Institut de Biologie Moléculaire et Cellulaire du C.N.R.S., Laboratoire de Biochimie, 15, rue René Descartes, F-67084 Strasbourg, France

Advanced Methods in Protein Microsequence Analysis
Ed. by B. Wittmann-Liebold et al.
© Springer-Verlag Berlin Heidelberg 1986

$$\text{Asp} \sim \text{AMP} + \text{R} - \text{NH}_2 \rightarrow \text{Asp } \alpha \text{ CO} - \text{NH} - \text{R}$$
$$\text{R} - \text{OH} \qquad\qquad - \text{O} - \text{R}$$
$$\text{R} - \text{SH} \qquad\qquad - \text{S} - \text{R}$$
$$\text{Nucleophilic acceptor} \qquad + \text{AMP}$$

Formation of amide (stable), ester or thioester (less stable) bonds with nucleophilic acceptors.

This new property of aspartyl-tRNA synthetase, linked to the unstability of the enzyme-adenylate complex, can be used to screen the peptides at the surface of the synthetase or of other proteins and so the three-dimensional structure of proteins in solution. So far such a phenomenon has not been observed for other synthetases and aspartylation reactions of proteins have not been described yet. Some structural and kinetic properties of the aspartylated synthetase have been studied and compared to those of the native enzyme [1, 2].

The covalent attachment of aspartic acid to the protein requires the presence of ATP and $MgCl_2$. It is specific to aspartic acid and to the aspartic system. In other words no other amino acids or aspartate analogs can replace aspartic acid in this reaction. The activated aspartic acid reacts with acceptors of the protein to form a stable bond; when incubated with tRNA^{Asp} this enzyme-bound aspartic acid cannot be transferred to the tRNA.

2 Experimental Procedures

2.1 Materials and Chemicals

The chemicals, products, and enzymes used in the aspartylation procedure are as described in [2]. Trypsin (treated with L-1 tosylamido-2-phenylethylchloromethyl ketone) was obtained from Worthington (Freehold, N.J., USA) and 4-N,N dimethyl-aminoazobenzene 4′ isothiocyanate (DABITC) from Fluka (Buchs, Switzerland). All chemicals used for the Edman-Chang degradation were analytical grade products of Pierce (Rockford, USA). Chromatography of peptides was performed on Sephadex G-50 (Pharmacia, Uppsala, Sweden), and on a HPLC apparatus (Waters Associates, USA) using a C18 Nova Pak column (Waters) (5 µM; 0.39 × 15 cm). 4-N,N dimethylaminoazobenzene 4′ thiohydantoin (DABTH) derivatives of the amino acids were identified by chromatography on micropolyamide thin layer sheets (F1700, Schleicher and Schüll, Dassel, Germany). Amino acid analyses were performed on a Durrum D500 analyzer (Durrum, Palo Alto, USA). All other chemicals were from Merck (Darmstadt, Germany), analytical grade.

2.2 Methods

2.2.1 Enzyme and tRNA

Aspartyl-tRNA synthetase (EC 6.1.1.12) from baker's yeast was prepared as described by Lorber et al. [4]. Brewer's yeast tRNA was prepared according to the procedure of Keith et al. [5] modified by Dock et al. [6].

2.2.2 Aspartyl-tRNA Synthetase Modification Reaction and Isolation of the Aspartylated Enzyme

Standard aspartylation mixtures contained 100 mM NaMes buffer, pH 6.0 or 100 mM Na Hepes buffer, pH 7.2 or 8.6, 2 mM ATP, 2 or 3 mM $MgCl_2$, 0.3 mM L-aspartic acid (either cold or ^{14}C labeled with a specific activity of $17,500 - 70,000$ cpm $nmol^{-1}$), 5 or 10 units of inorganic pyrophosphate ml^{-1} (in a negligible amount compared to aspartyl-tRNA synthetase), and 4.5 µM aspartyl-tRNA synthetase.

After 10 h of incubation at 30 °C, the medium was supplemented with fresh reactants ($1 - 2$ mM ATP, $1 - 2$ mM $MgCl_2$, and 100 µM L-aspartic acid), and the incubation was continued for another 14 h. A total of 200 µl of the incubation mixture was then filtered on a G-100 Sephadex column (25 cm × 1 cm) equilibrated with 20 mM Na Hepes buffer, pH 7.2 and 100 mM KCl. The eluted fractions were tested for radioactivity and tRNA charging activity. The aspartylated synthetase eluted in the dead volume of the column was separated from free reactants: $80 - 90\%$ of the synthetase activity was recovered. The extent of aspartylation of the synthetase was determined by transferring 5- or 10 µl aliquots to 3 MM Whatman paper discs; the discs were immersed and washed three times during 15 min in 5% TCA, then twice for 5 min in 95% ethanol and finally dried and counted by liquid scintillation.

2.2.3 Aspartylation of Proteins in a Two-Compartment Dialysis System

An equilibrium dialysis system containing eight pairs of compartments separated by a dialysis membrane was used. One compartment contained the aspartyl-adenylate-generating system; the other one, the protein to be labeled. To increase the diffusion rate through the membrane, the latter was made porous by $ZnCl_2$ treatment; membranes were moistened, immersed for 6 min in a $ZnCl_2$ solution (320 g of salt for 180 ml of H_2O), rinsed three times in 1 mM HCl, and extensively washed with water.

2.2.4 Protein Structure Analysis

Digestion of aspartyl-tRNA synthetase with trypsin, isolation, amino acid analysis, and sequencing of peptides were reported in [7].

3 Results

3.1 Identification of the Labeled Peptides of Aspartyl-tRNA Synthetase Modified by Aspartylation

The aspartylation reaction was carried out at three different pH values 6.0 (acidic), 7.2 (neutral), and 8.6 (basic). The aspartylated aspartyl tRNA-synthetase was digested with trypsin, which does not cleave at the lysine residues modified by aspartic acid, yielding overlapping peptides. The labeled peptides were purified by gel filtration on Sephadex G-50, followed by HPLC, and when necessary by electrochromatography, as described in [7]. These peptides were sequenced by the Edman-Chang method. The labeled aspartic acid was recovered at the first degradation step, together with the N-terminal amino acid of the peptide; in the case of polyaspartyl side chains bound to the same amino acid, labeled aspartic acid occurred in the next degradation steps.

The results are given in Table 1. In addition to aspartylated lysines, we found that some residues of threonine and serine were aspartylated too, thus the ester bond formed between threonine or serine and aspartic acid remains stable during the isolation procedure of the aspartylated peptides.

Another striking feature of the results listed in Table 1 is the increase in the number of aspartylated lysines, as the pH increases: 1 − 2 lysines (Lys 46 and Lys 371, yield 50%) at pH 6.0; 3 lysines (Lys 63, Lys 173 and Lys 403) at pH 7.2; and finally 3 residues (Lys 46, Lys 390 and Lys 540) at pH 8.6. Interestingly, the number of additional aspartic acid residues fixed per amino acid, especially di-asp, increases with the pH too: for instance one di-asp was bound to Ser 329 or Thr 331 at pH 6.0 in contrast to pH 8.6, where four di-asp were bound to Lys 46, Thr 268, 272, or His 271, Ser 329 or Thr 331 and Ser 367. It must be pointed out that under the experimental conditions used we could not always find out which Ser or Thr or Lys were aspartylated when they were simultaneously present within the same tryptic peptide.

It must also be stressed that aspartylation of some amino acids is strictly affected by the pH, for instance:

Ser 30   is aspartylated at pH 6.0 but not at pH 7.2 and 8.6
Thr 129 is aspartylated at pH 6.0 but not at pH 7.2 and 8.6
Lys 173 is aspartylated at pH 7.2 but not at pH 6.0 and 8.6
Lys 540 is aspartylated at pH 8.6 but not at pH 6.0 and 7.2

4 Discussion

4.1 Particular Behavior of Enzyme Synthesized Aminoacyl Adenylate in the Aspartic System

We describe here an aspartylation process of proteins; this phenomenon is used to map accessible domains at the surface of the proteins. It is based on the fact that, in contrast to other aminoacyl-tRNA synthetases-aminoacyl-adenylate complexes, the aminoacyl-tRNA synthetase-aspartyl-adenylate complex easily dissociates because of the poor affinity of aspartyl-adenylate for the synthetase. Then the activated aspartic acid can be transferred to nucleophilic acceptors, e.g, free amino acids or protein side chains. Concerning the location of some amino acid residues, a first conclusion could be drawn from kinetic studies showing that the maximal rates of tRNA charging and ATP-PPi exchange catalyzed by the synthetase are not significantly affected by the modification. Aspartylation es-

Table 1. Amino acids involved in the aspartylation process of aspartyl-tRNA synthetase at different pH's: acidic (6.0), neutral (7.2), and basic (8.6)

Peptide	Sequence	pH: 6 Amino acid aspartylated	Nb Asp fixed	pH: 7.2 Amino acid aspartylated	Nb Asp fixed	pH: 8.6 Amino acid aspartylated	Nb Asp fixed
28 – 31	Pro-Leu-Ser-Lys	Ser 30	1				
44 – 47	Gln-Arg-Lys-Lys	Lys 46	1			Lys 46	1
44 – 47	Gln-Arg-Lys-Lys					Lys 46	2
62 – 64	Glu-Lys-Lys			Lys 63	1		
78 – 84	Leu-Pro-Leu-Ile-Gln-Ser-Arg	Ser 83	1			Ser 83	1
129 – 130	Thr-Leu	Thr 129	1				
132 – 142	Gln-Gln-Ala-Ser-Leu-Ile-Gln-Gly-Leu-Val-Lys	Ser 135	1				
157 – 160	Ala-Gly-Ser-Leu			Ser 159	1		
170 – 174	Gly-Ile-Val-Lys-Lys			Lys 173	1		
194 – 204	Ile-Tyr-Thr-Ile-Ser-Glu-Thr-Pro-Glu-Ala-Leu					Ser or Thr	1
196 – 213	Thr-Ile-Ser-Glu-Thr-Pro-Glu-Ala-Leu-Pro-Ile-Leu-Leu-Glu-Asp-Ala-Ser-Arg					Ser or Thr	1
241 – 249	Thr-Val-Thr-Asn-Gln-Ala-Ile-Phe-Arg	Thr 241 or Thr 243	1				
263 – 265	Ala-Thr-Lys	Thr 264	1				
267 – 274	Phe-Thr-Glu-Val-His-Thr-Pro-Lys					Thr or His	2
326 – 333	Ala-Glu-Asn-Ser-Asn-Thr-His-Arg	Ser or Thr	2			Ser or Thr	2
362 – 372	Phe-Val-Phe-Ile-Phe-Ser-Glu-Leu-Pro-Lys-Arg	Ser 367	1			Ser 367	2
363 – 372	Val-Phe-Ile-Phe-Ser-Glu-Leu-Pro-Lys-Arg	Lys 371	0.5				
390 – 393	Lys-Leu-Pro-Lys					Lys 390	1
400 – 410	Leu-Thr-Tyr-Lys-Glu-Gly-Ile-Glu-Glu-Met-Leu-Arg			Lys 403	1		
476 – 485	Gly-Glu-Glu-Ile-Leu-Ser-Gly-Ala-Gln-Arg					Ser 481	1
532 – 544	Val-Val-Met-Phe-Tyr-Leu-Asp-Leu-Lys-Asn-Ile-Arg-Arg					Lys 540	1

sentially affects the tRNA binding sites; the strong decrease in the affinitiy of tRNA^{Asp} for the aspartylated synthetase indicates that modified amino-acid groups, most likely lysines, are in the tRNA site; the aspartylation also modifies the binding properties of ATP to the dimeric synthetase; but the binding of aspartic acid remains unaffected. (Obviously, aspartic acid covalently bound to the synthetase is not involved in tRNA charging).

4.2 Amino Acid Residues Modified in Aspartyl-tRNA Synthetase by Aspartylation

Theoretically cysteine residues also accept activates aspartic acid. We never found any labeled cysteine-containing peptide. This can be interpreted in two ways: (1) the thioester bond between aspartic acid and cysteine is very labile and does not survive during the course of peptide purification; (2) the cysteine-containing peptides are not located at the surface of aspartyl-tRNA synthetase and therefore are not accessible to activated aspartic acid. Indeed the three cysteine residues of the enzyme (positions 255, 512, 519; Fig. 2) lie within regions having a very low degree of hydrophilicity (Fig. 1). Tyrosine and histidine residues as well

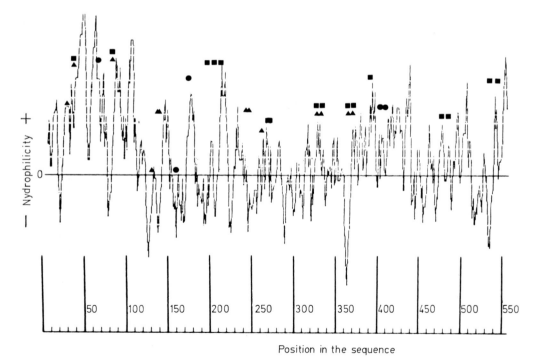

Position in the sequence

Fig. 1. Hydrophilicity curve of aspartyl-tRNA synthetase plotted according to the program of Hopp and Woods [13]. The various symbols for the aspartylated peptides are : ▲ pH 6.0; ● pH 7.2; ■ pH 8.6

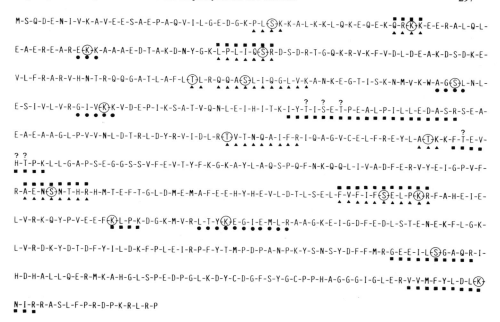

M-S-Q-D-E-N-I-V-K-A-V-E-E-S-A-E-P-A-Q-V-I-L-G-E-D-G-K-P-L-(S)-K-K-A-L-K-K-L-Q-K-E-Q-E-K-Q-R-(K)-K-E-E-R-A-L-Q-L-

E-A-E-R-E-A-R-E-(K)-K-A-A-A-E-D-T-A-K-D-N-Y-G-K-L-P-L-I-Q-(S)-R-D-S-D-R-T-G-Q-K-R-V-K-F-V-D-L-D-E-A-K-D-S-D-K-E-

V-L-F-R-A-R-V-H-N-T-R-Q-Q-G-A-T-L-A-F-L-(T)-L-R-Q-Q-A-(S)-L-I-Q-G-L-V-K-A-N-K-E-G-T-I-S-K-N-M-V-K-W-A-G-(S)-L-N-L-

E-S-I-V-L-V-R-G-I-V-(K)-K-V-D-E-P-I-K-S-A-T-V-Q-N-L-E-I-H-I-T-K-I-Y-T-I-S-E-T-P-E-A-L-P-I-L-L-E-D-A-S-R-S-E-A-

E-A-E-A-A-G-L-P-V-V-N-L-D-T-R-L-D-Y-R-V-I-D-L-R-(T)V-T-N-Q-A-I-F-R-I-Q-A-G-V-C-E-L-F-R-E-Y-L-A-(T)K-K-F-T-E-V-

H-T-P-K-L-L-G-A-P-S-E-G-G-S-S-V-F-E-V-T-Y-F-K-G-K-A-Y-L-A-Q-S-P-Q-F-N-K-Q-Q-L-I-V-A-D-F-E-R-V-Y-E-I-G-P-V-F-

R-A-E-N-(S)N-T-H-R-H-M-T-E-F-T-G-L-D-M-E-M-A-F-E-E-H-Y-H-E-V-L-D-T-L-S-E-L-F-V-F-I-F-(S)E-L-P-(K)R-F-A-H-E-I-E-

L-V-R-K-Q-Y-P-V-E-E-F-(K)L-P-K-D-G-K-M-V-R-L-T-Y-(K)E-G-I-E-M-L-R-A-A-G-K-E-I-G-D-F-E-D-L-S-T-E-N-E-K-F-L-G-K-

L-V-R-D-K-Y-D-T-D-F-Y-I-L-D-K-F-P-L-E-I-R-P-F-Y-T-M-P-D-P-A-N-P-K-Y-S-N-S-Y-D-F-F-M-R-G-E-E-I-L-(S)G-A-Q-R-I-

H-D-H-A-L-L-Q-E-R-M-K-A-H-G-L-S-P-E-D-P-G-L-K-D-Y-C-D-G-F-S-Y-G-C-P-P-H-A-G-G-G-I-G-L-E-R-V-V-M-F-Y-L-D-L-(K)

N-I-R-R-A-S-L-F-P-R-D-P-K-R-L-R-P

Fig. 2. Location of the aspartylated amino acids within the peptides of aspartyl-tRNA synthetase: ▲ pH 6.0; ● pH 7.2; ■ pH 8.6

are potential targets for aspartylation, but we failed to isolate labeled peptides containing these two amino acids. The only Tyr candidate (Table 1) would be in peptide 194 – 213 (Tyr 195) labeled with radioactive aspartic acid; but in the same experiment we isolated the labeled peptide 196 – 213, which rules out tyrosine as a possible target for aspartylation.

Nevertheless, His 271 (Table 1) could be di-aspartylated as well as Thr 268 or Thr 272. At present this possibility cannot be ruled out. On the other hand, His 332 (Table 1) is not aspartylated; indeed this His residue could be dinitrophenylated within the aspartylated peptide 326 – 333, which means a free imidazole ring.

4.3 Characteristics of the Aspartylated Peptides of Aspartyl-tRNA Synthetase

Interestingly, under the experimental conditions used, lysine residues are aspartylated as well as amino alcohol acid residues serine or threonine. As mentioned above, the attached labeled aspartic acid is obtained at the first degradation step of the Edman-Chang method, together with the N-terminal amino acid of the peptide; thus we could not definitely establish which of the three amino acids was aspartylated in the case of five peptides: peptide 194 – 213 (four possibilities: Thr 196, Ser 198, Thr 200, and Ser 212); peptide 241 – 249 (Thr 241 or Thr 243); peptide 267 – 274 (Thr 268 or Thr 272); peptide 326 – 333 (Ser 329 or Thr 331) and peptide 362 – 372 (ser 367 or Lys 371). The use of cleavage with carboxypeptidases A + B helped us to solve some ambiguities. Thus Ser 212 (peptide 194 – 213) seems to be free of aspartic acid, since this Ser is released by digestion with carboxypeptidases A + B. Using the same experimental approach, we could conclude that Thr 241 within peptide 241 – 249 and Ser 329 within peptide 326 – 333 are labeled. In the results we also stressed that the number of aspartylated lysine residues varies with the pH, from 1 – 2 residues at pH 6.0 to 3 at pH 8.6. These figures fit very well with the pK's of the NH_2 groups, being much more reactive at alkaline pH than at acidic pH. The formation of aspartic acid polymers on the synthetase also increases with alkaline pH, the acceptance ability of the α-NH_2 group of the aspartic acid being favored at alkaline pH. We essentially found di-Asp, i.e., a labeled aspartic acid was also recovered at the second step of the Edman-Chang degradation; at pH 8.6 there were 4 amino acids aspartylated by two successive Asp: Lys 46; Thr 268, His 271 or Thr 272; Ser 329 or Thr 331; and Ser 367 or Lys 371. Concerning Lys 46, it must be pointed out that at pH 6.0 this Lys was monoaspartylated; on the other hand, at pH 8.6 two possibilities exist: the mono- and the di-Asp are both present; the mono- and di-aspartylated peptides (Position 44 – 47) elute together on HPLC, they are separated by electrochromatography on a cellulose plate. It is obvious that the yield of aspartylation is not 100%; indeed in the experiment described above, we also isolated nonaspartylated peptide 44 – 47. Furthermore peptide 326 – 333 carrying a di-Asp elutes from the Sephadex column before the native one having a molecular ratio lower than the aspartylated form; on the other hand, the more acidic aspartylated peptide is not bound to the HPLC column resin, in contrast to the native one, which is weakly retained on the HPLC system. Peptide 363 – 372 bears two labeled aspartic acid residues as calculated from the amino acid composition. At pH 6.0 they react with Ser 367 and Lys 371; indeed we did not find any labeled aspartic acid at the second step of the Edman degradation; however, it looks as if Lys 371 is only aspartylated with a yield of 50% as compared to Ser 367. Surprisingly when aspartylation was carried out at pH 8.6, the two aspartic acid residues were both attached to Ser 367, since labeled aspartic acid was obtained at the first and second steps of Edman-Chang degradation. Furthermore Lys 371 could be dinitrophenylated, which means a free ε amino group. This result was confirmed by digestion with carboxypeptidases A + B generating a free lysine.

The effect of the pH of the aspartylation reaction is very important. As already discussed, lysine residues are more accessible at basic pH. On the other hand, Ser or Thr are covalently bound to activated aspartic acid at acidic as well as basic pH; therefore the targets are not necessarily the same at acidic or basic pH. Indeed only four peptides were found to be aspartylated under both conditions (acidic and basic pH): peptide 78 – 84, peptide 326 – 333, peptide 362 – 372 and peptide 44 – 47, (the latter being also mono- and di-aspartylated at pH 8.6).

4.4 Structural Features of the Aspartylated Enzyme

In conclusion we would like to emphasize some characteristic structural features. In our report on the primary structure of aspartyl-tRNA synthetase [3], we mentioned that peptide 362 – 372 could not be obtained by the different purification techniques used; this fact is probably related to its high degree of hydrophobicity strongly hindering its elution from the HPLC column [3]. In the aspartylation process the peptide covalently binds two radioactive aspartic acid residues; thus it becomes more acidic and can be eluted from the HPLC column. This result confirms the structure given for this peptide in [3].

The different amino acids involved in the aspartylation reaction are randomly distributed along the primary structure of aspartyl-tRNA synthetase (Fig. 2); the labeled peptides are located in the N-terminal region, as well as in the center or in the C-terminal part of the protein.

Finally it is worth noticing that our results fit very well with the hydrophilicity of aspartyl-tRNA synthetase plotted according to the program of Hopp and Woods [13]. The aspartylated peptides supposedly located at the surface of the aspartyl-tRNA synthetase, correspond to regions calculated to be highly hydrophilic (Fig. 1). There is a very good agreement between our experimental results concerning accessible domains of the aspartyl-tRNA synthetase and theoretical calculations about hydrophilicity, i.e., the possible location of some peptidic regions at the surface of the enzyme.

Aspartylation of proteins has not been described yet. As mentioned, other proteins can be labeled by aspartic acid. Thus the aspartylation process appears to be a useful tool to map peptides located at the surface of the three-dimensional structure of proteins.

References

1. Lorber B, Kern D, Giege R, Ebel JP (1982) FEBS Lett 146:59
2. Kern D, Lorber B, Boulanger Y, Giege R (1985) Biochemistry 24:1321
3. Amiri I, Mejdoub H, Hounwanou N, Boulanger Y, Reinbolt J (1985) Biochimie 67:607
4. Lorber B, Kern D, Dietrich A, Gangloff J, Ebel JP, Giege R (1983) Biochem Biophys Res Commun 117:259
5. Keith G, Gangloff J, Dirheimer G (1971) Biochimie 53:123
6. Dock C, Lorber B, Moras D, Pixa G, Thierry JC, Giege R (1984) Biochimie 66:179
7. Hounwanou N, Boulanger Y, Reinbolt J (1983) Biochimie 65:379
8. Kern D, Gangloff J (1981) Biochemistry 20:538
9. Giege R, Lorber B, Ebel JP, Moras D, Thierry JC, Jacrot B, Zaccai G (1982) Biochimie 64:357
10. Lorber B, Giege R (1983) FEBS Lett 156:209
11. Igloi GL, von der Haar F, Cramer F (1978) Biochemistry 17:3459
12. Lin SX, Baltzinger M, Remy P (1983) Biochemistry 22:681
13. Hopp T, Woods KR (1981) Proc Natl Acad Sci USA 78:3824

Chapter 7
Strategies and Specific Examples of Sequencing
Proteins and Peptides

7.1 Strategies of Biochemical Characterization of Hormonal Peptides

JOACHIM SPIESS[1]

Contents

1 Introduction

In the last 7 years, we have characterized a series of novel hormonal peptides exhibiting hypophysiotropic actions by complete amino acid sequence analysis [1 – 14, 29]. The size of these peptides varied from 9 – 134 amino acid residues.

---

1 The Clayton Foundation Laboratories for Peptide Biology, The Salk Institute, 10010 North Torrey Pines Road, La Jolla, CA, USA

Advanced Methods in Protein Microsequence Analysis
Ed. by B. Wittmann-Liebold et al.
© Springer-Verlag Berlin Heidelberg 1986

It is interesting to note that every characterization was accomplished with a different strategy adjusted specifically to the biochemical properties of the peptide to be analyzed and the pattern of contaminants. Despite this variety of strategies used, however, several general rules were followed in the approach of the biochemical microanalysis of a novel peptide hormone. On this basis, we want to propose here an experimental approach to the sequence analysis of an unknown hormone.

2 Purification

2.1 Establishment of an Assay

Usually, the chemical characterization of a hormonal peptide is preceded by the discovery of an interesting biological activity. It is of crucial importance that the biological observation can be translated into a highly specific and sensitive biological assay which can serve to monitor purification and characterization. Thus, in the purification of hypophysiotropic peptides we followed the hypophysiotropic activity by radioimmunoassays measuring the radioimmunoactivity of hypophysial hormones secreted from rat anterior pituitary cell cultures [15, 16]. Even without any knowledge of the primary structure of the active peptide, a bioassay can be used to develop a radioimmunoassay with monoclonal antibodies directed against the active peptide to be purified [17]. This procedure appears to be, for example, recommendable if the bioassay available is specific, but exhibits low sensitivity and thus requires significant amounts of peptide for testing.

2.2 Establishment of Peptidic Character

It is important to investigate as early as possible the peptidic character of the potential hormone to be characterized because this investigation will strongly influence the choice of strategies of purification and structure analysis. The peptidic character can be demonstrated by the biological inactivation (measured in the bioassay) of a partially purified fraction after exposure to a proteolytic enzyme. Since the enzyme used may interfere in the bioassay, it is advisable to use only enzymes which can be inactivated by specific inhibitors or chemical modification or which can be removed by precipitation or extraction prior to the bioassay. Thus, trypsin and one of its specific inhibitors could be useful to start this investigation with. The conditions of enzymatic digestion should be similar to those described below for fingerprinting experiments.

2.3 Tissue Homogenization and Purification

2.3.1 Homogenization and Extraction

The choice of the appropriate homogenization and extraction methods depends strongly on the stability, size, and solubility properties of the peptide of interest.

It is necessary to identify and optimize the methods of choice in a series of experiments carefully monitored with the (bio)assay. Because of interfering contaminants, it is not uncommon that the hormonal peptide under investigation has to be partially purified after homogenization and extraction in order to obtain information about the usefulness of the homogenization and extraction methods tested. It is advisable to carry these initial experiments out in the cold room at +4 °C to suppress enzymatic and nonenzymatic modifications.

Homogenization should first be attempted with mild methods such as a Potter Elvehjem homogenizer and, should these methods be without success, with rougher methods such as a Polytron homogenizer. It would be especially advantageous if the peptide under investigation could be extracted with organic solvents such as methanol or acetone or aqueous organic mixtures containing a high percentage of organic solvents. Such extraction procedures would remove most enzymatic activities which could modify the peptide of interest. Teleost gonadotropin-releasing hormone (GnRH), an N- and C-terminally blocked decapeptide, for example, was homogenized with a Polytron in a cold mixture of acetone and 1 M HCl (100:3) [9].

Alternative approaches would be extractions with aqueous buffers followed by precipitation with ammonium sulfate, other salts or polyethyleneglycol or by ion exchange chromatography or preparative reverse-phase high pressure liquid chromatography (HPLC). Such approaches would be appropriate with larger polypeptides such as for example the 32,000 dalton porcine inhibin [12].

2.3.2 Methods of Purification

The strategies of purification depend strongly on the size, the hydrophobic properties of the peptide to be analyzed and the purity requirements for sequence analysis.

It is often advisable to start the purification with gel permeation fast protein liquid chromatography (FPLC) to analyze for multiple active forms. FPLC should be carried out in 6 – 8 M guanidine hydrochloride or urea. Should the presence of the denaturing agents interfere with the bioassay, at least the sample application should be performed with 6 M urea.

If the peptide to be purified contains less than approximately 50 residues (corresponding to a molecular weight of less than 5000), FPLC could be followed by reverse-phase HPLC on C_{18} (C_4, C_8, CN) columns eluted with mixtures of 0.05 – 0.1% trifluoroacetic acid, acetronitrile or aqueous triethyl ammonium phosphate, acetonitrile or 0.1% heptafluorobutyric acid, acetonitrile [18 – 22] or others.

If the polypeptide contains 50 – 150 residues, the initial gel permeation could be followed by ion exchange chromatography (on FPLC), reverse-phase HPLC and possibly sodium dodecyl sulfate (SDS) polyacrylamide gel electrophoresis (PAGE).

Reverse-phase HPLC of large polypeptides (>50 residues) usually does not provide high resolution and thus cannot be recommended as a method of choice. However, the usefulness of reverse-phase HPLC as method for the purification of large polypeptides can be increased by chemical modification of the peptide of interest. Of course, this combination of reverse-phase HPLC and chemical modification may also be advantageous in the purification of polypeptides containing less than 50 residues. Thus, human hypothalamic growth hormone-releasing factor (hhGRF) was purified on reverse-phase HPLC after selective oxidation of its methionine to methionine sulfoxide (with H_2O_2) [7]. S-carboxy-methylation has frequently been combined with reverse-phase HPLC to achieve an efficient purification. Examples of this approach are the characterizations of ovine hypothalamic somatostatin-28 [3], ovin ovarian oxytocin [29] and porcine follicular inhibin [14].

Combination of chemical modification with reverse-phase HPLC should be reserved to the final steps of purification because the modified peptide may have lost its biological activity and hence may escape detection by bioassay.

For large polypeptides which are difficult to purify, SDS PAGE may be the purification method of choice, especially, when combined with electroblotting ([23], for details see also Sects. 3.4 and 3.5).

3 Sequence Analysis

3.1 General Considerations

Before a strategy of sequence analysis can be designed, it is important to get to know the amount of purified peptide available, its purity, subunit structure and the reactivity of the N-terminal amino group (which may be free or blocked). The size of the intact molecule should be already known at this stage from the gel permeation experiments constituting part of the purification (see above).

3.2 Peptide Determination by Amino Acid Analysis

The amount of available peptide may most reliably be determined by amino acid analysis. The methods of amino acid analysis described to date differ mainly in the indicator reaction chosen, column packing, and chronologic order of chromatography and coupling (post- versus pre-column coupling). Ninhydrin, ortho-phthalaldehyde (OPA), fluorescamine, phenylisothiocyanate (PITC), or other agents have been used for coupling. None of these methods is fully satisfying at this time because they are either relatively specific, but not very sensitive (like the methods employing ninhydrin) or relatively unspecific, but sensitive (like the methods with fluorescent detection methods).

To date we have been using in our laboratory mainly anion exchange chromatography combined with a post-column ninhydrin detection system (Beckman 121 MB amino acid analyzer with a 20×0.28 cm column packed with Beckman AA-10 resin). We hydrolyze peptides routinely with the following procedure [3]: Approximately 0.3 µg of peptide is transferred to a hydrolysis tube (6×50 mm, Corning, Cat. No. 9820). After addition of 1.0 nmol of norleucine (in aqueous

3.0 M acetic acid) as internal standard, the glass tube is dried in a Savant Speed Vac Concentrator. Subsequently, 25 µl of methane sulfonic acid containing 0.2% tryptamine (Pierce Cat. No. 25600) is added. The tube is evacuated with a high vacuum pump until the pressure in the tube is at the most 100 mm Hg and sealed.

The mixture is then incubated either for 3 h at 140°C or for 22–24 h at 110°C. After this incubation, 25 µl of aqueous 3.9 M NaOH is added. An aliquot of the mixture (normally 80%) is applied to the analyzer. It is usually possible to determine all frequently occurring amino acids including cysteine, tryptophan, and proline after hydrolysis with methane sulfonic acid and 0.2% tryptamine, whereas it may be difficult to obtain reliable data about the relative content of cysteine and tryptophan after hydrolysis with HCl.

3.3 Purity

The purity of the peptide under investigation can be assessed by Edman degradation, fast atomic bombardment (FAB) mass spectrometry, SDS PAGE and amino acid analysis.

3.3.1 Amino Acid Analysis as Purity Test

Thus, amino acid analysis is not only used to determine the amount of peptide available, but may also be helpful to study the purity of a peptide. Of the methods mentioned above, amino acid analysis usually provides the least information about the purity of a peptide except possibly for the analysis of small peptides (containing less than 20 residues). In this instance, insignificant deviations of all amino acid ratios from integer numbers and possibly complete absence of some amino acids are suggestive for high purity of the hydrolyzed peptide. Since even amino acid analysis of small peptides can be misleading, when small amounts of peptide (<100 pmol) are applied [24], one should not derive important decisions of analytical strategies from such studies alone.

It is usually not advisable at this stage of the characterization to attempt to establish the amino acid composition of the peptide of interest. In view of the reliability of the methodology available today, amino acid compositions are most often established by direct or indirect sequence analysis.

3.3.2 Edman Degradation

After amino acid analysis has been performed, one can continue with any of the other methods listed above. In most of our characterization projects, we subjected the peptide to Edman degradation next.

Edman degradation can be carried out manually or automatically with gas phase, spinning cup and solid phase sequencers. Each of these approaches has its own merits which are discussed in Chapters 2 and 3 of this book. At this time, we

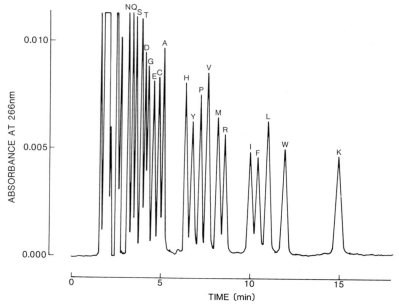

Fig. 1. PTH-amino acid analysis on reverse-phase HPLC. A mixture of PTH-amino acids (each representing 100 pmol) in 20 µl of methanol is applied to a (25 × 0.46 cm) Zorbax PTH column and eluted with a mixture (18:16:66) of acetonitrile, tetrahydrofuran (THF) and 3.2 mM sodium phosphate (adjusted with phosphoric acid to pH 3.15) in the presence of 12 µM dithiothreitol at a flow rate of 1.5 ml min^{-1}, a column temperature of 36 °C and a wavelength of 266 nm (for details see [25]). The chromatography was carried out with a Hewlett-Packard 1084B liquid chromatograph

are mainly working with a gas phase sequencer (470 A Protein Sequencer, Applied Biosystems) and standard sequencer programs provided by the manufacturer. Only one minor modification has been introduced: an additional nitrogen gas line has been placed in the fraction collector to dry the amino acid phenylthiohydantoin (PTH) derivatives after collection. Thus, some of the PTH-amino acids, especially PTH-carboxymethyl-cysteine are stabilized. In our laboratory, PTH-amino acid analysis is carried out with a Hewlett Packard 1084B liquid chromatograph equipped with a variable wave length detector or with a Perkin-Elmer series 4 liquid chromatograph equipped with a Kratos Spectroflow 773 Monitor on a (0.46 × 25 cm) PTH column (Dupont) (Fig. 1). Serine residues are identified as PTH-serine and a product which elutes immediately after PTH-alanine and may be an adduct of dithiothreitol with a PTH-serine derivative. The same product is formed from cysteine residues (W. Woo and J. Spiess, unpublished data).

This isocratic chromatographic procedure of PTH-amino acid analysis is compatible with the gas phase sequencer, but has not been very useful in the analysis of Edman cycles obtained from our spinning cup sequencer (W. Woo and J. Spiess, in preparation). For those fractions, we use an ODS column and acetonitrile gradients as described earlier [26]. We like to apply 100 – 300 pmol of peptide to the polybrene-containing glass fiber filter to determine 30 – 50

residues. However, even after the application of only 30 – 50 pmol of peptide we
have recognized approximately 25 – 30 residues (for example in the sequence
analysis of the A chain of porcine 18,000 dalton inhibin [14]).

By Edman degradation, two important aspects of the purity analysis can be
satisfied. It can be determined by comparison of the yields of the PTH derivatives of the first few residues with the peptide amount applied to the sequencer
that the predominant peptide of the analyzed fraction is not N-terminally blocked or an N-terminal block can be suspected on the basis fo the available data.
High purity of the predominant peptide of the analyzed fraction is indicated by
the complete or nearly complete absence of PTH-amino acids which cannot be
referred to the predominant peptide.

3.3.3 FAB Mass Spectrometry and SDS PAGE

If significant amounts of PTH-amino acids cannot be detected although a significant amount of peptide was applied to the sequencer, the purity of the peptide of
interest may be assessed by FAB mass spectrometry or SDS PAGE. Both
methods complement one another in the analysis of purity. Whereas FAB mass
spectrometry would show the presence of intact molecular species possessing
molecular weights of less than 6000 (see C. Eckhart, Chap. 3.1), the normally
used experimental protocol of SDS PAGE [27] should not be applied to polypeptides with molecular weights of less than 10,000. If it is suspected on the basis
of the amino acid analysis that the peptide of interest contains cysteine residues,
SDS PAGE should be performed with and without reduction of the sample.

3.3.4 Conclusion

Purity analysis may lead to a wide variety of observations, only a few of which
will be considered here.

If the peptide of interest found to be highly purified is suspected to contain
cysteine residues and if additional Edman degradation is intended, it is recommended to perform S-carboxymethylation. Although discrimination between
serine and cysteine residues can usually be achieved with a gas-phase sequencer
and compatible PTH-amino acid analysis (see above), discrimination may
become difficult after exposure of serine and cysteine residues to more than
15 – 20 Edman cycles. S-carboxymethylation may also improve the retention of
the peptide in gas phase and spinning cup sequencers by the introduction of
carboxylic groups especially when Edman degradation is carried out in the
presence of the positively charged Polybrene.

If the predominant peptide of the analyzed fraction is highly purified
(>85%), N-terminally unblocked, and probably contains less than 50 residues,
complete sequence analysis should be pursued with Edman degradation and C-
terminal analysis (see below).

Sequence analysis of proteins of high purity should be performed with a combined approach of recombinant DNA technology and partial sequence determination by Edman degradation.

If the degree of purity of the peptide of interest is observed to be 5 – 50%, further purification should be attempted. Alternatively, it should be considered to obtain partial sequence information of the major components of the fraction and to select for the sequence of interest by biological immunoneutralization experiments.

3.4 S-Carboxymethylation

We are currently using the following procedure for S-carboxymethylation [3]: Approximately, 0.1 – 2.0 nmol of peptide is dried in a microfuge 1.5 ml tube (Eppendorf Catalogue No. 2236411-1) by a Savant Speed Vac Concentrator. *Immediately* after drying, the residue is dissolved with 50 μl of reducing buffer which consists of 8 M urea (Ultra pure, Schwartz/Mann Catalogue No. 821527), 6.0 M disodium ethylene diamine tetraacetate (Na_2 EDTA), 0.173 M dithiothreitol (DTT) and 0.58 M tris (hydroxymethyl) amino methane (TRIS) adjusted to pH 9.0 – 9.1 with 1 M HCl. The mixture is incubated for 60 min at 50°C under a protective argon gas layer, cooled to 23°C and treated with 25 μl 0.66 M iodoacetic acid (Sigma Catalogue No. I-6250) in reducing buffer (without DTT) adjusted to pH 8 – 9 with aqueous 3.5 N NaOH (sigma Catalogue No. 505-8). After incubation for 15 min at room temperature, 7.5 μl of glacial acetic acid is added to the mixture, which is subsequently subjected to reverse-phase HPLC. The conditions of reverse-phase HPLC should be similar to those applied to the purification of the nonmodified peptide. Usually, we employ (25 × 0.46 cm) C_{18}, C_8 or C_4 columns with pore sizes of 100 – 330 Å for peptides containing less than 50 residues and C_4 columns with a pore size of 330 Å for larger polypeptides. We often carry HPLC out with mixtures of 0.05% trifluoroacetic acid and acetonitrile at flow rate of 0.8 – 1.0 ml min^{-1}, temperatures of 23° – 37°C and flat linear gradients of increasing acetonitrile concentrations (0.7 – 1.0%/min). Fractions are manually collected to achieve optimal separation and immediately frozen at – 80 or – 20°C.

3.5 Complete Sequence Determination by Edman Degradation and C-Terminal Analysis

3.5.1 Edman Degradation

Highly purified peptides (containing less than 50 residues) which are N-terminally unblocked and free of cysteine or carboxymethylated, are subjected in one or two experiments to automated Edman degradation carried out with 100 – 300 pmol of peptide per experiment. A second experiment is performed when the first experiment has provided incomplete information and optimization of the sequencer program appears promising. Especially increased carry-over generated by proline residues can be reduced by the introduction of a second cleavage in the affected cycles. Often, the C-terminal 1 – 3 residues of a peptide (especially if they are hydrophobic amino acids) escape detection because of untimely extractions (wash out). However, even if a peptide is completely degraded, the sequence data do not allow discrimination between a C-terminal amide and a free carboxylic group. Since the biological activity of hormonal peptides often depends on their C-terminal sequence, especially the presence of a C-terminal amide group, Edman degradation of (hormonal) peptides must always be supplemented by C-terminal analysis.

3.5.2 C-Terminal Analysis

Determination of the molecular weight by FAB mass spectrometry may already fulfil this requirement of supplementation of Edman data. Alternatively, a C-terminal fragment cleaved from the peptide with chemical or enzymatic procedures should be analyzed by mass spectrometry or amino acid analysis, automated Edman degradation, and reverse-phase HPLC. We favor the use of enzymes for fragmentations of 0.1 – 2.0 nmol of peptide, and especially like to employ trypsin, clostripain, thermolysin, and staphylococcal protease for the cleavage at basic, hydrophobic, and acidic residues, respectively. Usually, we use enzyme to substrate weight ratios of 1 : 20 to 1 : 50, incubate for 1 – 5 h and subject the digests to reverse-phase HPLC. Details of the optimization of enzymatic fragmentation (as well as other methods of C-terminal analysis) are presented elsewhere [28]. Discrimination between a C-terminal amide and a free carboxylic group can be achieved by chromatographic comparison of the C-terminal fragment, its synthetic replicate and synthetic analogs. Should the natural product represent an oligopeptide, the comparative chromatographic studies can be carried out without prior fragmentation (Fig. 2).

3.5.3 Comparison of the Natural Product with its Synthetic Replicate

After the complete chemical characterization of the purified peptide, it has not been established that the analyzed peptide represents the natural product of interest. It is therefore obligatory to demonstrate that the analyzed peptide exhibits the observed activity. This objective can be achieved with immunoneutralization experiments using antibodies against the synthetic replicate of the analyzed peptide or by biological and chromatographic comparison of the natural product with a synthetic replicate and possibly synthetic analogs of the analyzed peptide. The peptides required for these experiments may be synthesized with solid phase methods [30], purified on reverse-phase HPLC [18 – 22] and characterized similarly as natural peptides [31]. Comparative studies of the natural product and the synthetic replicate of the analyzed peptide must include the determination of the intrinsic biological activity and potency in the bioassay used for the purification of the natural product. Besides this biological comparison, it is recommended to compare the chromatographic behavior on reverse-phase HPLC. For the assessment of the biological significance of the analyzed product it could be crucial to obtain information about the biological action of the synthetic replicate in vivo. However, this important aspect is not nessarily an obligatory part of the complete sequence analysis.

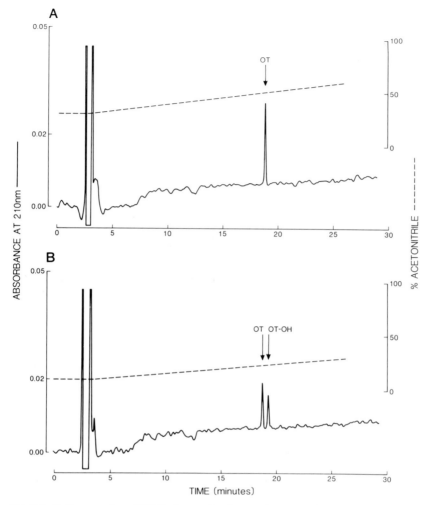

Fig. 2 A, B. Reverse-phase HPLC of natural ovine ovarian oxytocin (OT), its synthetic replicate and desamidated analog. Samples were applied to a Vydac C_{18} column (25×0.46 cm; particle size, 5 μm; pore size, 330 Å) and eluted at 32 °C and 1.0 ml min^{-1} with a mixture of 0.05% trifluoroacetic acid and acetonitrile. A Hewlett-Packard 1084B liquid chromatograph was used. Approximately, 130 pmol of natural OT and either 250 pmol of synthetic OT (**A**) or 130 pmol of synthetic OT-OH (**B**) were mixed before chromatography (Watkins et al. [29])

If the natural product elutes before and is less active than the synthetic peptide, the possibility should be considered that the natural product might have been oxidized at its methionine residues. Such an oxidation generated by air oxygen represents a common event during the purification of peptides. The possibility of oxidation of methionine residues (usually to methionine sulfoxide) can be tested by repeating the comparative experiments with synthetic peptides containing methionine sulfoxide instead of methionine. These peptides could be obtained by de novo synthesis or mild oxidation of methionine containing peptides with H_2O_2 in an acidic aqueous medium [31].

3.6 Complete Sequence Determination of Small N-Terminally Blocked Peptides

If a peptide of high purity and small size (<50 residues) is found to be N-terminally blocked, one should consider the possibility of the presense of an N-terminal pyroglutamyl residue and attempt to deblock the peptide by digestion with L-pyroglutamyl peptide hydrolase [9]. Should this deblocking procedure be successful, the characterization should continue as outlined in Section 3.5. Otherwise, the blocked peptide should be cleaved in two separate experiments at different sites and overlapping sequence analysis by Edman degradation should be performed. In this instance, the N-terminal fragments could be subjected to mass spectrometry. C-terminal analysis, immunologic experiments, and comparative studies with synthetic peptides should be carried out as described in Section 3.5.

3.7 Partial Sequence Determination by Edman Degradation

3.7.1 Analysis of Sequences with N-Terminal Proline

If peptides with free N-termini are only partially purified and complete purification turns out to be difficult, partial sequence information can be obtained by exposing the peptide mixture to orthophthalaldehyde (OPA) in cycles showing N-terminal proline residues. OPA will block all N-terminal residues representing primary amines and leave N-terminal proline residues as secondary amines unblocked. Thus, quasi-homogeneous conditions are generated in the sequencer and unambiguous sequence data are provided. This approach can be repeated several times by adding OPA in different proline cycles. The sequence of the hormonal peptide of interest can be found by neutralization of the biological activity with antibodies directed against the analyzed sequences. These antibodies are to be raised against peptides synthesized on the basis of the sequence data of Edman degradation after blocking with OPA.

The approach has been developed on the basis of observations [32, 33] that fluorescamine and OPA can be successfully used to suppress background during Edman degradation. OPA blocking has been accomplished in solid phase [33], spinning cup [8, 34] and gas phase [35] sequencers.

In the characterization of rat hypothalamic corticotropin releasing factor (CRF), the OPA strategy was employed for the first time to the analysis of a natural product. Because of the characteristic N-terminal sequence of CRF, X-X-X-Pro-Pro, exposure of CRF containing peptide mixtures to OPA in cycles 4 and 5 is expected to leave mainly CRF unblocked. Thus, the OPA strategy is attractive for the partial sequence analysis of CRF purified from various sources (Fig. 3).

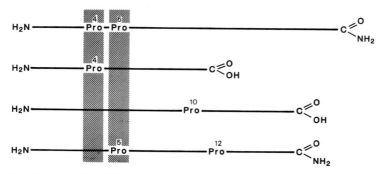

Fig. 3. Application of the orthophthalaldehyde (OPA) strategy to CRF. *Shadowed columns* illustrate the effect of OPA on CRF and contaminating peptides

3.7.2 Edman Degradation as Search Method of Recombinant DNA Methods

The primary structure of large polypeptides, especially of molecular weights >10,000, should mainly be deduced from the nucleotide sequences of the corresponding cDNA's obtained from cDNA clone banks. The identification of a cDNA clone can be accomplished with hybridization probes synthesized on the basis of amino acid sequence information derived from data of Edman degradation.

If the carboxymethylated polypeptide (or protein) is found to be of high purity (>90%), the N-terminal sequence (of at least 20 – 30 residues) should be determined by Edman degradation of 100 – 300 pmol of peptide. Should subsequent sequence information of other parts of the molecule be desired or should the polypeptide be found to be N-terminally blocked, fragmentation followed by reverse-phase HPLC as described in Section 3.5.2 should be performed. It is advantageous to generate relatively small peptides by enzymatic cleavage (for example with clostripain as shown in Fig. 4) to facilitate their purification on reverse-phase HPLC. If the cloning strategy requires the identification and Edman degradation of the C-terminal fragment, C-terminal tritiation [28, 38] marking the C-terminus should be carried out prior to fragmentation.

If the polypeptide of interest is significantly contaminated, the OPA strategy outlined in Section 3.7.1 could be employed.

It is of crucial importance to demonstrate that the recognized amino acid sequences are part of the polypeptide exhibiting the observed biological activity. This can be accomplished by neutralization of the biological activity of the purified peptide with antibodies directed against the analyzed sequences. For the production of antibodies, peptides have to be designed and synthesized on the basis of the data of Edman degradation.

Although strategies combining recombinant DNA technology and Edman degradation are attractive because of their efficiency and sensitivity, it should be kept in mind that one may miss post-translational modifications in the prediction of amino acid sequences of peptides on the basis of DNA structures. Errors on this basis may be especially disturbing for the characterization of hormonal

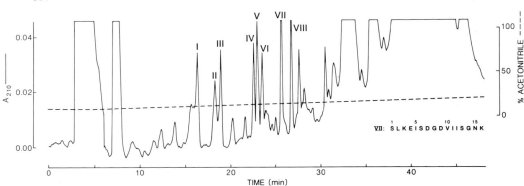

Fig. 4. Fragmentation of the epidermal growth factor-related peptide (ERRP) and fragment analysis. *Digestion:* 77 µg (1.0 nmol) of ERRP was digested with clostripain (Boehringer, Mannheim) (Substrate to enzyme weight ratio, 20:1) in 50 µl of aqueous 0.05 M Quadrol, 5 mM $CaCl_2$ 1 mM dithiothreitol and 6% (v/v) 1-propanol (adjusted to pH 8.0 with trifluoroacetic acid) for 5 h at 37 °C. *Reverse-phase HPLC:* The digest was applied to a Vydac C_4 column (25 × 0.46 cm; particle size, 5 µm; pore size, 330 Å) and eluted at 31 °C and a flow rate of 1.0 ml/min with a mixture of 0.055% (v/v) trifluoroacetic acid and acetonitrile. *Product analysis:* Products I – VIII were subjected to amino acid analysis followed by Edman degradation of Product VII in a Wittmann-Liebold [36] modified Beckman 890C spinning-cup sequencer. (In collaboration with Weber and Gill [37])

peptides, the activity of which often depends on post-translational modifications.

3.8 Trends

On the basis of our experiences and observations, it appears highly probable that reverse-phase HPLC and Edman degradation will remain the two most important procedures for the purification and sequence analysis of small hormonal peptides, whereas DNA cloning techniques and Edman degradation as search method (possibly in connection with electroblotting) may serve as optimal procedures for the sequence analysis of large polypeptides. It is expected that FAB mass spectrometry will be increasingly used to supplement data of Edman degradation of small peptides and fragments from large polypeptides. It may be especially useful in the discovery and possibly characterization of post-translational modifications.

Acknowledgments. This research was supported by national Institutes of Health grants AM-26741, AM-20971, HD-13527, AM-26378 and AM-16921, and the Clayton Foundation for Research − California Division. JS is a Clayton Foundation Investigator. The excellent technical assistance of W. Woo, D. Karr and T. Richmond is gratefully acknowledged. We thank B. Connor for manuscript preparation.

References

1. Spiess J, Rivier J, Rodkey J, Bennett C, Vale W (1979) Proc Natl Acad Sci USA 76(6): 2974 – 2978
2. Noe B, Spiess J, Rivier J, Vale W (1979) Endocrinology 105:1410 – 1415

3. Spiess J, Villarreal J, Vale W (1981) Biochemistry 20:1982–1988
4. Spiess J, Rivier J, Rivier C, Vale W (1981) Proc Natl Acad Sci USA 78:6517–6521
5. Spiess J, Rivier J, Thorner M, Vale W (1982) Biochemistry 24:6037–6040
6. Spiess J, Rivier J, Thorner M, Vale W (1983) In: Dumont JE, Nunez J, Denton RM (eds) Hormones and cell regulation. Elsevier/North-Holland Biomedical Press, Amsterdam, pp 231–242
7. Spiess J, Rivier J, Vale W (1983) Nature 303:532–535
8. Spiess J, Rivier J, Vale W (1983) Biochemistry 22:4341–4346
9. Sherwood N, Eiden L, Brownstein J, Spiess J, Rivier J, Vale W (1983) Proc Natl Acad Sci USA 80:2794–2798
10. Spiess J, Noe BD (1985) Proc Natl Acad Sci USA 82:277–281
11. Spiess J, Noe BD (1985) In: Patel Y, Tannenbaum G (eds) Somatostatin. Plenum, New York, pp 141–154
12. Rivier J, Spiess J, McClintock R, Vaughan J, Vale W (1985) Biochem Biophys Res Commun 133:120–127
13. Vale W, Rivier J, Vaughan J, McClintock R, Corrigan A, Woo W, Karr D, Spiess J (1986) Nature 321:776–779
14. Mayo KE, Cerelli GM, Spiess J, Rivier J, Rosenfeld MG, Vale W, Evans RM (1986) Proc Natl Acad Sci USA 83:5849–5853
15. Vale W, Vaughan J, Jolley D, Yamamoto G, Bruhn T, Seifert H, Perrin M, Thorner M, Rivier J (1986) In: Conn PM (ed) Methods in enzymology: Neuroendocrine peptides, vol 124. Academic Press, Orlando, FL, pp 389–401
16. Vale W, Vaughan J, Yamamoto G, Bruhn T, Douglas C, Dalton D, Rivier C, Rivier J (1983) In: Conn PM (ed) Methods in enzymology: Neuroendocrine peptides, vol 103. Academic Press, London New York, pp 565–577
17. Schönherr OT, Houwink EH (1984) Antonie van Leeuwenhoek J Microbiol Ser 50:597–623
18. Rivier J (1978) J Liq Chromatogr 1:343–367
19. Rivier J, McClintock T (1983) J Chromatogr 268:112–119
20. Rivier J, McClintock R, Galyean R, Anderson H (1984) J Chromatogr 288:303–328
21. Bennett HP, Hudson HM, McMartin C, Purdon GE (1977) Biochem J 168:9
22. Bennett HPJ, Browne CA, Solomon S (1980) J Liq Chromatogr 3:1353–1365
23. Aebersold RH, Teplow DB, Hood LE, Kent SBH (1986) J Biol Chem 261:4229–4238
24. Spiess J, L'Italien J (1986) In: L'Italien J (ed) Modern methods in protein chemistry, Plenum, New York (in press)
25. Glajch JL, Gluckman JC, Charikofsky JG, Minor JM, Kirkland JJ (1985) J Chromatogr 318:23
26. Spiess J, Rivier J, Rivier C, Vale W (1982) In: Elzinga M (ed) Methods in protein sequence analyses. Humana, Clifton, NJ, pp 131–138
27. Laemmli UK (1970) Nature 227:680–685
28. Spiess J (1986) In: Shively JE (ed) Microstructural studies in peptide and protein chemistry. Humana, Clifton, NJ, pp 363–377
29. Watkins WB, Choy VJ, Chaiken IM, Spiess J (1986) Neuropeptides (in press)
30. Barany G, Merrifield RB (1979) In: Gross E, Meienhofer J (eds) The peptides. Academic Press, London New York, pp 3–284
31. Märki W, Spiess J, Taché Y, Brown M, Rivier JE (1981) J Am Chem Soc 103:3178–3185
32. Bhown AS, Bennett JC, Morgan PH, Mole JE (1981) Anal Biochem 112:158–162
33. Machleidt W, Hofner H (1982) In: Elzinga M (ed) Methods in protein sequence analysis. Humana, Clifton, NJ, pp 173–180
34. Spiess J (1985) In: Hruby V, Rich D (eds) Peptides: Structure and function. Pierce Chemical Co, Rockford, IL, pp 711–714
35. Margolies MN, Brauer AW, Matsueda GR (1986) In: L'Italien J (ed) Modern methods in protein chemistry. Plenum, New York (in press)
36. Wittmann-Liebold B (1980) In: Beers Jr, RF, Basset EG (eds) Polypeptide hormones. Raven, New York, pp 87–120
37. Weber W, Gill GN, Spiess J (1984) Science 224:294–297
38. Matsuo H, Narita K (1975) In: Needleman SB (ed) Protein sequence determination. Springer, New York Berlin Heidelberg, pp 104–113

7.2 Use of a Metalloproteinase Specific for the Amino Side of Asp in Protein Sequencing

Herwig Ponstingl, Gernot Maier, Melvyn Little, and Erika Krauhs[1]

Contents

1 Introduction

A larger panel of proteolytic enzymes analogous to DNA restriction enzymes would be useful for analyzing protein structures. As we are sequencing larger polypeptides in smaller quantities, generation of a moderate number of fragments would, in most cases, facilitate their separation and overlapping. In addition, such proteinases would be helpful in the elucidation of domains with biological activity, in generating peptide maps [1] and in the recent attempts to sequence end-labeled proteins on gels [2, 3].

Numerous highly specialized proteases exist which modulate the biological activity of proteins, e.g., viral proteases, which cleave a large precursor into mature viral proteins [4, 5], pro-hormone and propheromone convertases, such as the propheromone convertase Y from yeast, which hydrolyzes the bond between two adjacent basic residues [6], proteases from the blood-clotting cascade, of which thrombin, preferentially splitting the C-terminal side of a Pro-Arg sequence preceded by a hydrophobic dipeptide, is already in fairly widespread use [7], or collagenases, cleaving the bond between Pro-X/Gly-Pro [8, 9]. Their specificity may not always be exactly the same as in the native substrate, which frequently is a unique protein, and it may be necessary to denature substrates before digestion, but as a rule highly specific sequences are recognized.

In this context we report here on the use of a metalloprotease which is strictly specific for the amino side of aspartyl and (if the substrate has been oxidized) cysteic acid residues. Like most of the other highly specific proteases, it has not received adequate attention, probably because it is not yet commercially avail-

---

1 Projektgruppe Molekulare Biologie der Mitose, Institut für Zell- und Tumorbiologie, Deutsches Krebsforschungszentrum, Im Neuenheimer Feld 280, D-6900 Heidelberg, FRG

Advanced Methods in Protein Microsequence Analysis
Ed. by B. Wittmann-Liebold et al.
© Springer-Verlag Berlin Heidelberg 1986

able, and vice versa. While other metalloproteinases, such as thermolysin, cleave peptide bonds at the amino side of large hydrophobic residues, the single protease secreted by the wild-type *Pseudomonas fragi* prefers small and hydrophilic residues [10]. It has been mutated by G. R. Drapeau and selected by growth on elastin as the sole carbon source to yield an enzyme strictly specific for Asp and CysSO$_3$H, and since Cys can be alkylated, it can be made specific for Asp alone [11].

2 Cleavage Conditions

We have digested several polypeptides with this enzyme. The cleavage conditions are given in Table 1. For an illustration of its usefulness in sequencing, Table 2 lists the 13 sites that have been cleaved in α-tubulin, together with the remaining 14 out of the 27 Asp-containing sequences that remained uncleaved [12]. Each site was either digested quantitatively or not at all, no site appears to have been partially hydrolyzed. It may be premature to draw general conclusions on the causes of unrecognized aspartyl residues, but a valine preceding the aspartyl residue appears to be unfavorable, as well as a potential of the sequence in question to form turns, at least if calculated by the Robson prediction method [13]. Of the 12 cysteines of α-tubulin, after alkylation none gave rise to a cleavage site.

The resulting 14 peptides were purified by high pressure chromatography and most of them yielded valuable sequences overlapping tryptic and CNBr-derived

Table 1. Alkylation of the substrate polypeptides and cleavage conditions

1. Deionize a solution of 8 M urea with Amberlite AG (Serva, Heidelberg) by stirring for 20 min at room temperature.
2. In this solution, dissolve 121.1 mg ml^{-1} tris, 1 mg EDTA ml^{-1}, add 8.4 µl 2-mercaptoethanol ml^{-1}, bring to pH 8.6 with conc. HCl, adjust volume with water.
3. Dissolve dialyzed and lyophilized polypeptide in a small volume of this buffer, reduce overnight at room temperature.
4. To 9 volumes of the reduced sample add dropwise one volume of 1.2 M iodoacetate (232 mg ml^{-1}; Serva, Heidelberg) in freshly deionized 8 M urea in the dark for 15 min under stirring.
   Note: It is essential that the iodoacetate be colorless. If it is yellow it contains free iodine which will also react with the protein. Traces of iodine can be extracted by grinding iodoacetate crystals in a mortar with diethyl ether. Store the reagent dark and frozen. At lower pH, iodoacetate may also alkylate His and Met. Therefore, check the pH of the solution while adding the reagent dropwise.
5. Terminate the reaction with 150 µl 2-mercaptoethanol per ml of alkylating reagent.
6. Dialyze the sample against four batches of 0.01 – 0.1 M ammonium bicarbonate in freshly deionized 2 M urea, or desalt on a Sephadex G25 column in the dark. As the *P. fragi* enzyme is a metalloprotease, EDTA should be removed quantitatively from the reducing solution.
7. Add *P. fragi* protease at an enzyme to substrate ratio of 1 : 100 (w/w), incubate at 37 °C for 24 h.
   Note: The enzyme was a gift of Dr. Gabriel Drapeau, Department of Microbiology, University of Montreal. It is stored at − 20 °C in 10 mM Tris-HCl, pH 7.5, containing 0.02% sodium azide. It will be damaged by lyophilization.
8. To terminate digestion, add 1 mg EDTA ml^{-1}. Fractionate the digest (gel filtration, reversed phase or ion exchange chromatography).

Table 2. Asp-containing sequences in porcine α-tubulin

Cleaved			Uncleaved		
by the Asp-specific metalloproteinase from *Pseudomonas fragi.*					
		Asp position			Asp position
LEHS	DCAF	199	GGGD	DSFN	47
TGKE	DAAK	98	QMPS	DKTI	39
EARE	DMAA	424	ALEK	DYEE	431
VPGG	DLAK	367	GIQP	DGQM	32
LYRG	DVVP	322	IGGG	DDSF	46
MVKC	DPRH	306	RKLA	DQCT	126
RLSV	DYGK	160	ALNV	DLTE	251
PTVI	DEVR	76	EVGV	DSVE	438
RRNL	DIER	218	AFMV	DNEA	205
WARL	DHKF	392	IQFV	DWCP	345
EAIY	DICR	211	AVFV	DLEP	69
DHKF	DLMY	396	KEII	DLVL	116
VVPK	DVNA	327	DLVL	DRIR	120
			SLRF	DGAL	245

fragments [12]. They were also used for localizing binding sites of microtubule-associated proteins within the tubulin sequence by an overlay assay [14].

Surprisingly, we are aware of only two other groups that have since used this proteinase in sequencing histocompatibility antigens [15, 16] and the phosphorylation sites in enolase and lactate dehydrogenase [17]. In its Asp-content of 6%, tubulin is close to the 5.5% of average proteins [18]. Its cleavage sites with the *P. fragi* protease comprise 3% of the residues. Thus, a similar frequency of cleavage sites may be expected for many other proteins.

3 Summary

A mutated proteinase from *Pseudomonas fragi* specifically cleaves peptide bonds at the amino side of aspartyl residues. It has been used as a new tool in protein sequencing.

Acknowledgments. We thank Dr. G. R. Drapeau, Dept. of Microbiology, University of Montreal, for the gift of the P. fragi protease, and J. Kretschmer and H. Scherer for their skilled technical assistance. This work was supported by the Deutsche Forschungsgemeinschaft.

References

1. Cleveland DW, Fischer SG, Kirschner MW, Laemmli UK (1977) J Biol Chem 252:1102
2. Jue RA, Doolittle FR (1985) Biochemistry 24:162
3. Jay DG (1984) J Biol Chem 259:15572
4. Klump W, Marquardt O, Hofschneider PH (1984) Proc Natl Acad Sci USA 81:3351
5. Yoshinaka Y, Katoh I, Copeland TD, Oroszlan S (1985) Proc Natl Acad Sci USA 82:1618
6. Mizuno K, Matsuo H (1984) Nature 309:558
7. Chang J-Y (1985) Eur J Biochem 151:217
8. Coletti-Previero MA, Cavadore J-C, Tonnelle C (1975) Immunochemistry 12:93
9. Henschen A, Lottspeich F, Hessel B (1979) Hoppe-Seyler's Z Physiol Chem 360:1951
10. Noreau J, Drapeau GR (1979) J Bacteriol 140:911
11. Drapeau GR (1980) J Biol Chem 255:839
12. Ponstingl H, Little M, Krauhs E, Kempf T (1981) Proc Natl Acad Sci USA 78:2757
13. Robson B, Suzuki E (1976) J Mol Biol 107:327
14. Littauer UZ, Giveon D, Thierauf M, Ginzburg I, Ponstingl H (1986) In: Dustin P and Porter KR (eds) 3rd Int. Symp. on Microtubules and Microtubule Inhibitors. Elsevier, Amsterdam, p 171
15. Guild BC, Strominger JL (1984) J Biol Chem 259:9235
16. Krangel MS, Pious D, Strominger J (1984) J Immunol 132:2984
17. Cooper JA, Esch FS, Taylor SS, Hunter T (1984) J Biol Chem 259:7835
18. Dayhoff MO (1978) Atlas of protein sequence and structure, vol 5, Suppl 3. Nat Biomed Res Found, Washington, DC, p 363

7.3 Primary Structure of Rabbit Apolipoprotein A – I High Performance Liquid Chromatography, PTC-Amino Acid Analysis, and Microsequencing

CHAO-YUH YANG, TSEMING YANG, HENRY J. POWNALL, and
ANTONIO M. GOTTO, Jr.[1]

Contents

1 Introduction

Apolipoprotein A – I [apo A – I], the major protein of high density lipoproteins [HDL] [1], is an activator of the plasma enzyme lecithin, cholesterol acyltransferase [LCAT]. This enzyme transfers fatty acids from lecithin to cholesterol in high density lipoproteins and is responsible for the formation of nearly all plasma cholesteryl esters [2]. It has been suggested that the activation of LCAT by apo A – I is due in part to its amphiphilic character. Synthetic models of apolipopeptides with a primary structure different from that of apo A – I, but having a similar amphiphilic character and helical potential, also activate LCAT

---

1 Baylor College of Medicine and The Methodist Hospital, Houston, TX 77030, USA

Advanced Methods in Protein Microsequence Analysis
Ed. by B. Wittmann-Liebold et al.
© Springer-Verlag Berlin Heidelberg 1986

[3, 4]. The role of apo A – I in the mechanism of LCAT activation is important but poorly understood.

The application of reverse-phase high performance liquid chromatography [RP – HPLC] to studies of protein chemistry has increased in recent years because this method affords both rapid separation of components and high sensitivity in detection and quantitation of peptides [5]. Picomole amounts of material can be quantitated (WATERS Pico-Tag system, [6]) using HPLC for amino-acid analysis. Sequencing of proteins at picomole levels by automatic sequencers and manual methods [7, 8] is possible using HPLC for the separation and identification of the phenylthiohydantoin [PTH] and diaminoazobenzenthiohydantoin [DABTH] amino acids, respectively. Furthermore, an HPLC with a photodiode array spectrophotometer permits the detection of tryptophan containing peptides in complex mixtures [9]. In this chapter, we describe the elucidation of the primary structure of rabbit apo A – I [10] by microsequencing and HPLC methods.

2 Materials

2.1 Rabbit Apo A – I Isolation

Pooled rabbit plasma was shipped from Pel-Freeze on wet ice. Apo A – I was purified [11] and subsequently dialyzed against 2 liters of 0.1 M ammonium bicarbonate, pH 8.0, for two days.

2.2 Enzymatic Hydrolysis

According to the sequence of human apo A – I [12] and the amino acid analysis data of the rabbit apo A – I, the protein contains no cysteine residues. Thus enzymatic hydrolysis can be performed without reduction and alkylation to protect the -SH groups.

2.2.1 Digestion with Trypsin. Tryptic hydrolysis of apo A – I (6.5 mg ml^{-1} in 0.1 M ammonium bicarbonate, pH 8.0) was performed with TPCK-treated trypsin (Worthington Biochemicals). To obtain the tryptic hydrolysate, 1 mg of trypsin was used for every 50 mg of protein (enzyme : substrate ratio = 1 : 50) and the mixture was left at room temperature for 5 h. The tryptic hydrolysate was then subjected to HPLC separation.

2.2.2 Digestion with Staphylococcus aureus V8 protease (Miles Biochemicals). A second digestion of apo A – I (6.5 mg l^{-1} in 0.1 M NH$_4$HCO$_3$, pH 8.0) was performed with *Staphylococcus aureus* protease [13]. The same enzyme : substrate ratio was used; the mixture was left at room temperature for 24 h before HPLC purification.

3 Peptide Purification

3.1 Peptide Separation by HPLC

Separations were performed either on a WATERS HPLC system (WATERS, MA, USA) equipped with a Knauer (Berlin, W. Germany) variable wavelength detector or a Hewlett Packard (Palo Alto, CA, USA) 1090 LC system, equipped with a photodiode array spectrophotometer.

3.1.1 Primary Separation

A Vydac C18 HPLC column (250×4.6 mm), heated to $50\,°C$, was utilized to isolate the peptides at a flow rate of 1.5 ml min^{-1}. To obtain a primary separation of the tryptic and staphylococcal digests, a trifluoroacetic acid [TFA] buffer system [9] (buffer A: 0.1% TFA in water, v/v; buffer B: 0.08% TFA in 95% acetonitrile and 5% water, v/v/v) was used.

A gradient of buffer B was linearly increased from 0 to 45% B at a rate of 1% B per min. Peptides were monitored at 220 nm. The eluent under each peak was collected manually in tubes and dried by a Speedvac (Savant Instruments, New York, USA). The purity of each peak fraction was checked by the manual sequencing method (see Sect. 4.2). The pure peptides were then sequenced continuously until no more DABTH-amino acid could be detected by the thin layer chromatography [TLC] method.

3.1.2 Rechromatography

An ammonium acetate buffer system (A: 0.025 M ammonium acetate, pH 6.5; B: 30% 0.05 M ammonium acetate, pH 6.5, 70% acetonitrile, v/v) was used for rechromatography [14, 15]. For this purpose, the remaining partially purified fractions were then dried in a Speedvac and reconstituted in 200 µl of buffer A or a 1:1 mixture of buffer A and B; the latter solution was used to ensure that peptide mixtures eluting at a high percentage of B could be dissolved and transferred from the sample tube to the column. Twenty µl (10%) of the sample was then used to test and find the conditions for an optimum separation. When the rechromatography conditions for the particular sample had been established, the remainder of the sample was injected into the system and the eluting peptides were collected in tubes. To maximize recovery of the peptides from the impure fraction, the original sample tube was washed with 50 to 100 µl of 50% formic acid and this solution was applied on the HPLC column. The purity of the collected fractions was again checked with the manual sequencing method by using a small aliquot of the sample. For pure peptides, amino-acid analysis and sequencing of the whole peptide were performed; otherwise, the fractions were subjected to a second rechromatography using a sodium phosphate buffer system [14].

3.1.3 Examples

Figure 1 shows the separation of 50 µg of staphylococcal digest of rabbit apo A – I by the HP 1090 HPLC system on a Vydac C18 column with the TFA system. The gradient runs from 0% to 45% in 45 min. Manual sequence determination showed that we had isolated most peptides in pure form. Rechromatography of less pure fractions was performed with the ammonium acetate buffer system on a WATERS HPLC system. Figure 2 shows the rechromatography of the

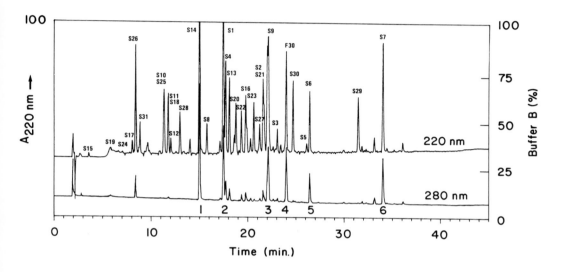

Fig. 1. Separation of peptides derived by *Staphylococcus aureus* V8 protease digestion of rabbit apo A – I. *Column:* Vydac C18, 4.6 × 250 mm, 50 °C. *Elution system:* (A) 0.1% TFA in water; (B) 0.08% TFA in 95% acetonitrile + 5% water. *Flow rate:* 1.5 ml min^{-1}. Gradient from 0% B to 45% B in 45 min

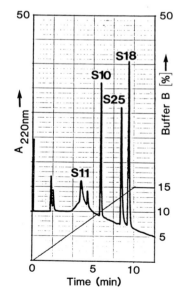

Fig. 2. Rechromatography of peptides S10, S11, S18, S25 with an ammonium acetate buffer system. *Column:* Vydac C18, 4.6 × 250 mm, 50 °C. *Elution system:* (A) 0.025 M AmAc, pH 6.5; (B) 30% 0.05 M AmAc, pH 6.5, + 70% acetonitrile. *Flow rate:* 1.5 ml min^{-1}. Gradient from 0% B to 15% B in 10 min

staphylococcal peptides on the same Vydac column with ammonium acetate buffer using a gradient from 0% to 15% B in 15 min. By this procedure, peptides S10, S11, S18, and S25 were obtained in pure form.

3.1.4 Comments

1. Since both TFA and ammonium acetate systems are volatile, peptides dissolved in these solutions are free of salt after lyophilization. This is important for further gas-phase sequencing and Pico-Tag amino-acid analysis (see Sect. 5).
2. Since ammonium acetate absorbs at 220 nm and buffer B (only 30% 0.05 M NH_4Ac) contains less salt than buffer A, the baseline of the chromatogram moves downward when increasing the percentage of B. To avoid this, the salt concentration of buffer B can be adjusted to 0.025 M. Should it be necessary to rechromatograph peptides eluting at a very high percentage of acetonitrile, the acetonitrile concentration of buffer B can be increased to 95% with an adequately increased ammonium acetate content to maintain the buffer at 0.025 M.
3. A phosphate buffer system (A: 0.005 M KH_2PO_4/K_2HPO_4, pH 6.0; B: 30% A + 70% CH_3CN), which was not used in this sequence study, is also a good system for peptide purification [14]. It is a nonvolatile buffer system.
4. This HPLC procedure can be applied in general to the separation of peptide fragments derived from different kinds of proteins [10, 14 – 17]. To achieve varying separation patterns, the pH of the solution may be changed.
5. Preparative columns are recommended for the purification of large amounts of material. Despite the greater injection volumes and the higher flow rates needed for larger columns, the separation pattern is the same as regular size columns. Instead of the regular size Vydac C18 column, we employed a preparative column with a diameter of 10 mm (normal diameter = 4.6 mm), increased the flow rate from 1.5 to 4.5 ml min^{-1}, and injected 5 mg of sample. The other conditions remained the same. As a result, we achieved the same separation pattern with significant time savings.

3.2 HPLC for Peptide Analysis

The characteristic absorption spectra of aromatic amino acids between 240 and 310 nm can be used to identify tryptophan-, tyrosine-, and phenylalanine-containing peptides. In an acidic solution, the absorption spectra of these amino acids exhibit minima or maxima at 255, 270, and 286 nm. Based on these characteristics, the content of the aromatic amino acids in peptides can be estimated [9].

3.2.1 Identification of Tryptophan-Containing Peptides

The peptide mixture was separated on a Vydac C18 column with a Hewlett Packard 1090 LC system. The exact conditions are described in Section 3.1.

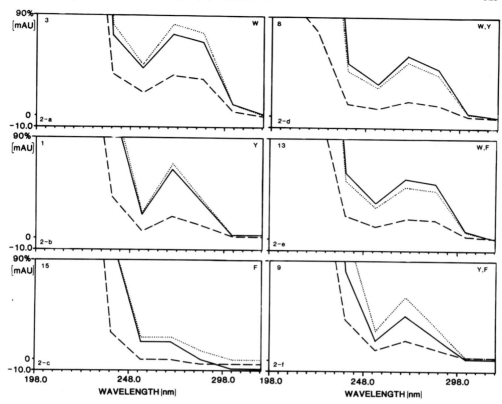

Fig. 3. Absorption spectra of aromatic amino acids containing peptides (range from 220 to 320 nm). The aromatic amino acids in each peptide are indicated on the right-hand side of the spectra. Numbers on the left-hand side of the spectra are the peak numbers according to Fig. 1 from [8]. Four spectra were stored by the system: at peak apex, at up and down slopes, and at the baseline after a peak. The base line spectrum was used as a reference for the other three spectra (apex · · · · · · , up ———, and down – – –)

Peptides were monitored by a photodiode array spectrophotometer at the center wavelengths of 220 nm and 280 nm (bandwidth of 60 nm), simultaneously using a Hewlett Packard Hook program. The peak detector used peak recognition techniques to store spectra automatically at certain points on chromatographic peaks. These points were peak apex, the inflection points on the ascending and descending portions of the peak, and the baseline immediately following a peak. Spectra were scanned from 200 nm to 320 nm. The valleys of the peaks were used as references for the spectra.

The spectra of the fractions which show positive absorption at 280 nm in Fig. 1 were recorded. The forms of the spectra differ between peptides containing different aromatic amino acids. Namely, tryptophan has two maximum absorption wavelengths $[\lambda_{max} A]$ at 270 nm and 286 nm, and one minimum absorption wavelength $[\lambda_{min} A]$ at 255 nm; tyrosine has one $\lambda_{max} A$ at 270 nm and one $\lambda_{min} A$ at 255 nm and phenylalanine has two $\lambda_{min} A$ at 255 nm and 286 nm and one $\lambda_{max} A$ at 270 nm. We found that they all have a common character, namely that the

spectra have either maxima or minima at 255, 270, and 286 nm [7]. As an
example, Fig. 3 shows the characteristic spectra of peptides containing different
aromatic amino acids [9]. According to the results of the spectra, six peptides,
shown in Fig. 1, contained tryptophan. Sequence analysis confirmed this finding
as shown in Fig. 5 except that peak 4 (or F30), which covers the same residue
positions as peptide S9, differs from S9 by an exchange of residues 69 and 70.

3.2.2 Comments

The method described here offers a rapid, direct, and sensitive procedure for the
identification of aromatic amino acid-containing peptides, especially the tryptophan-containing peptides. Since tryptophan is unstable under the 6 N HCl hydrolytic conditions, a special treatment for its analysis is often required to isolate
tryptophan-containing peptides [18, 19]; these procedures are usually cumber-
some and require substantial amounts of material. Using a photo diode array
detector, the tryptophan-containing peptide can be identified directly after
HPLC separation; therefore, it can be selectively isolated from complex mixtures.

4 Sequencing

4.1 Automatic Sequence Analysis

The sequence of longer peptides (more than 25 residues) was determined with a
gas-phase sequenator from Applied Biosystems. Additionally, the intact protein
was subjected to sequencing by a Beckman liquid-phase sequenator. PTH-amino
acids were identified on an IBM cyano column (250 × 4.6 mm) with the Beckman
HPLC system connected to a WATERS WISP auto sampler (model 710B) [7].

4.2 Manual Sequence Analysis

Sequence determination of shorter peptides (less than 25 residues) was performed
according to the method of Chang et al. [20] using a modified Edman reagent, di-
methylaminoazobenzeneisothiocyanate [DABITC]. DABTH-amino acids were
identified either by thin-layer chromatography [TLC] [21, 22] or HPLC method
[8].
    The modified Edman sequencing technique has increased sensitivity through
the chromophore of the reagent, which releases red-colored DABTH-amino acid
derivatives. The color reagent is advantageous for manual sequence analysis,
since the amount of peptide can be estimated during the whole process by its
color intensity. Using several modifications, we have applied this method to the
simultaneous sequencing of 20 or more polypeptides in amounts of 0.5 to 1 nmol.
The reagents and solvents used for the procedure are pyridine, phenylisothio-

cyanate [PITC], trifluoroacetic acid [TFA], dimethylaminoazobenzene-isothio-cyanate [DABITC], n-heptane, butylacetate, and acetone; all reagents, including water, should be of HPLC grade purity. The standard DABTH-amino acid mixture was prepared according to Chang et al. [22]. A Multi-Block Heater (Lab-Line Inst., Melrose Park, IL, USA) was used to maintain a constant reaction temperature.

4.2.1 Modified Edman Degradation

Reagent quantities given in the procedure apply for 20 samples; if more than 20 samples are sequenced, correspondingly more reagent should be prepared by the following procedure:

1. Prepare a DABITC solution by dissolving 28.2 mg of the reagent in 10 ml acetone. This solution need only be freshly prepared weekly.
2. Take one volume (1000 µl for 20 samples) of the DABITC solution and dry it in a Speedvac.
3. Dissolve the DABITC first in two volumes (2000 µl) of pyridine and then add one volume (1000 µl) of water.
4. Add 120 µl of this solution to each sample tube and mix well. The tubes (8 × 75 mm) have a conical bottom and can be capped with a plastic stopper (Radnoti Glass Tech. CA, USA).
5. Flush each sample with nitrogen for 5 to 10 s and cap the tubes.
6. Mix the samples again and incubate the samples for 50 min at 54 °C; mix the samples occasionally during this time.
7. After 50 min have elapsed, uncap the tubes, add 10 µl of PITC to each sample, and mix well.
8. Flush the samples again with nitrogen and cap the tubes.
9. Mix samples and incubate them for 30 min at 54°C, while mixing occasionally.
10. After incubation, extract excess reagent from the samples by adding 500 µl of a 2:1 solution of n-heptane: ethyl acetate to each tube.
11. Mix samples well and spin in Speedvac for about 2 min.
12. Discard upper phase (yellow) of each sample by using an aspirator.
13. Repeat steps (10) to (12) four to five times until upper phase is clear.
14. Dry lower phase of samples in a Speedvac.
15. Add 50 µl of anhydrous TFA to each sample, flush with nitrogen, and quickly cap sample tubes.
16. Mix samples well and incubate them for 15 min at 54 °C. Mix samples occasionally.
17. Dry samples in Speedvac for no more than 5 min and add 50 µl of water and 200 µl of n-butyl acetate to each sample; then mix well.
18. Spin tubes in Speedvac for 3 min and pipet the upper phase (the organic phase containing the derivatized amino acid) into a small Pyrex glass tube (6.5 × 50 mm).
19. Dry the samples and the organic phase in a Speedvac. The degraded sample, in the original glass tube, is now ready for the next Edman degradation step [beginning at step (2) if DABITC solution is already made up].

20. Add 50 µl of 30% TFA to each amino acid derivative; then flush with nitrogen and cap amino-acid sample tube.
21. Mix amino-acid sample well and incubate for 20 min at 54 °C.
22. Dry amino-acid sample in Speedvac. The amino acid derivative is now ready for identification by TLC.

4.2.2 DABTH-Amino Acid Identification

Two-dimensional TLC using polyamide sheets (cut to 3×3 cm squares, Schleichert & Schuell, Dassel, W. Germany) is run for the identification of DABTH-amino acids. The solvent systems are acetic acid: water (1 : 2, v/v) for the first dimension and toluene: n-hexane: acetic acid (2 : 1 : 1, v/v/v) for the second dimension. Use of a light box is recommended.

1. Apply amino acid standard mixture (about 20 pmol each) on the lower right corner of the sheet, about 0.5 cm from each side. Attempt to keep the spot at minimum size, as larger spots prevent good resolution of individual DABTH-amino acids in the chromatogram.
2. Turn the plate on its backside. Then, using a light box, apply the sample (dissolved in several µl of EtOH) directly on top of the DABTH-amino acid standard spot shining through from the frontside. Again, try to keep the spot as small as possible and reapply the sample until enough amounts (1 to 5 times of the sample depending on concentration) of the DABTH-amino acid are on the polyamide sheet. Then turn the plate so that the sample spot is now in the lower right corner and mark the top left corner for future reference with the sample number by using an unsharpened pencil (be careful not to damage the surface of the sheet).
3. Attach the polyamide sheet on the top side to a sheet holder (Fig. 4).
4. Fill a container with the first-dimension solution (33% acetic acid) to a height of no more than 0.4 cm and put in the polyamide sheet, Make sure that all of the lower side is wet and touches the liquid. Then quickly place an airtight cover on the container (Fig. 5).
5. Let the liquid rise almost to the top of the plate (3 to 5 min) and remove the sheets; be sure not to invert the plate while doing so.
6. Dry the bottom edge of the plate with tissue paper and blow-dry the plate for 15 to 20 min.
7. Reorient the plates so that the previously marked top left corner becomes the top right corner on the sheet holder.
8. Fill a second container with the second-dimension solution to a height of no more than 0.4 cm; then quickly put in the plates, making sure that all of the bottom edge is immersed in liquid. The container should be covered quickly after this. Since this solution is very volatile and small changes in concentration of the individual components of the solution due to evaporation can affect the separation of the DABTH-amino acids adversely, this part of the procedure should be carried out quickly.
9. Again, let the liquid move almost to the top (2 to 3 min). Then take out the sheets and dry the bottom edge with tissue. Blow-dry sheets for 15 min.

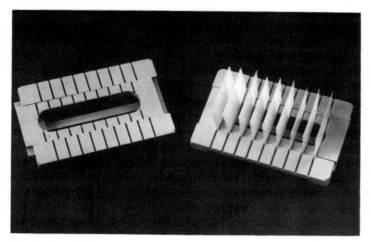

Fig. 4. Sheet holder for polyacrylamide thin layer chromatography (TLC)

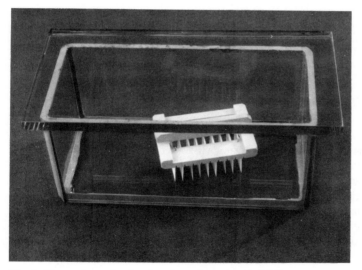

Fig. 5. Chromatography chamber for polyacrylamide TLC

10. Expose plates to concentrated HCl vapors for 2 to 3 s. This makes the DABTH-amino acid spots visible (pink).

The individual sample spots can then be identified using the DABTH-amino acid standard on the reverse side. To make TLC easier and more accurate, a sheet holder (made by Max-Planck-Institute, Goettingen, W.Germany) was utilized to keep nine sheets in a vertical position so that the known standard mixture on one side and the unknown on the other side could be run symmetrically, and thus the

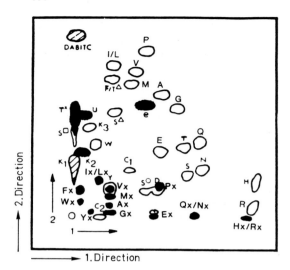

Fig. 6. Location of DABTH-amino acids (red spots) and their by-products (blue spots) on a polyamide sheet (3 × 3 cm) after two-dimensional thin-layer chromatography [20]. The solvent systems used for separation are described in the text. The positions of the derivatives are *circled* (represented by red), whereas the *solid* (represented by blue) and *hatched areas* (represented by purple) are the by-products. DABTH-residues are identified with the single-letter code of the corresponding amino acid. Characteristic side spots (blue) derived from the DABTH-amino acids are designated with an x; these usually appear with the main DABTH-derivative. Other by-products are: e = DABTC-diethylamine; C_1 = DABTH-carboxymethylcysteine; C_2 = DABTH-cysteic acid; U = N-dimethylaminoazobenzene-N-phenylthiourea; S°, $S^{\square}$, $S^{\triangle}$ = by-products of DABTH-Ser; T^x and $T^{\triangle}$ = by-products of DABTH-Thr; K_1 = α-DABTH-ε-DABTC-Lys; K_2 = α-PTH-ε-DABTC-Lys; K_3 = α-DABTH-ε-PTC-Lys

identification of the unknown residue was easily accomplished. Most DABTH-amino acid residues derived from this double-coupling method are accompanied by side spots on the chromatogram (which appear in blue). The chemistry of these side spots is still unknown; however, their location on the polyamide sheet after two-dimensional TLC was very characteristic. Through their positions, the interpretation of the DABTH results was simplified. A chromatogram with all 21 DABTH-amino acids and their side spots is shown in Fig. 6. DABTH-Ile and DABTH-Leu can be separated either by a one-dimensional separation on polyamide using a formic acid:ethanol:water (1:9:9, v/v/v) mixture [21] or by HPLC method [8].

4.2.3 Comments

1. This method is well suited for manual peptide sequencing, since the reagent chromophore, DABITC, gives the DABTH-amino acid derivatives a visible red color that can be followed easily during the whole manual process.
2. Using our procedure and the multi-sheet holder, a technician can work with 20 samples simultaneously for 1.5 cycles every day; thus, 30 residues of sequence can be determined each day.

3. The manual sequencing procedure can be interrupted only after a drying step. Thus, for samples to be stored, they should be removed from the sequencing cycle immediately following either step (14) or (22), where steps (1) to (14) and (14) to (22) comprise a 1/2 cycle each.
4. Except for arginine, the C-terminus of a peptide is difficult to identify using this manual method, due to loss of the derivatized hydrophobic residue during the extraction step. However, the C-terminus residue of a peptide can be deduced by amino-acid composition; by treating with carboxypeptidase A, B, and Y; or by hydrazinolysis.
5. The first-dimension solvent (acetic acid solution) does not need to be exchanged for up to 3 days. However, the solution for the second dimension should be replaced for each run and should be kept in a tightly capped bottle to achieve a constant two-dimensional chromatogram.

5 PTC-Amino Acid Analysis

Two systems, an LKB amino-acid analyzer and a WATERS Pico-Tag system, were available in our laboratory for determination of the composition of proteins and peptides. Rabbit apo A – I was analyzed by the LKB amino-acid analyzer using ninhydrin as the reagent, whereas peptides were analyzed by the WATERS Pico-Tag system using PITC for the derivatization of an amino-acid mixture before HPLC analysis. Because it was more sensitive and faster than the LKB system, we primarily used the latter. To achieve optimum results for routine analysis, we modified the procedure by using a Hewlett Packard 1090 HPLC system. The 22 PTC-amino acid derivatives can be identified in 22 min on a Spherisorb ODS II column (250×4.6 mm) [6].

The Pico-Tag technique involves three steps: (a) hydrolysis of the protein or peptide sample to yield free amino acids, (b) pre-column derivatization of the sample, and (c) analysis by reverse-phase HPLC. The procedures are as follows:

5.1 Hydrolysis of the Sample

1. The samples and 2 nmol of a standard amino-acid mixture (Pierce Chemicals) are dried in small Pyrex tubes (6.5×50 mm) and placed in a reaction vial with 200 µl of 5.7 N HCl and 20 µl of 0.1% phenol. As a check, a blank tube should also be included.
2. Seal the reaction vial under vacuum and hydrolyze at 150°C for 1 h or 110°C for 20 h.
3. Open the vial immediately after hydrolysis and remove the acid by vacuum.

5.2 Pre-Column Derivatization

1. Add 20 μl of redrying solution (1:1:1, ethanol:water:triethylamine) to each sample; then dry under vacuum.
2. Add 20 μl of derivatization reagent (7:1:2:1, ethanol:water:triethylamine:PITC) to each redried sample; then mix by vortexing for a few seconds, incubate at room temperature for 20 min, and place the samples under vacuum.
3. To redissolve the samples for injection, use a diluent having the following composition: 0.5 M Na_2HPO_4, pH 7.4 (titrated with 10% H_3PO_4) and 5% acetonitrile.

5.3 Reverse-Phase HPLC Analysis

The 2 nmol PTC-amino acid standard is reconstituted in 200 μl of sample diluent; of this solution, 3 μl (30 pmol) are injected into the column for calibration purposes, whereas 50% of the samples and the blank are applied for analysis (reconstituted in 30 μl and injection of 15 μl). The separation of the PTC-amino acids can either be achieved by a WATERS HPLC system (WATERS Pico-Tag operation manual, WATERS, MA) or a modified system, which has been developed in our lab using our Hewlett Packard HPLC system and a Sperisorb ODS II column. Our HPLC conditions are:

Column: Spherisorb ODS II 3 μm (150×4.6 mm).
Eluent: A: 0.153 M sodium acetate, 0.05% TEA, and 6% acetonitrile
                  (pH 6.4).
                  B: 40% water + 60% acetonitrile.
Flow rate: 0.8 ml/min.
Oven temp: 47°C.
Gradient: 0% to 12% B, 10.0 min; 12% to 48% B, 10.0 min; 48% to 100%
                  B, 0.1 min; isocratic at 100% B, 5.0 min; 100% to 0% B,
                  5.0 min.
Detection wavelength: 254 nm.

Figure 7 shows the separation of 30 pmol of PTC-amino-acid standard on Spherisorb ODS II colum by the HP-HPLC system with the conditions described above. All amino acid derivatives separated well under those conditions in 21 min.

5.4 Comments

a) Advantages of the PTC-amino acid analysis method include shorter analysis time, analysis of samples at the picomole level, and significant reduction of solvent consumption due to a slower flow rate.
b) Microscale analysis requires preparation of clean PTC-amino acid derivatives. Therefore, the reagent, the vial, the glass tubes, and the instrument itself should be kept scrupulously clean.

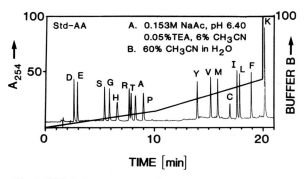

Fig. 7. HPLC chromatogram of a standard PTC-amino acid mixture (30 pmol each) on a Spherisorb ODS II 3 μm colum. Chromatographic conditions are described in the text. The solvent gradient is indicated in the figure. *D* aspartic acid; *E* glutamic acid; *S* serine; *G* glycine; *H* histidine; *T* threonine; *A* alanine; *P* proline; *R* arginine; *Y* tyrosine; *V* valine; *M* methionine; *C* cysteine; *I* isoleucine; *L* leucine; *F* phenylalanine; *K* lysine

c) The retention time of PTC-arginine can be increased by decreasing the salt concentration of buffer A. Hence, the position of the PTC-Arg on the chromatogram can be changed as desired in order to achieve a better separation. Using the Spherisorb ODS column and the conditions as described before, the PTC-Arg peak can be moved from its position between PTC-His and PTC-Thr to a position behind PTC-Pro by altering the sodium acetate concentration from 0.153 to 0.030 M, while the order of the other PTC-amino acids remains the same.

6 Results

The complete amino-acid sequence of rabbit apo A – I was obtained through alignment of staphylococcal and tryptic peptides (Fig. 8). All of the staphylococcal peptides of rabbit apo A – I were isolated and sequenced except the peptide from position 33 to 37. A second digestion with trypsin was employed to find overlapping and missing peptides.

6.1 Staphylococcal Peptides

Separation of the staphylococcal peptides S1 to S31 is shown in Fig. 1. The TFA buffer system provided a good separation of most of the liberated peptides. The fractions which were only partially purified and were subjected to rechromatography with the ammonium acetate buffer were S2, S10, S11, S17, S18, S21, S24, and S25. One chromatogram obtained by a second chromatography (Fig. 2) illustrates the separation of S10, S11, S18, and S25. We sequenced and analyzed peptides S5 and S7 as shown in Fig. 7, although these peptides resulted not from a cleavage after an aspartic acid or glutamic acid residue, but after a glycine residue.

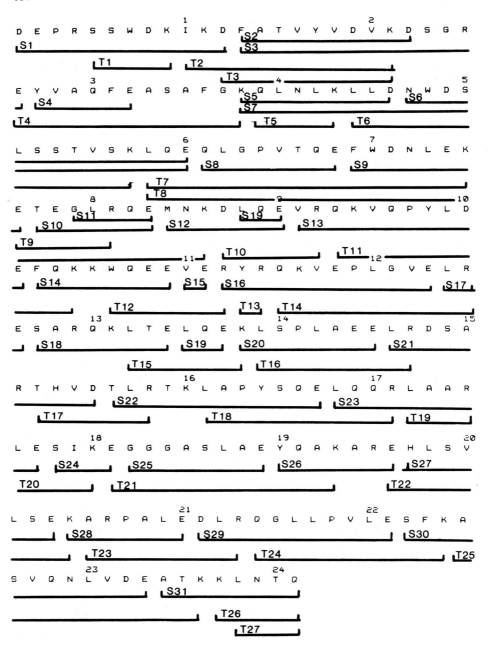

Fig. 8. Complete amino-acid sequence of apolipoprotein A – I from rabbit. Abbreviations: *S* digestion by *Staphylococcus aureus* protease; *T* tryptic digestion

6.2 Tryptic Peptides

The 27 major tryptic peptides, most of which were homogeneous, were numbered T1 – T27, as shown in Fig. 8. Since the tryptic peptides were used for overlapping and filling the rabbit apo A – I sequence, not all of the tryptic peptides from rabbit apo A – I were isolated and purified. However, sequence analysis of the isolated tryptic peptides permitted alignment of the staphylococcal peptides for the complete amino acid sequence of rabbit apo A – I.

6.3 Alignment of Peptides

As indicated in Fig. 8, the amino-acid sequence was found by aligning overlapping peptides of both digestions. The N-terminus of rabbit apo A – I was confirmed by sequencing the intact protein to the first 20 residues with a Beckman liquid-phase sequenator. The C-terminus of rabbit apo A – I was identified through its homology with that of human apo A – I reported by Brewer et al. [24]. The primary structure of rabbit apo A – I has now been identified and contains 241 residues with the sequence shown in Fig. 8.

7 Discussion

The sequencing of rabbit apo A – I was complicated by microheterogeneity that may have been caused by our use of pooled plasma. These microheterogenities were identified through yield calculations. The differing residues [$Tyr_{98} \rightarrow Phe_{98}$, $Val_{122} \rightarrow Ala_{122}$, and the inversion of Phe_{69} and Trp_{70}] were not included in Fig. 8. Like human apo A – I, rabbit apo A – I contains very little tryptophan (4), histidine (2), and methionine (1); it also contains only two residues of isoleucine. A comparison of rabbit apo A – I with human, canine [25], and rat apo A – I [26] shows that all apo A – I sequences exhibit extensive homologies to each other (Fig. 9). Rabbit apo A – I shares at a 78% homology to the sequence of human apo A – I [24]. Its homology with canine apo A – I is 80%, while the homology with rat apo A – I is a surprisingly low 61%. However, the overall homology of all 4 apo A – I proteins with respect to each other is very high.

All four tryptophan and all six tyrosine residues are conserved in all apo A – I sequences except rat apo A – I, which has one additional tryptophan at position 22 and lacks a tyrosine at position 106. The four tryptophan residues are located within the N-terminus half of apo A – I, whereas histidine residues are found in the C-terminus half of apo A – I. The leucine residues, which are known to be strong helix formers [27] on the basis of their circular dichroism [CD] [28], occur more frequently in apo A – I than any other amino acid. Over 90% of the leucine residues are conserved in specific positions of apo A – I.

Pownall et al. [3] and Yokoyama et al. [4] have reported the activation of LCAT by short amphiphilic synthetic peptides designed to have minimal amino-acid sequence homology with any segment of apo A – I. They concluded that the

TOTAL PRIMARY STRUCTURE OF RABBIT APOLIPOPROTEIN A-I

```
                        1                     2                     3
Rabbit  D E P R S S - W D K I K D F A T V Y V D - V K D S G R E Y V
Human           P Q P     R V     L         V L             D
Canine          - Q S P   R V     L         A V             D
Rat             - Q   Q   R V     F         A V             D
        + + - * + + * - + + - + + - - - - - + - - + + + 0 + + - - -
                          4                     5                     6
Rabbit  A Q F E A S A F G K Q L N L K L L D N W D S L S S T V S K L
Human   S       Q G     L                   V T       F S
Canine  A       E A     L                   L S       V T
Rat     S       E S   T L       N           T L G     V G R
        * + - + * - - - 0 + + - + - + - - + + - + * - * + - - * + -
                        7                     8                     9
Rabbit  Q E Q L G P V T Q E F W D N L E K E T E G L R Q E M N K D L
Human   R     L                             G               S
Canine  R     I                             V               S
Rat     Q     L               A           D W       N       N
        + + + - 0 - - - + + - - * + - + + + - + * _ + + + - + + + -
                        10                    11                    12
Rabbit  Q E V R Q K V Q P Y L D E F Q K K W Q E E V E R Y R Q K V E
Human   E     K A             D               M   L               E
Canine  E     K Q             D               V   L               A
Rat     E N   K Q   M     H   E   E       N       V   A       L E
        + + - + * + - + - - - + + - + + + - + + + - + * - + + + - *
                        13                    14                    15
Rabbit  P L G A E L R E S A R Q K L T E L Q E K L S P L A E E L R D
Human       R A     Q   G         H               G Q Q M
Canine      G S     R   G         Q               G E E L
Rat     E   G T     H K N - -   - - M   R H   K V V A E E F
        - - * * + - * + + - + + + - * + - + + + - + - - * + + - + +
                        16                    17                    18
Rabbit  S A R T H V D T L R T K L A P Y S N E L Q Q R L A A R L E S
Human   R     A H V     A   T H         D E   R Q           E A
Canine  R     T H V   A     A Q         D D   R E           Q A
Rat     R M   V N A   A     A K F G L   D Q M R E N     Q   T E
        + - + - * - + - - + - * - * - - + + - + + + - - * + - * *
                        19                    20                    21
Rabbit  I K E G G G A S L A E Y Q A K A R E H L S V L S E K A R P A
Human   L   N       R       H   K T   H     T S         K
Canine  L   G       S       H   R   S Q     A G         R
Rat     I R - N H P T -   I     H T K G D H   R T   G   K
        - + + * * * - + - - + - * - + - * + * - + - - * + + - + - -
                        22                    23                    24
Rabbit  L E D L R Q G L L P V L E S F K A S V Q N L V D E A T K K L N T Q
Human       E                     K V     F K S A K E       T         T
Canine      Q                     K V     L L A A I D       A         A
Rat         D       G       M     A W K A K I M S M I D     A K       A -
        - + + - * + 0 - - - - - + * - + - + - * * - * + + - * + - + - +
```

Fig. 9. Comparison of the amino-acid sequence of rabbit apo A–I with the published sequence of human [24], canine [25] and rat [26] apo A–I. Only sequences different from that of rabbit apo A–I are indicated. The hydrophobic-hydrophilic character [31] of the amino-acid sequence is represented as follows: hydrophobic (−), hydrophilic (+), neutral (○), and uncertain (*)

PEPTIDE	SEQUENCE AND HYDROPHOBICITY
1	V S S L L S S L K E Y W S S L K E S F S – + + – – + + – + + – – + + – + + + – +
2	P K L E E L K E K L K E L L E K L K E K L A – + – + + – + + + – + + – – + + – + + + – –

Fig. 10. Simplication of Peptide 1 [3] and Peptide 2 [4] amino-acid sequence with hydrophobic (–) and hydrophilic (2) characters

Table 1. Five regions of highest homology in apo A – 1 derived from the comparison of apo A – 1 with the synthetic peptides 1 and 2. The homologies, in parentheses, are expressed as a percentage where 100% corresponds to identical structures

Peptide	Position and percentage of homology				
	1	2	3	4	5
1	90 – 109 (85%)	68 – 87 (80%)	72 – 91 (80%)	79 – 98 (75%)	101 – 120 (80%)
2	88 – 109 (82%)	99 – 120 (82%)	77 – 98 (77%)	17 – 38 (77%)	66 – 87 (73%)

presence of a sufficiently large amphiphilic structure is sufficient for LCAT activation. At the same time, amphiphilic helices are also the same functional domains that are responsible for lipid binding [29]. All of the different specific apo A – I proteins that have been isolated and studied to date are known to activate LCAT [30]. Consequently, the functional domain must be fairly well conserved in the species studied.

To simplify analysis of similar structures between the synthetic peptides and apo A – I, we classified the amino acids as either hydrophobic, hydrophilic, neutral, or uncertain. After this simplification (Figs. 9 and 10), we found via a computer fit that these two synthetic peptides (reported by Pownall et al. [3] and Yokoyama et al. [4]) and apo A – I share similarities in the sequential order of hydrophilic-hydrophobic amino acids [31]. Table 1 shows the five segments of apo A – I that are most similar to the synthetic peptides. Of the five, four were in the same region of the protein. All of these segments are located within the N-terminus half of the protein. It is possible that, in a protein as large as apo A – I, more than one amphipathic section can activate LCAT. We suggest that the N-terminus region contains the major portion of these activating regions.

8 Outlook

It has been demonstrated that HPLC can play an important role in the purification of peptides, identification of amino-acid derivatives, and quantitation of amino-acid composition for protein microsequencing. The trifluoroacetic acid (TFA), ammonium acetate, and phosphate buffer systems are well suited for peptide separations. The combination of the TFA system with the ammonium acetate or phosphate system facilitates the isolation of pure peptides in a shorter period of time. In order to increase detection sensitivity and to save solvent and time, use of the microbore column has increased. Commercial columns are available from a number of different companies. Since the column flow rate is proportional to the product of the linear velocity and the column cross-sectional area, solvent consumption can be reduced proportionally to the square of the column radius [32]. Thus the microbore column reduces solvent consumption and increases mass sensitivity. The use of microbore columns for analysis of amino-acid derivatives could increase detection sensitivities to the fentomol level, so that the amount of material necessary for sequencing studies could be further reduced.

Acknowledgments. We thank Billy Touchstone for automatic sequence analysis, Susan Kelly for the artwork, and Marjorie Sampel for her assistance in the preparation of this manuscript. This work was supported in part by the Specialized Center of Research in Arteriosclerosis Grant (HL-27341) and a grant from the American Heart Association, Texas Affiliate (85G-202).

References

1. Schaefer EJ, Eisenberg S, Levy RJ (1978) J Lipid Res 19:667 – 687
2. Fielding CJ, Shore VG, Fielding PE (1972) Biochem Biophys Res Commun 46:1493 – 1498
3. Pownall HJ, Hu A, Gotto AM Jr, Albers JJ, Sparrow JT (1980) Proc Natl Acad Sci USA 77:3154 – 3158
4. Yokoyama S, Fukushima D, Kupferberg JP, Kezdy FJ, Kaiser ET (1980) J Biol Chem 255:7333 – 7339
5. Elzinga M (ed) (1982) Methods in protein sequence analysis. Humana, Clifton, NJ
6. Yang CY, Sepulveda FI (1985) J Chromatogr 346:413 – 416
7. Hewick RM, Hunkapillar MW, Hood LE, Dryer WJ (1981) J Biol Chem 256:7990 – 8025
8. Yang CY, Wakil SJ (1984) Anal Biochem 137:54 – 57
9. Yang CY, Pownall HJ, Gotto AM Jr (1985) Anal Biochem 145:67 – 72
10. Yang CY, Yang TM, Pownall HJ, Gotto AM Jr (1986) The complete primary structure of apolipoprotein A – I from rabbit high density lipoprotein. Eur J Biochem (in press)
11. Jackson RL, Gotto AM Jr (1972) Biochim Biophys Acta 285:36 – 47
12. Baker HN, Gotto AM Jr, Jackson RL (1975) J Biol Chem 250:2725 – 2738
13. Drapeau GR, Bioly Y, Houmard J (1972) J Biol Chem 247:6720 – 6726
14. Yang CY, Pauly E, Kratzin H, Hilschmann N (1981) Hoppe-Seyler's Z Physiol Chem 362:1131 – 1146
15. Kratzin H, Yang CY, Krusone J, Hilschmann N (1980) Hoppe-Seyler's Z Physiol Chem 361:1591 – 1598
16. Kratzin H, Yang CY, Goetz H, Pauly E, Koelbel S, Egert G, Thinnes FP, Wernet P, Altevogt P, Hilschmann N (1981) Hoppe-Seyler's Z Physiol Chem 362:1665 – 1669
17. Yang CY, Kratzin H, Goetz H, Thinnes FP, Kruse T, Egert G, Pauly E, Koelbel S, Wernet P, Hilschmann N (1982) Hoppe-Seyler's Z Physiol Chem 363:671 – 676
18. Lundhlad RL, Noyes CM (1984) Anal Biochem 136:93 – 100
19. Sasagawa T, Titanik K, Walsh KA (1983) Anal Biochem 134:224 – 229
20. Chang JY, Brauer D, Wittmann-Liebold B (1978) FEBS Lett 93:205 – 214
21. Yang CY (1979) Hoppe-Seyler's Z Physiol Chem 360:1673 – 1675

22. Chang JY, Creaser EH, Bentley KW (1976) Biochem J 153:607 – 611
23. Heinrikson RL, Meredith SC (1984) Anal Biochem 136:65 – 74
24. Brewer HB Jr, Fairwell T, LaRue A, Ronan R, Hauser A, Bonzert TJ (1978) Biochem Biophys Res Commun 80:623 – 630
25. Chung H, Randolph A, Reardon I, Heinrikson RL (1982) J Biol Chem 257:2961 – 2967
26. Poncin JE, Martial JA, Gillen JE (1984) Eur J Biochem 140:493 – 498
27. Chou PY, Wells M, Fasman GD (1973) Biochemistry 11:3028 – 3043
28. Chou PY, Fasman GD (1973) J Mol Biol 74:263 – 281
29. Segrest JP, Jackson RL, Morrisett JD, Gotto AM Jr (1974) FEBS Lett 38:247 – 253
30. Chen CH, Albers JJ (1983) Biochim Biophys Acta 753:40 – 46
31. Levit M (1976) J Mol Biol 104:59 – 107
32. Scott RPW (ed) (1984) Small bore liquid chromatography columns. Wiley, New York

7.4 Sequence Analysis of Complex Membrane Proteins (Cytochrome c Oxidase)

G. Buse, G. J. Steffens, G. C. M. Steffens, L. Meinecke, S. Hensel, and J. Reumkens[1]

Contents

1 Introduction

In recent years, amino-acid sequence analysis has been extended to hydrophobic proteins, especially those which are constituents of biological membranes and are known as integral or intrinsic membrane proteins. Both expressions refer to the fact that parts of the structures of such proteins extend into and more often penetrate the lipid bilayer of biomembranes, giving rise to a firm association between the hydrocarbon chains of the fatty acids or cholesterol and protein domains made up from about 12 mostly hydrophobic amino acids out of the 20 known to build the protein structure. Integration of these residues into the membrane − most often in ~20-membered helical conformation − is governed by the free energy $(-\Delta G)$ of their transfer from a random coil configuration in the aqueous phase into an ordered (helical) conformation in the lipid phase [1].

---

1 Fachgebiet Molekulare Biologie der Proteine, Abteilung Physiologische Chemie, RWTH Aachen, Klinikum Pauwelsstraße, D-5100 Aachen, FRG

Advanced Methods in Protein Microsequence Analysis
Ed. by B. Wittmann-Liebold et al.
© Springer-Verlag Berlin Heidelberg 1986

Table 1. Protein components of bovine heart cytochrome c oxidase

Poly-peptide	M_r	Syn-thesis	Hydrophobic domains	Stoichio-metry	N-terminal sequences
I	56993[a]	Mit.	ca. 12, core	1	Formyl-Met-Phe-Ile-Asn-
II	26049	Mit.	2	1	Formyl-Met-Ala-Tyr-Pro-
III	29918[a]	Mit.	ca. 6, core	1	(Met)-Thr-His-Gln-
IV	17153	Cyt.	1	1	Ala-His-Gly-Ser-
V	12436	Cyt.	None	1	Ser-His-Gly-Ser-
VIa	10668	Cyt.	None	1	Ala-Ser-Gly-Gly-
VIb	9419	Cyt.	1	1	Ala-Ser-Ala-Ala-
VIc	8480	Cyt.	None	1	Ser-Thr-Ala-Leu-
VII	10063	Cyt.	None	1	Acetyl-Ala-Glu-Asp-Ile-
VIIIa	5441	Cyt.	1	1	Ser-His-Tyr-Glu-
VIIIb	4962	Cyt.	1	2	Ile-Thr-Ala-Lys-
VIIIc	6243	Cyt.	1	1	Phe-Glu-Asn-Arg-

[a] From mtDNA sequence [5].

Difficulties encountered with the proteinchemical investigation of these membrane proteins generally start with their isolation and purification, which require detergent solubilization steps and often forbid the use of effective separation methods such as ion-exchange chromatography or electrophoresis. These handicaps pertain directly to the sequence analysis of the proteins itself.

One of the first and most significant examples of the protein chemical investigation of a membrane protein was that of monomeric bacteriorhodopsin [2] from the purple membrane of *Halobacterium*. Our investigations have been mainly concerned with cytochrome c oxidase, a complex oligomeric enzyme from the inner membrane of mitochondria [3], which is taken as an example here (Table 1). This enzyme, known as complex IV of the respiratory chain, is made up of 12 different protein components with M_r ranging from 5,000 to 57,000 [4]. The three largest and most hydrophobic of these are coded for by mitochondrial genes.

2 Preparative Isolation of a Membrane Enzyme Complex and Purification of the Subunits (Polypeptides)

2.1 Preparation of Membranous Enzyme Complexes

No attempt will be made here to describe the preparation of a specific membrane enzyme. It is, however, well known that preparative isolation and purification of membrane proteins require solubilization procedures, i.e., the disruption of the imbedding biological membranes by detergent action (e.g., with Triton X-100, cholate, desoxycholate, laurylmaltoside). In the final preparation these detergents replace most, but not all of the natural lipids of the membrane at the surface of principally intact proteins. Solubilization and the following purification steps, for instance repeated differential fractionation in ammonium sulfate again in the presence of detergents, is without principal difficulty only in the case of monomeric proteins. However, in the case of enzyme complexes, such as those constituting the respiratory chain of mitochondira and the ATP synthase, the immediate question arises whether on disruption of the membrane and the subsequent preparatory steps using varying detergents, the biochemist is left with a faithful representation of what originally existed. Thus the number of subunits in the functional monomer of bovine heart cytochrome c oxidase has been investigated in some hundred papers [6], values from 2 [7] to 15 [8] have been reported. Enzymatic activity is but one criterion to answer the above question. Activity is often difficult to determine if it needs the reconstitution of an intact membrane, e.g., to measure a membrane-transport process. There is also the risk of missing regulatory subunits, which could have been lost together with parts of the membrane in the course of an overdone "purification". It is generally accepted that a complex functional association in situ is stoichiomeric, and this has to be preserved in the final preparation. The unequivocal protein chemical characterization and demonstration of an integral subunit stoichiometry, as well as the enzymatic activity, are both mandatory criteria for an intact preparation. In fact, these criteria have not been met or been demonstrated in many membrane enzyme complex preparations. These preparations are seldom monodisperse, which may be one reason for the many futile experiments aiming at the three-dimensional crystallization of complex membrane proteins.

2.2 Preparative Isolation and Purification of Protein Components of an Oligomeric Protein

Isolation of the protein components of an oligomeric complex may aim either at obtaining subunits in a native state for functional studies or at providing polypeptides in a denatured form for further protein chemical investigation.

In cytochrome c oxidase, the isolation of functional, e.g., copper or heme-binding or H^+ translocating subunits, has so far not been achieved. Isolation of the constituting polypeptides is possible by (preparative) sodium dodecyl sulfate gel electrophoresis. For large-scale purification of the hydrophobic polypeptides of cytochrome c oxidase and other membrane polypeptides, we have introduced the chromatography on Bio-Gel columns in the presence of unbuffered neutral aqueous 2 – 5% sodium dodecyl sulfate solutions. The polyacrylamide matrix, together with the under these conditions enhanced critical micellar concentration [9] and therefore higher deodecyl sulfate monomer activity, provide a sufficient means for the dissociation even of very hydrophobic polypeptides. Pre-incubation of the sample in the SDS solution (with or without β-mercaptoethanol depending on the cysteine status) is recommended. Elevated temperatures should, however, be avoided, as these often lead to irreversible aggregation of hydrophobic proteins. By choosing an appropriate separation range of the gel, at least a suitable prefractionation should be possible even for rather complex membrane proteins. In the case of cytochrome c oxidase preparations with low contents of lipids and detergents we have used short (5 ×35 cm) Bio-Gel P-60 and P-100 columns for one-step purification in micromole quantities of the four

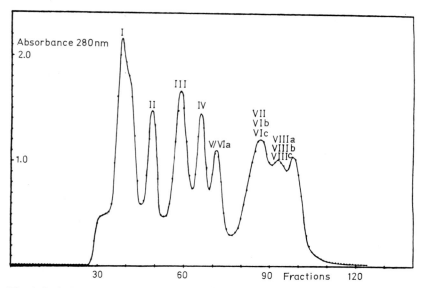

Fig. 1. Isolation of polypeptides of beef heart cytochrome c oxidase on a Bio-Gel P-100 column (5×35 cm, minus 400 mesh) in 3% SDS. Preparation of the sample (100 mg) and the chromatography were performed as described in Table 1. *Roman numerals* refer to polypeptides as given in Section 1

Table 2. Isolation of hydrophobic polypeptides from oligomeric membrane proteins by chromatography on Bio-Gel

---

1. Preparation of Bio-Gel columns: weigh 40 g of Bio-Gel P-100 (−400 mesh, Bio-Rad Laboratories) or, according to the molecular weight of the polypeptides to be separated, P-60 or P-150 and swell in about 2 l of bidistilled water. After most of the gel material has been settled, the upper layer remains cloudy (fines) and is taken off by suction. Add again 2 l of water, stir gently and take off the upper layer after the gel material has settled. Repeat this procedure about four times, till a clear upper layer is achieved. Take off the clear upper layer and replace by 2 l of 3% SDS. Repeat this procedure once. The total procedure takes about 8 h.

2. Take off the upper layer and pour the gel slurry into the column (5×35 cm). Adjust a flow rate of 0.5 ml min^{-1}; a hydrostatic pressure of about 80 cm is appropriate. Equilibrate overnight under these conditions. Attention must be paid to prevent a too high hydrostatic pressure; do not use a peristaltic pump.

3. Dissolve the membrane protein (ca. 100 mg or less) in 5 ml of 3% SDS, containing, if necessary, 0.1 ml of β-mercaptoethanol. Incubate 1 h at room temperature.

4. Stop the column flow. Layer the protein solution with a teflon tube on top of the gel and start the chromatography. Collect fractions of 4 ml and read absorbance at 280 nm. Pool fractions corresponding to a specific polypeptide and lyophilize. Check homogeneity by SDS-polyacrylamide gel electrophoresis.

---

largest components, among them the extremely hydrophobic polypeptides I and III (Fig. 1 and Table 2).

As with SDS-PAGE of hydrophobic proteins [10], the apparent M_r's obtained from the column experiments are, as compared to the more hydrophilic proteins or the usual test proteins, a gross underestimation of the exact values obtained from the primary structures: polypeptides I: app. $M_r \approx 36{,}000$, $M_r \approx 56{,}993$; and III: app. $M_r \approx 22{,}000$; $M_r \approx 29{,}918$ (see Sect. 1).

2.3 Removal of Sodium Dodecyl Sulfate from the Membrane Proteins

Using the procedures described above, separated polypeptides are obtained in relatively highly concentrated (2 – 5%) solutions of SDS. Removal of most of this detergent is necessary for further protein chemical investigation. Again here procedures described in the literature [11] for hydrophilic water-soluble proteins cannot be applied to hydrophobic proteins as these are insoluble in aqueous media. The way which we have chosen (Table 3) starts with lyophilization of the pooled fractions. The protein is thus obtained as a powder of a few mg in grams of SDS. A first gross removal of SDS can then be achieved by chromatography over a column of Bio-Gel P-4 equilibrated with unbuffered 0.01% SDS. The unbound and probably also some bound SDS is thus removed by molecular sieving, the protein remains in a solubilized form. Lyophilization of the column eluate yields a material with about equal amounts of protein and SDS. Further removal of SDS can be done by extracting once or twice with cold ethanol. Caution should be taken at this step, since losses may occur with some small hydrophobic proteins, such as the so-called proteolipid (subunit) IX of ATP synthase. This should be controlled by inspection of the UV absorption of the extract. The protein is left behind as a precipitate suitable for amino acid or sequence analysis. It is favorably suited for enzymatic cleavages probably because a few molecules of SDS remain bound to hydrophobic protein segments.

Table 3. Removal of excess SDS from polypeptides isolated in 3% SDS

1. Weigh about 85 g of Bio-Gel P-4 or P-6 (200 – 400 mesh, Bio-Rad Laboratories) and swell in about 2 l of 0.01% SDS.

2. Pour the gel slurry into the column (5 × 35 cm) and equilibrate overnight with 0.01% SDS at a flow rate of 2 ml min^{-1}.

3. Dissolve the lyophilized material obtained from the chromatography on the Bio-Gel P-100 column in 4 – 5 ml bidistilled water.

4. Stop the column flow and layer the dense solution of the isolated polypeptide on top of the gel. Start the chromatography, the flow rate should be 2 ml min^{-1}.

5. Collect fractions of 4 ml and read absorbance at 280 nm. In the case of drop counting, reduction of the fraction volume indicates the elution of the bulk SDS from the column. Alternatively a conductivity meter equipped with a flow cell may be used to indicate the elution of the SDS not bound to the protein.

6. Lyophilize the pooled fractions.

7. Extract, if desired, the lyophilizate with 2 × 5 ml of cold ethanol. Spin at 5000 r.p.m. Check eventual losses of protein material by spectrophotometric control of the supernatant.

8. The pellet can either be lyophilized or directly used for protein chemical investigations.

3 Sequence Analysis of Hydrophobic Proteins

3.1 Cleavage of Membrane Proteins

In general all fragmentation methods used for sequence analysis of proteins may also be used with smaller, weakly hydrophobic membrane proteins, e.g., all the nuclear coded polypeptides of cytochrome c oxidase which have no more than one membrane domain. The amphiphilic subunits of the respiratory enzyme complexes (e.g., subunit II of the oxidase) are also accessible by these methods.

Difficulties occur, however, with large hydrophobic proteins having numerous membrane-bound and extended inner-protein domains. The most successful methods for preparing complete sets of fragments of suitable length for automated sequencing, without leaving behind an unfragmented core, are those methods which cleave at hydrophobic residues and thereby fragment hydrophobic domains. The well-known chemical cleavage with CNBr at methionyl residues [12] alternatively in combination with the cleavage at tryptophan [13] and digestion with chymotrypsin (cleavage preferentially at Tyr, Trp, Phe) have been most successful in our hands. The latter proved especially effective in the presence of 0.001% sodium dodecyl sulfate with proteins obtained from the procedures described in Sections 2.2 and 2.3. Chymotryptic cleavage under these conditions also avoids the formation of fragments too large for an effective separation on reverse-phase HPLC columns.

3.2 Separation of Fragments Obtained from Membrane Proteins

If the number of peptides to be theoretically expected from the cleavage of the protein is not too large, a separation can be achieved over Bio-Gel columns (P-10 or P-30) with 10% or 70% acetic acid as eluant. The smaller nuclear coded polypeptides of cytochrome c oxidase can also be isolated with this method [14]. Successive chromatographies in these eluents with those peptides that are soluble in the respective solvent may also be useful. A further step may be the use of up to 90% formic acid; however, the Bio-Gel material is hardly stable under these conditions and break-down products appear in the samples for amino acid and sequence analysis. Also formic acid may lead to blocking of the peptide N-termini.

High performance liquid chromatography is the method of choice if a large number of peptides is to be isolated, as in the case of the 57 kD subunit I of cytochrome c oxidase (Fig. 2 and Table 4). The numerous chymotryptic fragments of this polypeptide are reproducibly resolved into about 90 peaks (Fig. 3), giving peptides covering the entire chain of this protein. In both this and the above-mentioned separation procedure, if the peptide mixture obtained on cleavage is not readily soluble in the chromatographic solvent, a few minutes of sonication can be helpful. Note that the use of organic acid eluants forbids registration at the main peptide bond absorption. At 228 nm (LKB Uvicord SII) absorbance results from the peptide bond as well as from contributions of Trp, Try, His, Met, Cys and Phe. Therefore even small peaks in the peptide pattern may be important.

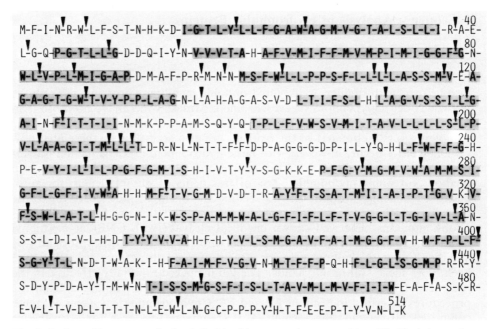

Fig. 2. Amino-acid sequence of subunit I of beef heart cytochrome c oxidase [5]. *Shaded areas* indicate the main hydrophobic domains. *Arrows* show main sites of chymotryptic cleavage

Table 4. Chymotryptic cleavage of hydrophobic proteins and separation of the fragments by high performance liquid chromatography

1. Dissolve about 10 mg of the polypeptide from which the bulk SDS has been removed by chromatography on Bio-Gel P-4 (see Table 3), in 10 ml 0.1 M ammonium hydrogen carbonate. Due to the presence of residual SDS the protein in most cases readily dissolves.

2. Dilute to 100 ml with 0.1 M ammonium hydrogen carbonate and adjust the pH to 8.5.

3. Dissolve 1 mg chymotrypsin in 1 ml 0.1 M ammonium hydrogen carbonate and add 0.3 ml to the protein solution. The cleavage is performed at 37 °C under gentle shaking. After 5 h another 0.3 mg of chymotrypsin are added. The cleavage reaction is stopped by lyophilization after a total reaction time of 10 h.

4. Dissolve the lyophilized, still SDS-containing, material in 2 ml 50 mM ammonium acetate pH 7.0. Sonicate several minutes and pass through a 0.2 μ filter. 200 μl are chromatographed on a HPLC column filled with wide pore material (Vydac TPRP, 300 Å, 10 μ, 0.4 × 25 cm) at a flow rate of 0.7 ml min^{-1}. Use a gradient of 0 – 60% acetonitrile in 50 mM ammonium acetate pH 7.0. Other systems such as 0.1% trifluoracetic acid/acetonitrile and 0.1 M ammonium hydrogen carbonate/acetonitrile pH 8.5 may be used as well.

5. Read the absorbance at 228 nm and collect fractions manually. Be aware of a delay time between detection and actual collection of the fragments.

6. Rechromatograph those fractions, which appear to be heterogeneous in 0.1% trifluoroacetic acid or 0.1 M ammonium hydrogen carbonate pH 8.5 on the same column.

7. Very hydrophobic fragments, which may not dissolve in either of the mentioned systems, are dissolved in 80% formic acid and chromatographed in 0.1% trifluoroacetic acid/0 – 60% 1-propanol.

3.3 Sequencing

For automated sequencing [15] we have used the liquid phase machine from Beckman; either the 890 C version with samples from 30 – 300 nmol with hand conversion of the 2-anilino 5-thiazolinone amino acids [16] or the 890 M version with programmer, single pump cup vacuum and Beckman autoconversion (samples 300 pmol to 30 nmol). The conversion in the latter is performed in the fraction collector at 55 °C by addition of ≈ 100 μl 25% TFA to the dried thiazolinone. The dry PTH-amino acids remain at ~ 30°C under vacuum (N_2) without problems for 1 or 2 days. Only minor amounts of PTH-serine and PTH-threonine are generally observed. However, with the (Polybrene) program provided by the manufacturer for peptide sequencing and addition of dithioerythritol to the solvents ethylacetate (S_2) and 1-chlorobutane (S_4) and the reagents heptafluorobutyric acid (R_3) and 25% trifluoroacetic acid (R_5), both amino acids are detected as derivatized PTH's in the HPLC identification (adopted from [17]), with two peaks from the allo-Thr and Thr isomers.

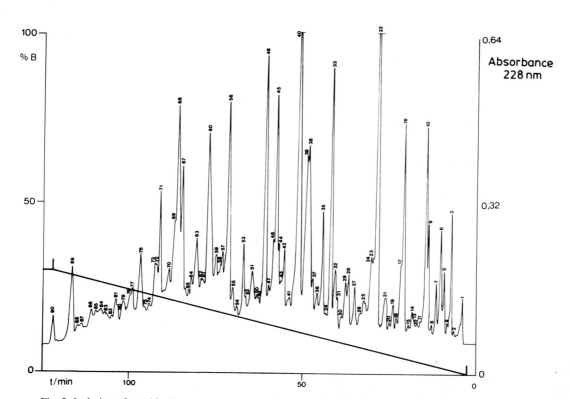

Fig. 3. Isolation of peptides from a chymotryptic digest of subunit I of bovine cytochrome c oxidase (see Table 4). The chromatography was performed at room temperature on a Vydac TPRP column (10 μ, 0.4 × 25 cm) using the system (A) 50 mM ammonium acetate, 0.1% trifluoracetic acid, pH 7.0; (B) acetonitrile and the indicated gradient. Flow rate: 0.7 ml min^{-1}

We have made much use of a peptide program published originally by Crewther and Inglis [18], which we adapted to the Beckman liquid phase machine in order to prevent losses of hydrophobic peptides during the ethylacetate (S_2) wash. The essential of this program, which uses 0.1 M Quadrol (R_2 or R_4) and no Polybrene, is the substitution of the ethylacetate (S_2) wash by a wash with 1-chlorobutane $+0.1\%$ acetic acid (S_3 or S_4 depending on program configuration). The wash time is adjusted so that the reaction cup above the undercut is finally clean of Quadrol, but a residual amount of Quadrol remains in the cup to stabilize the peptide. Quadrol is partially lost also with the 1-chlorobutane (S_4 or S_3) extraction. This can be seen from a small Quadrol induced peak following the PTH-amino acids in the HPLC chromatogram (Fig. 4). Correspondingly the program uses a low $\rightarrow$ high vacuum, N_2, drying step for heptafluorobutyric acid (R_3) at the end of the cleavage, instead of a restricted $\rightarrow$ low vacuum drying. If flow rates and step times are properly adjusted to the individual machine, sequencing through $30-40$ cycles to the C-terminal residue is generally possible with either peptide program. If, however, the 890 M has to perform with peptide or protein amounts of only a few 100 pmol, the repetitive yield falls to $\approx 92\%$ or less in our experience.

4 Determination of the Stoichiometry of Subunits in Oligomeric Protein Complexes

As outlined in Section 2.2, this question is of special significance for a proper characterization of oligomeric membrane proteins. Biosynthetic labeling with radioactive amino acids [19] is possible only with microorganisms. Quantitation of stain in SDS-gels [20] or C-terminal amino-acid detection methods such as hydrazinolysis [21] or carboxypeptidase treatment [22], either need quantitative separation of all the proteins of a complex or/and are not easily quantitatively performed.

4.1 Determination of Stoichiometry by Direct Edman Degradation of an Oligomeric Enzyme

After the gel chromatographic separation of all protein components of an oligomeric complex has been achieved and all the polypeptides have been characterized by their N-terminal sequence, a direct determination of their stoichiometry becomes possible by Edman degradation of the entire enzyme complex. The amount of protein introduced into the sequencer cup is determined. Quantitation of the HPLC separated PTH-amino acids is obtained over several cycles from their known extinction coefficients at 269 or 254 nm [23], and corrected for the repetitive yield of the degradation. Difficulties with quantitation of PTH-serine are thus circumvented. A practically quantitative release of the N-terminal residues in the first cycle is necessary. Here there may be an advantage in the use of a liquid phase sequencer instead of the gas-phase or solid-

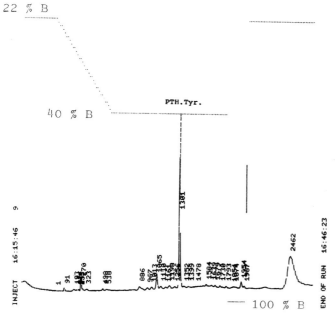

Fig. 4. Typical HPLC chromatogram (e.g., of PTH-tyrosine) with Quadrol peak from Edman degradation in a Beckman 890 c Sequencer, using a 1-chlorobutane + 0.1% acetic acid instead of ethylacetate wash. 2-Anilino-5-thiazolinone derivatives were converted in 20% trifluoroacetic acid at 55 °C during 25 min. PTH-amino acids were extracted with 2 × 1 ml ethylacetate and identified on a HPLC column (Zorbax ODS, 5 μ, 0.4 × 25 cm) at 62 °C using the system (A) 10 mM sodium acetate pH 4.5; (B) acetonitrile with indicated gradient

phase machines presently on the market. Without any subunit separation the result of this experiment gives a clearcut answer to the question of the stoichiometry of the subunits in a protein complex. Figure 5 gives an example taken from four degradation cycles of cytochrome c oxidase.

4.2 Determination of Stoichiometry from UV Absorption of Individual Polypeptides

Finally these data can be supplemented: On the basis of a complete set of primary structural data for all the polypeptides of the enzyme a direct calculation of the UV-absorption at 280 nm with tryptophan $\varepsilon_M^{280} = 5.559$ and tyrosine $\varepsilon_M^{280} = 1.197$ [24] of the individual polypeptides is possible. The absorption of these residues can be summed up and directly compared to the peak area obtained in the separations done on Bio-Gel columns under denaturing 3% SDS conditions. The stoichiometry of N-terminal blocked subunits in cytochrome c oxidase was thus obtained as well (see Sect. 1).

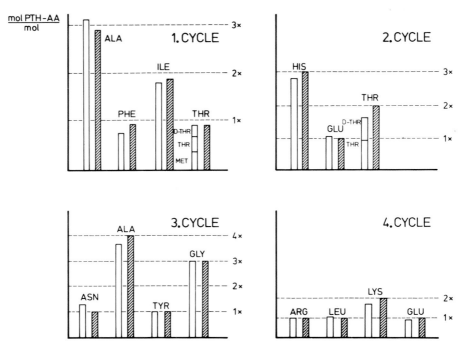

Fig. 5. Edman degradation of beef heart cytochrome c oxidase. Quantitation of significant PTH-amino acids (mol PTH-amino acid/mol oxidase) of the first four cycles from HPLC chromatogram. (For the system see legend to Fig. 4). *Open bars* represent the relative amounts of the PTH-amino acids released in the respective cycles, whereas the amounts calculated from the initial amount and the repetitive yield are shown as *shaded bars*

Acknowledgment. This work has been supported by the Sonderforschungsbereich 160 Biologische Membranen of the Deutsche Forschungsgemeinschaft.

References

1. Heijne G von (1981) Eur J Biochem 116:419–422
2. Ovchinnikov Y, Abdulaev W, Feigina M, Kiselev A, Lobanow N (1979) FEBS Lett 100:219–224
3. Capaldi RA, Malatesta F, Darley-Usmar VM (1983) Biochim Biophys Acta 726:135–148
4. Buse G, Steffens GJ (1978) Hoppe-Seyler's Z Physiol Chem 359:1005–1009
5. Anderson S, de Bruijn MHL, Coulson AR, Eperon IC, Sanger F, Young IG (1982) J Mol Biol 156:683–717
6. Kadenbach B, Ungibauer M, Jarausch J, Büge U, Kuhn-Nentwig L (1983) Trends Biochem Sci 8:398–400
7. Komai H, Capaldi RA (1973) FEBS Lett 30:273–276
8. Griffin DC, Landon M (1981) Biochem J 197:333–344
9. Helenius A, Simons K (1975) Biochim Biophys Acta 415:27–79
10. Nobrega FG, Tzagoloff A (1980) J Biol Chem 255Ü9828–9837
11. Weber K, Osborn M (1975) In: Neurath H, Hill R (eds) The proteins. Academic Press, London New York, pp 180–221
12. Gross E, Witkop B (1979) J Am Chem Soc 83:1510–1514

13. Ozols J, Gerard G (1977) J Biol Chem 252:5986 – 5989
14. Steffens GCM, Steffens GJ, Buse G, Witte L, Nau H (1979) Hoppe-Seyler's Z Physiol Chem 360:1633 – 1640
15. Edman P, Begg G (1967) Eur J Biochem 1:80 – 91
16. Biewald R, Buse G (1982) Hoppe-Seyler's Z Physiol Chem 363:1141 – 1153
17. Lottspeich F (1980) Hoppe-Seyler's Z Physiol Chem 361:1829 – 1834
18. Crewther WG, Inglis AS (1975) Anal Biochem 68:572 – 585
19. Sebald W, Machleidt W, Otto J (1973) Eur J Biochem 38:311 – 324
20. Merle P, Kadenbach B (1980) Eur J Biochem 105:499 – 507
21. Akabori S, Ohno K, Hanaka T, Olanda Y, Hanafusa H, Jaruan I, Tsugita A, Sugai K, Matsushima T (1956) Bull Chem Soc Jpn 29:507 – 514
22. Ambler RP (1967) In: Hirs CHW (ed) Methods in enzymology, Vol XI. Academic Press, London New York, pp 155 – 166
23. Edman P (1970) In: Needleman SB (ed) Protein sequence determination. Springer, Berlin Heidelberg New York pp 211 – 255
24. Sober AK (1979) In: Handbook of biochemistry, 2nd ed. CRC, Cleveland, Ohio, B-75

7.5 Sequence Studies on the α-, β-, and γ-Chains of Elongation Factor 1 from *Artemia*: Some Practical Notes

R. AMONS[1]

Contents

1 Introduction

In eukaryotic organisms, elongation factor 1 encloses three polypeptide chains, designated as EF-1α, EF-1β and EF-1γ. EF-1α, a 51,000 MW protein, is the eukaryotic equivalent of the prokaryotic, aminoacyl tRNA-carrying enzyme EF-Tu. Both EF-1α and EF-Tu can bind aminoacyl tRNA and GDP or GTP. EF-1β, having a MW of 26,000, corresponds to prokaryotic EF-Ts, the enzyme that exchanges in the EF-Tu. GDP-complex, GDP for GTP. The function of EF-1γ, which has no prokaryotic counterpart, is still unknown. This protein, which has a MW of 46,000, is normally tightly associated with the β-chain, forming the complex EF-1βγ. In the brine shrimp *Artemia,* a high molecular weight complex between the three polypeptide chains can be isolated from dormant cysts; in the free-swimming nauplii, however, free EF-1α is found predominantly. The significance of the complexing of EF-1α with the other two chains in dormant cysts is still unclear (for a recent review see [1]).

Recently, the primary structure of *Artemia* EF-1α has been elucidated in our laboratory, simultaneously via direct protein sequencing and via cDNA sequenc-

---

1 Sylvius Laboratories, Faculteit der Geneeskunde, Rijksuniversiteit Leiden, P.O. Box 9503, NL-2300 RA Leiden, The Netherlands

Advanced Methods in Protein Microsequence Analysis
Ed. by B. Wittmann-Liebold et al.
© Springer-Verlag Berlin Heidelberg 1986

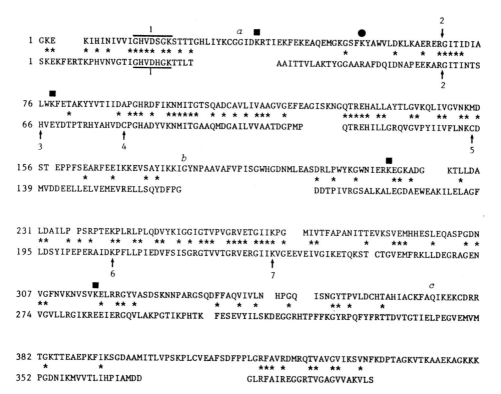

Fig. 1. Comparison of EF-1α from *Artemia* (*upper line*) with EF-Tu from *E. coli* (*lower line*). Homologous residues are represented by * [for the original papers see [2] (*Artemia*), and [3, 4] (*E. coli*)]. *1* Proposed phosphate binding site for nucleotides [6, 7]; *2* Site of limited tryptic digestion (Arg-68/Gly-69 in *Artemia* EF-1α [8], Arg-58/Gly-59 in *E. coli* EF-Tu [9, 10]); *3* His-66 (EF-Tu) is the target of the reaction with bromoacetyllysyl tRNA, whereas *4* modification of Cys-81 (EF-Tu) abolishes binding of aminoacyl tRNA; *5* modification of Cys-137 (EF-Tu) abolishes binding of GDP, GTP and EF-Ts (for a review see [11]); *6* Lys-208 and Lys-237 (both EF-Tu) are reported to be located in the vicinity of tRNA binding sites [12, 13]. ■ and ● are methylated amino acids in EF-1α from *Artemia* (see Sect. 4 and Fig. 6). cDNA sequencing (see [2]) revealed three differences compared with the sequence of EF-1α from *Artemia* as given here; *a* Ser, *b* Asp, *c* Glu (cf. also [19])

ing [2]. In Fig. 1, its primary structure has been schematically compared with that of of EF-Tu from *E. coli* which protein was previously sequenced [3 – 5]. The direct homology between these proteins amounts to about 25%. It is seen that regions in *E. coli* EF-Tu which have been correlated with certain enzymatic functions have been conserved in the eukaryotic protein as well (see legend of Fig. 1 for details). This conclusion could be confirmed when sequences of other "prokaryotic" [14, 15] and eukaryotic [16 – 19] EF-Tu/EF-1α proteins became available.

Current research in our laboratory is focused on the mapping of the nucleotide-binding regions of *Artemia* EF-1α, for which our experience with the elucidation of its primary structure is of great advantage, as well on the crystallization of the protein.

Less detailed information is available for the prokaryotic factor EF-Ts, although the amino-acid sequence of the *E. coli* protein has been deduced from its cDNA sequence [20].

In our laboratory, the primary structures of the proteins EF-1β and EF-1γ from *Artemia* are currently being investigated via cDNA sequencing techniques, and partially via direct protein sequencing. In addition, attention will be paid to localizing the binding sites between EF-1β and EF-1γ, and between EF-1β and EF-1α, using chemical modification techniques. Our familiarity with primary structure determinations will be of great help in these studies.

The present paper will be confined to practical notes on purification and sequencing methods, routinely used in this laboratory, in as much as these techniques could be also useful to other investigators.

2 Separation Procedures for Peptides Obtained from Elongation Factors (and Other Proteins)

2.1 Gel Filtration in 0.5% Trifluoroacetic Acid

Gel filtration on Sephadex G75 superfine, with 0.5% (v/v) trifluoroacetic acid as an eluent, was found to be very useful for the separation of BrCN-generated peptides and other (large) protein fragments. An example of the separation of the BrCN-generated peptides of S-pyridylethyl-EF-1α from *Artemia* is given in Fig. 2 (see [8] for another application). 0.5% (v/v) trifluoroacetic acid is an excellent solvent for many peptides, and it allows monitoring at 220 nm; in addition, it is

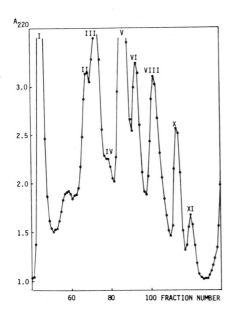

Fig. 2. Elution profile of the BrCN-generated peptides from S-pyridylethyl-EF-1α, soluble in 0.5% (v/v) trifluoroacetic acid. The column of Sephadex G75 superfine (1.6 cm × 110 cm) was eluted at 4 ml h^{-1} and fractions of 40 drops were collected [8]

Table 1. Gel filtration in 0.5% (v/v) trifluoroacetic acid

---

1. Swell Sephadex G 75 superfine in H_2O, remove fines as usual.

2. Add trifluoroacetic acid in small portions, while stirring with a glass rod, to make a final concentration of 0.5% (v/v). Deaerate the suspension.

3. Pour a column as usual. As the gel material becomes a little softer in the solvent used, it is important to keep the hydrostatic pressure lower than 120 cm of H_2O during the column preparation. Also, the diameter of the column should not be smaller than 0.8 cm.

4. Elute the column at a flow rate of $2-4$ cm h^{-1}, at room temperature or at 4 °C. Monitor the effluent at 220 nm.

5. As Sephadex G 75 superfine is not stable if kept for longer periods in 0.5% (v/v) trifluoroacetic acid, empty the column immediately after the run, and bring the gel suspension to pH 7 with dilute NaOH. It can then be stored indefinitely at 4 °C.

---

completely lyophilizable. A convenient procedure for column preparation and elution is given in Table 1.

2.2 Gel Filtration in the Presence of SDS

If peptides or proteins are found to be insoluble in aqueous solvents or form an aggregate, it is sometimes useful to apply gel filtration in the presence of SDS. Gel filtration media like Sephadex (G50 or G75, superfine), Sephacryl (S200, S300 or S400) all could be used, keeping in mind that the apparent molecular weight of the peptides to be chromatographed is increased at least two- or three-fold, due to binding of SDS. Applications can be found in [21]. Recently, we use high performance gel filtration on a Zorbax GF 250 column, in the presence of SDS, as a rapid and convenient procedure for the separation of the EF-1βγ complex of *Artemia* into the constituent protein chains (see Fig. 3) (see Sect. 2.5 for removal of SDS).

2.3 Reverse-Phase High Performance Liquid Chromatography

During the past 5 years, this technique has become the most important peptide separation procedure in virtually every laboratory; it needs therefore no further introduction. In Fig. 4, a few applications concerning the primary structure determination of the elongation factors are given.

2.4 Electroelution

In certain instances, e.g., for sequencing purposes, the need becomes apparent to elute proteins or peptides from polyacrylamide gels. It should be stressed that in these cases, (heavy) staining of the gels should be avoided as far as possible. Proteins separated on isoelectric focusing gels can often be visualized by immersing the gels for 30 min in 10% (w/v) trichloroacetic acid. The proteins precipitate in

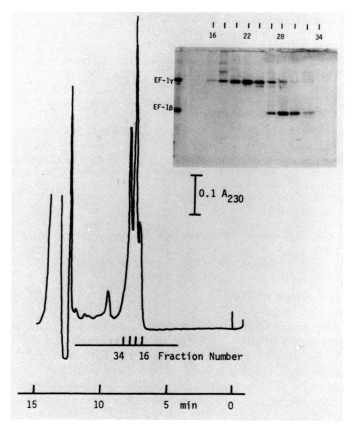

Fig. 3. High performance gel filtration on a Zorbax GF 250 column Dupont). About 20 μg of EF-1βγ
was chromatographed in 0.1 M sodium phosphate (pH 6.8) containing 1% (w/v) SDS. The flow rate
was 1 ml min^{-1}. *Inset:* Polyacrylamide gel electrophoresis [22] of the protein fractions indicated

the form of white discs, which are easily localized and cut from the gels. This
procedure does not work satisfactorily for SDS-containing gels. In that case, proteins can usually be visualized by immersing them for 15 min in a solution of
10 mg 8-anilinonapthalene sulfonic acid in 100 ml 0.1 M HCl; the fluorescent
protein bands can easily be localized under ultraviolet (366 nm) light (cf. [24]).
Finally, proteins to be separated on SDS-containing gels can be labeled with
fluorescamine prior to electrophoresis [25]; as the fluorescamine treatment does
not interfere in the determination of amino-acid composition and sequence, this
technique could be very useful, especially for peptide purification.

The excised gel slices can be stored at − 20 °C, or are processed immediately.
In that case, they are cut into small cubes, which are transferred into an Eppendorf tube, and incubated at room temperature with soaking buffer [0.5 M ammonium carbonate (pH 8.0), 2% (w/v) SDS and 0.2% (v/v) 2-mercaptoethanol]
for 2 h. During this incubation, the pH sould occasionally be checked, and be

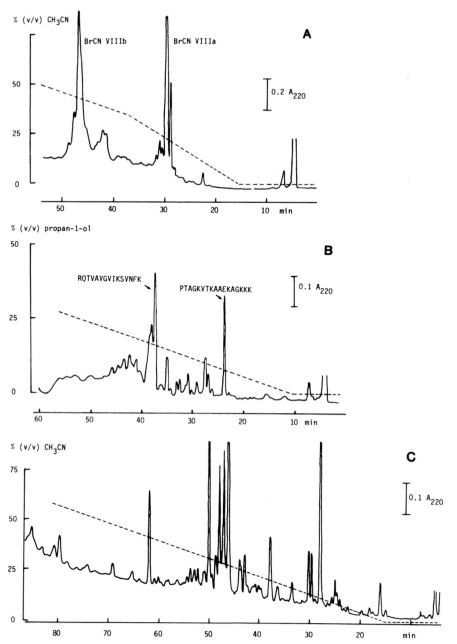

Fig. 4. A Separation of BrCN-peptide VIIIa from BrCN-peptide VIIIb with reverse-phase high performance liquid chromatography (RP-HPLC). Peak VIII (Fig. 2) was rechromatographed on a μBondapak C18 column (Waters), using a linear gradient of 0–75% (v/v) CH_3CN in 0.1% (v/v) trifluoroacetic acid, as indicated, at a flow rate of 1.0 ml min^{-1}. **B** RP-HPLC of a digest, obtained by partial acid hydrolysis [23] of BrCN-peptide VIIIa (see **A**), on a μBondapak C18 column, using a gradient of 0–50% (v/v) propan-1-ol in 0.1% (v/v) trifluoroacetic acid, as indicated, at a flow rate of 1.0 ml min^{-1}. **C** RP-HPLC of a tryptic digest of about 300 μg of EF-1β, on a μBondapak C18 column, using a linear gradient of 0–75% (v/v) trifluoroacetic acid, as indicated, at a flow rate of 1.0 ml min^{-1}

Table 2. Procedure for electroelution

---

1. Introduce into one end of the bent glass tube (see Fig. 5) a 3 mm-piece of Tygon tubing (1/4'' × 1/8''), meant as a holder for a filter paper disc (see below).

2. Slip a 3 cm-piece of silicone rubber tubing (6 mm × 9 mm) for half its length over the same tube end.

3. Close the other end of the bent tube by a piece of presoaked dialysis tubing (Spectrapor No. 3, MW cut off 3500 can generally be used), clenched into the tube by a 6-mm-piece of Tygon tubing (1/4'' × 1/8''). Keep the dialysis tubing wet by immersing the bent tube in a small beaker filled with water or electrophoresis buffer.[a]

4. Introduce into the bottom end of the straight tube (see Fig. 5) a 3 mm piece of Tygon tubing (1/4'' × 1/8'').

5. Clench a piece of dialysis tubing into the side arm of the tube as described above. (This dialysis membrane will allow the sample to be electrophoresed, free from possible proteinaceous impurities, originating from the buffer compartment etc., which would otherwise be concentrated in the protein solution during the electroelution).

6. Fit the straight glass tube into the upper compartment of the electrophoresis apparatus, and connect the bottom of the tube with the bent tube (see Fig. 5), which has been filled with electrophoresis buffer, and prechecked for leaks. A disc of filter paper (5 mm ∅), placed between the two glass tubes, prevents the gel pieces to be introduced into the straight tube (see below) from getting into the bent tube. Air bubbles in the connection region should be removed.

7. Remove any electrophoresis buffer, present above the paper filter disc, with a long Pasteur pipet.

8. Introduce the gel pieces, presoaked in soaking buffer[b], through the funnel and push them to the bottom of the tube by means of a glass rod. Transfer any remaining liquid, in which the gel pieces were presoaked, into the tube as well.

9. Overlay the contents carefully with electrophoresis buffer, until the tube is filled above the side arm.

10. Fill the lower and upper buffer compartments with the electrophoresis buffer. The level of buffer in the upper compartment should be just below the upper end of the straight tube.

11. Remove any remaining air bubbles on the dialysis membranes, and perform the electroelution at 100 – 150 V for 10 h at room temperature, the anode being at the bottom.

12. Reverse the current for 5 min at the end of the run, remove the contents of the bent tube carefully and lyophilize the solution. Note: If the gel pieces to be eluted contain urea, or another uncharged substance in considerable amount, this may contaminate the protein sample by diffusing into the bent tube. In that case, the device can be used the other way round, i.e., the gel pieces are loaded into the bent tube, and the current is reversed (anode at top). The protein accumulates during the electroelution in the straight tube, and can be collected there.

---

[a] The electrophoresis buffer consists of 50 mM ammonium acetate (pH 6), containing 0.02% SDS.
[b] The soaking buffer consists of 0.5 M ammonium hydrogen carbonate, containing 2% SDS and 0.2% 2-mercaptoethanol.

adjusted to approximately 8 by addition of 2 M ammonium hydroxide. Then, the gel pieces and the soaking solution are transferred into the device (Fig. 5), as described in Table 2.

2.5 Separation of Proteins and Peptides from SDS

Small amounts of SDS usually do not interfere with automatic protein sequencing or digestion with proteases etc. (see, however, [26]), and there is no need for

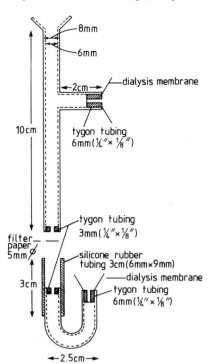

Fig. 5. Device for the electroelution of proteins and peptides separated on polyacrylamide gels. See Table 2 for further description

removal. In many cases, however, e.g., when peptides have been isolated by gel filtration in the presence of SDS, removal is necessary.

We have found a very simple method for the complete removal of SDS from protein or peptide (chain length >20 residues)-containing solutions [27]. Briefly, the lyophilized sample (which also may contain salts, urea, etc.) is dissolved in a mixture of propionic acid, formic acid and H_2O (2:1:2; v/v/v). Under these conditions, protein-SDS (or peptide-SDS) complexes, as well as SDS micelles, are dissociated. The protein or peptide can then be separated from the (monomeric) SDS by gel filtration in the same solvent. The procedure is given in Table 3.

3 Preparation of Protein Samples to be Digested or Sequenced

3.1 Sulfhydryl Group Modification in Small Amounts of Protein

Before fragmentation of proteins by chemical and proteolytic cleavage, or before sequencing the protein, it is necessary to block the sulfhydryl groups. S-carboxymethylation [29] and S-pyridylethylation [30] are both useful for peptides to be cleaved; a procedure for S-carboxymethylation is given in Table 4. If S-carboxymethylated proteins are sequenced, it should be noted that the elution position of the phenylthiohydantion of S-carboxymethylcysteine coincides with that of aspartic acid in the system of Zimmermann et al. [31]; both components, howev-

Table 3. Removal of SDS from proteins and peptides (see [27])

---

1. Swell Sephadex G25 superfine in a mixture of propionic acid, formic acid and water (2:1:2; v/v/v). Pour a column (0.7 cm × 10 cm). (This column can be stored in the solvent at $2° - 4°C$ for several years).

2. Apply the sample to the column; the sample volume should not exceed $0.7 - 0.8$ ml, using a column of these dimensions. The sample may contain $10 \mu g - 10$ mg of protein and up to 50 mg of SDS.

3. Collect fractions of $0.10 - 0.20$ ml. 3 µl of each fraction is tested for the presence of SDS (see below).

4. Pool all fractions collected before the appearance of SDS, and remove the solvent by rotatory evaporation and lyophilization.

5. The appearance of SDS in the effluent is tested as follows (cf. [28]). Dilute an aliquot of 3 µl of each fraction to be tested with 0.80 ml H_2O. Add 0.20 ml of 0.4 mM pararosaniline hydrochloride (purified as described [28]); vortex the solution with 1.0 ml of a mixture of chloroform and ethyl acetate (1:1; v/v). If SDS is present, the dye will appear in the organic (lower) phase, due to ion pair formation. Under these conditions, the lower limit of visual detection is well below an amount of 0.5 µg of SDS.

Table 4. Blocking of sulfhydryl groups in small amounts of protein with iodoacetate

---

1. Add to an amount of $100 - 200$ µg of protein, dissolved in 150 µl of H_2O (or dilute buffer), preferably in a 2-ml Eppendorf test tube: 50 µl of 2 M Tris HCl (pH 7.5), 100 mg urea and 20 µl of 10% (v/v) dithiothreitol. Mix until all contents are dissolved, flush with nitrogen, and incubate for 2 h at 37°C.

2. Add 4 mg iodoacetic acid, dissolved in 25 µl of 1 M Tris base, flush with nitrogen, and place the tube in the dark for 30 min at room temperature.

3. Add 1.0 ml H_2O and 120 µl 100% (v/v) trichloroacetic acid, leave on ice for 15 min, centrifuge 10 min at 10,000 g.

4. Remove and discard the supernatant. Wash the precipitate with 500 µl of a mixture of ethanol and hexane (1/1; v/v), centrifuge.

5. Remove and discard the supernatant. Dry the protein precipitate in a stream of nitrogen.

---

er, are easily separated when the elution buffer is diluted with an equal volume of water and the column temperature is lowered to 30°C.

3.2 Sulfhydryl Group Modification in the Spinning Cup Prior to Sequencing

For proteins to be sequenced, it is very convenient to modify the sulfhydryl groups directly in the spinning cup itself. We have found that the modification procedure using 4-vinylpyridine and tributyl phosphine [30] is superior for this purpose. The procedure is outlined in Table 5. Although the original protocol uses a pH of 7.5, the reaction proceeds well at pH 9.0, the pH normally used in the Edman degradation. It is therefore not necessary to adjust the pH of the quadrol/propan-1-ol buffer, used for the modification, to 7.5 before addition of the reagents. The method gives a high modification yield, even with proteins and peptides that are difficult to reduce, e.g., basic protease inhibitor and insuline).

Table 5. Blocking of sulfhydryl groups in proteins to be sequenced in the spinning-cup sequenator with 4-vinylpyridine

---

1. Adjust the temperature of the reaction cup at 37 °C.

2. Apply 50 – 1000 μg of protein and 3 mg Polybrene to the cup. Dry the sample.

3. Start the normal sequence program, but without addition of phenylisothiocyanate, till the quadrol/propan-l-ol buffer has been delivered. Set the programmer on hold.

4. Open the reaction cup, apply 10 μl tributylphosphine and 10 μl 4-vinylpyridine (redistilled) at high speed to the bottom of the cup.

5. Close the reaction cup, flush with nitrogen, allow the modification reaction to proceed for 1 h at 37 °C at high speed.

6. Continue the normal program, while raising the temperature to 57 °C. Excess of reagents, and reaction products are effectively removed during the subsequent solvent extraction steps.

The phenylthiohydantoin of S-pyridylethylcysteine is easily detected in the procedure of Zimmermann et al. [31]. However, like other positively charge phenylthiohydantoin amino acids (e.g., those of Arg and His), the elution position of the phenylthiohydantoin of S-pyridylethylcysteine is sensitive to small changes in buffer composition. It is therefore necessary to check its chromatographic position frequently.

4 Methylated Amino Acids in EF-1α

EF-1α from cysts of *Artemia* is an interesting protein with respect to post-translational modifications. At four positions, exclusively ε-trimethyllysine is found, and at one site presumably ε-monomethyllysine (see Fig. 6 and [2, 8]). The determination of these amino acids could only be accomplished by back hydrolysis of the corresponding phenylthiohydantoin derivatives (see legend of Fig. 6 for details).

```
        30          ■        40
H L I Y K C G G I D K R T I E K F E K E A

   70            ■   80
R G I T I D I A L W K F E T A K Y Y V T I

   210           ■   220
P W Y K G W N I E R K E G K A D G K T L L

   310           ■   320
V G F N V K N V S V K E L R R G Y V A S D

      50          ●      60
E A Q E M G K G S F K Y A W V L D K L K A
```

Fig. 6. The methylated amino acids in EF-1 α from *Artemia* cysts. The ε-trimethylated lysine residues are indicated by ■, the (presumably) ε-monomethylated lysine residue by ● . Unexpectedly the phenylthiohydantoins of ε-trimethyllysine as well as of the putative ε-monomethyl-lysine did not show a peak when analyzed according to Zimmermann et al. [31]; the sample obtained after conversion of the chlorobutane extract by acid treatment as usual, did however, contain the phenylthiohydantoin amino acids, as was shown by back hydrolysis [32]. The reason for this behavior is not known; it cannot be excluded that the very polar cationic derivatives are strongly bound to certain (negative) groups of the column matrix. ε-trimethyllysine is eluted some 3 min before lysine on the amino-acid analyzer used, which was equipped with the one-column system; the identity of the component was confirmed with authentic sample of this amino acid. The phenylthiohydantoin of the putative ε-monomethyllysine yielded on back hydrolysis a product that coincided on the amino-acid analyzer with lysine, and ε-mono- and ε-dimethyllysine, and it could therefore not be identified with certainty as yet

Acknowledgments. The author thanks Professor Dr. W. Möller and Dr. J. A. Maassen for critical reading of the manuscript, and Mr. W. J. M. Pluijms and Mr. H. J. G. M. de Bont for expert technical assistance.

References

1. Möller W, Amons R, Lenstra JA, Maassen JA (1985) In: Jaspers E, Sorgeloos P, Moens L, Decleir W (eds) Proc 2nd Int Symp Brine Shrimp *Artemia*. Universa Press, Wetteren, 1986, in press
2. Van Hemert FJ, Amons R, Pluijms W, Van Ormondt H, Möller W (1984) EMBO J 5:1109−1113
3. Jones MD, Petersen TE, Nielsen KM, Magnusson S, Sottrup-Jensen L, Gausing K, Clark BFC (1980) Eur J Biochem 108:507−526
4. Laursen RA, L'Italien JJ, Nagarkatti S, Miller DL (1981) J Biol Chem 256:8102−8109
5. Nakamura S, Nakayama N, Takahashi K, Kaziro Y (1982) J Biochem 91:1047−1063
6. Halliday KR (1984) J Cycl Nucleotide Protein Phosphoryl Res 9:435−448
7. Möller W, Amons R (1985) FEBS Lett 186:1−7
8. Amons R, Pluijms W, Roobol K, Möller W (1983) FEBS Lett 153:37−42
9. Jacobson GR, Rosenbush JP (1976) Biochemistry 15:5105−5109
10. Wittinghofer A, Frank R, Leberman R (1980) Eur J Biochem 108:423−431
11. Liljas A (1982) Prog Biophys Mol Biol 40:161−228
12. Van Noort JM, Kraal B, Bosch L, La Cour TFM, Nyborg J, Clark BFC (1984) Proc Natl Acad Sic USA 81:3969−3972
13. Van Noort JM, Kraal B, Bosch L (1985) Proc Natl Acad Sci USA 82:3212−3216
14. Nagata S, Tsunetsugu-Yokota Y, Naito A, Kaziro Y (1983) Proc Natl Acad Sci USA 80:6192−6196
15. Montandon P-E, Stutz E (1983) Nucleic Acids Res 11:5878−5892
16. Nagata S, Nagashima K, Tsunetsugu-Yokota Y, Fujimura K, Miyazaki M, Kaziro Y (1984) EMBO J 3:1825−1830
17. Cottrelle P, Thiele D, Price VL, Memet S, Micouin J-Y, Marck C, Buhler J-M, Sentenac A, Fromageot P (1985) J Biol Chem 260:3090−3096
18. Schirmaier F, Philipsen P (1984) EMBO J 3:3311−3315
19. Brands JHGM, Maassen JA, Van Hemert FJ, Amons R, Möller W (1986) Eur J Biochem 155:167−171
20. An G, Bendiak DS, Mamelak LA, Friesen JD (1981) Nucleic Acids Res 9:4163−4172
21. Amons R, Pluijms W, Kriek J, Möller W (1982) FEBS Lett 153:37−42
22. Laemmli UK (1970) Nature 227:680
23. Inglis A, McKern N, Roxburgh C, Strike P (1980) In: Birr C (ed) Methods in peptide and protein sequence analysis. Elsevier Biomed Amsterdam New York, pp 329−343
24. Talbot DN, Yphantis DA (1971) Anal Biochem 44:246−253
25. Vandekerckhove J, Van Montagu M (1974) Eur J Biochem 44:279−288
26. Walker AI, Anderson CW (1985) Anal Biochem 146:108−110
27. Amons R, Schrier PI (1981) Anal Biochem 116:439−443
28. Pitt-Rivers R, Impiombato FSA (1968) Biochem J 109:825−830
29. Crestfield AM, Moore S, Stein WH (1963) J Biol Chem 238:622−627
30. Friedman M, Zahnley JC, Wagner JR (1980) Anal Biochem 106:27−34
31. Zimmermann CL, Appella E, Pisano JJ (1978) Anal Biochem 77:569−573
32. Smithies O, Gibson D, Fanning EM, Goldfliesh RM, Gilman JG, Ballantyne DL (1971) Biochemistry 10:4912−4921

Chapter 8
Alternative Sequencing Techniques for Elucidation of Peptide and Protein Structures

8.1 RNA Sequencing

Martin Digweed, Tomas Pieler, and Volker A. Erdmann[1]

Contents

1 Introduction

This chapter was written to illustrate the fact that protein sequencing may also be performed indirectly by sequence determinations of the corresponding mRNA's or DNA's. The consideration of nucleic acid sequencing as an alternative to protein sequencing may become important when only small amounts of a protein are available, or when the sequencing of a protein is hindered by amino acid modifications in the protein.

In this chapter we describe procedures for RNA sequencing, one of which, the chemical sequencing method, is very similar to the Maxam and Gilbert procedure for DNA sequencing. For more details on DNA sequencing and required cloning procedures the reader is referred to the more specialized literature [1, 2].

---

1 Institut für Biochemie, Freie Universität Berlin, Otto-Hahn-Bau, Thielallee 63, D-1000 Berlin 33

Advanced Methods in Protein Microsequence Analysis
Ed. by B. Wittmann-Liebold et al.
© Springer-Verlag Berlin Heidelberg 1986

2 Isolation of RNA

2.1 Extraction of RNA

Most procedures for RNA extraction consist of tissue disruption followed by phenol extraction to separate nucleic acids from protein. The method of physical disruption will depend largely upon the type of tissue involved; while animal cells can be opened with relatively mild methods, plant and bacterial cells require stronger forces to damage their cell walls. Most disruption methods are of the liquid shear type in which the forces are exerted onto cells and tissue in liquid suspension. The Waring blender uses a specially designed cup and blades driven at high speeds to procedure strong shearing forces, while hand-held or motor-driven homogenizers such as the Dounce and Potter Elvejham generate shearing forces between the wall of the glass tube a closely fitting glass or Teflon pestle. The choice of suspension medium will also depend upon the material in question and many recipes are in circulation. Buffered salt solution may be supplemented with sucrose to prevent bursting of cell organelles, particularly lysosomes, a non-ionic detergent (e.g, Triton X-100) to promote break-down of subcellular membranes, or the ionic detergent SDS to inhibit nuclease activity and disrupt nucleo-protein complexes. Concentrations of EDTA, 2-mercaptoethanol and salts vary considerably from one method to another. Depending upon the nature of the material, powdering by liquid nitrogen treatment may be helpful. In all cases precautions must be taken against RNases present in the material itself and possibly contaminating glassware and other apparatus. All procedures should be

Table 1. Cetyltrimethylammonium bromide (CTAB) extraction of plant DNA and RNA

1.	Freeze the plant material in liquid nitrogen and lyophilize it. Grind the dried material to a fine powder with pestle and mortar.
2.	Resuspend the powder in 50 mM Tris-HCl pH 8. 0.1% CTAB, 10 mM EDTA, 700 mM NaCl, 1% λ-mercaptoethanol (use 20 ml per g dry weight of powder).
3.	Incubate the mixture at 50°C for 30 min with occasional stiring to mix.
4.	Remove the mixture to room temperature and add an equal volume of chloroform: isoamyl alcohol (24:1) shake to mix phases.
5.	Centrifuge at 13,000 g for 10 min.
6.	Remove the aqueous phase to a fresh tube or bottle and add 1/10th volume of 10% CTAB. Add an equal volume of chloroform: isoamyl alcohol, shake and centrifuge as above.
7.	Remove the upper aqueous phase and add an equal volume of precipitation buffer (50 mM Tris-HCl ph 8.0, 1% CTAB, 10 mM EDTA, 1% λ-mercaptoethanol) mix thoroughly and leave for 30 min at room temperature to ensure complete precipitation.
8.	Collect the nucleic acid/CTAB precipitate by centrifugation at 4000 g for 5 min.
9.	Resuspend the pellet in 6 – 18 ml 50 mM Tris-HCl pH 8.0, 5 mM EDTA, 50 mM NaCL, 1 M CsCl and layer onto 2 ml of 5.7 M CsCl in 50 mM Tris-HCl pH 8.0, 5 mM EDTA, 50 mM NaCl in a Beckmann SW 50.1 tube. Centrifuge for 12 h at 120,000 g.
10.[a]	Drain the RNA pellet and resuspend in 10 mM Tris-HCl pH 7.6, 1 mM EDTA and precipitate by addition of 2 vol of ethanol and leaving at − 20°C for several hours.
11.	Collect the RNA pellet by centrifugation, dry briefly under vacuum and resuspend in a small volume of distilled water.

[a] DNA bands at or below the 1 M/5.7 M CsCl interface and may be collected with a Pasteur pipet and rebanded on CsCl density gradients.

Table 2. Isolation of RNA from animal tissue

1. (A) Freeze the tissue in liquid nitrogen and homogenize in a Waring blender at 0 °C for several minutes, resuspend the homogenate in 10 mM Tris-HCl pH 8.0, 50 mM KCl, 10 mM Mg-acetate or, (B) Homogenize the tissue in 10 mM Tris-HCl pH 8.0, 50 mM KCl, 10 mM Mg-acetate in a Dounce homogenizer at 4 °C
2. Add SDS to an end concentration of 1% and EDTA to 2 mM.
3. Add two volumes of phenol saturated with buffer and shake vigorously for 10 min
4. Centrifuge at 10,000 g for 10 min at room temperature
5. Remove this upper aqueous phase and place on ice.
6. Re-extract the phenol phase by adding an equal volume of buffer and vigorous shaking.
7. Recentrifuge as above and add the second aqueous phase to the first.
8. Re-extract the aqueous phase by addition of an equal volume of phenol-chloroform (1:1, vol:vol) shaking and centrifugation as above.
9. If a large amount of white coagulated protein is still present at the interface the phenol extraction of step 8 may be repeated. The final aqueous phase is made 0.3 M Na-acetate and two volumes of ethanol added.
10. Allow precipitation of RNA to proceed at -20 °C for several hours.
11. Collect the precipitate by centrifugation at 10,000 g for 10 min, remove the supernatant and add a small volume of 70% ethanol (precooled to -20 °C) and recentrifuge.
12. Dry the pellet under vacuum and resuspend the RNA in a small volume of distilled water. Store at -80 °C prior to further purification.

Table 3. Isolation of ribosomes from bacterial cells

1. Place 40 g (wet weight) of frozen bacterial cells in an ice-cooled mortar, add 80 g of cold aluminum oxide (Alcoa No. 305).
2. With a pestle, grind the cells to a paste and then continue grinding until a viscous fluid is obtained.
3. Resuspend the cell/Alcoa mix in 200 ml cold 10 M Tris-HCl pH 7.8. 30 mM NH_4Cl, 10 mM $MgCl_2$, 6 mM 2-mercaptoethanol containing 1 µg ml^{-1} RNase-free DNase.
4. Centrifuge the suspension at 10,000 rpm for 15 min, decant the supernatant and place on ice.
5. Resuspend the pellet of cell debris and Alcoa in 50 ml of cold buffer and recentrifuge as above.
6. Add the second supernatant to the first and centrifuge for 45 min at 15,000 rpm.
7. The clear supernatant is a crude ribosome fraction. The ribosomes are collected by centrifugation overnight at 40,000 rpm and resuspended in a small volume of the buffer described above without DNase.

conducted at around 4 °C, glassware must be scrupulously clean and 'finger-RNase' contamination avoided by wearing latex gloves.

Extraction from plant tissues presents further problems, due to the large amounts of polysaccharides normally present and the relatively high content of RNases typical of plant material. However, a technique has been developed [3] which exploits the formation of a specific nucleic acid/cetyltrimethylammonium bromide complex. Coupled with the powdering of freeze-dried plant material, this provides an efficient method for plant RNA and DNA preparation.

Tables 1 and 2 detail basic procedures for extraction of RNA from plant and animal sources. However, considering the enormous varieties of tissue available, these methods can only be viewed as starting points for the development of specialized procedures. We recommend that the reader turn to the literature when considering extraction of a particular tissue.

As an alternative to direct phenolization, mRNA and rRNA can be isolated by first preparing polysomes, ribosomes, post-ribosomal supernatant, or other subcellular fractions. A procedure for the isolation of a crude ribosome extract from bacteria cells is presented in Table 3, where the disruption procedure is of the solid-shear type in which cells are ground with abrasive particles (in this case aluminium oxide). The further purification of the ribosomes is presented in the following section.

2.2 Purification to Single-Species RNA

The purity of the RNA preparation to be analyzed is of crucial importance for the quality of the final results. Since eukaryotic cells in particular contain a complex composition of different small and large RNA's, it is often advisable to include a subcellular fractionation step prior to the removal of protein by phenolization. In general RNA is found associated with specific proteins, forming either RNP's with discrete biological functions (e.g., ribosomes, prosomes and splicosomes), or representing cellular storage particles (e.g., 7S and 42S RNP's during amphibian oogenesis).

A first fractionation of either cellular or nuclear extracts is often achieved by high-speed centrifugation (Table 3, S100) which is then followed by density gradient centrifugation. As a classical example of this procedure we describe the

Table 4. Separation of ribosomal subunits by density gradient centrifugation

---

1. Prepare 34 ml linear gradients of 10% to 30% sucrose in 10 mM Tris-HCl pH 7.8, 30 mM NH_4Cl, 0.3 mM $MgCl_2$ in SW28 (Beckman) centrifugation tubes or similar. Load each gradient with 300 yA$_{260}$ units of 70S ribosomes.
2. Centrifuge the gradients in the appropriate rotor at 20,000 rpm for 18 h at 4 °C.
3. Fractionate the gradients by puncturing the tubes at the base or inserting a glass capillary tube into the gradient and starting flow by light suction. Measure the absorption of the fractions at 260 nm.
4. Pool the fractions containing the 30S and the 50S subunits and collect them by centrifugation at 40,000 rpm overnight.

Table 5. Isolation of 7S RNP particles from *Xenopus laevis* ovaries

---

1. Homogenize one mature ovary with an equal volume of 20 mM Tris-HCl pH 7.5, 70 mM KCl, 10 mM $MgCl_2$, 3 mM DTT in a Dounce homgenizer using 3 – 5 strokes of a loosely fitting pestle. Carry out the homogenization on ice.
2. Centrifuge the homogenate at 13,000 g for 10 min.
3. Remove the clear supernatant, avoiding the yellow lipid layer on the surface, and dialyze against 50 mM Tris-HCl pH 6.8, 5 mM $MgCl_2$ at 6 °C.
4. Remove any precipitate by centrifugation at 13,000 g for 20 min.
5. Load the clear supernatant onto a Whatman DE 52 anion exchange column (4 × 30 cm) equilibrated in 50 mM Tris-HCl pH 6.0, 5 mM $MgCl_2$ at 6 °C.
6. Elute the column with a linear 0 – 0.6 M NaCl gradient in 50 mM Tris-HCl pH 6.8, 5 mM $MgCl_2$: see Fig. 1.

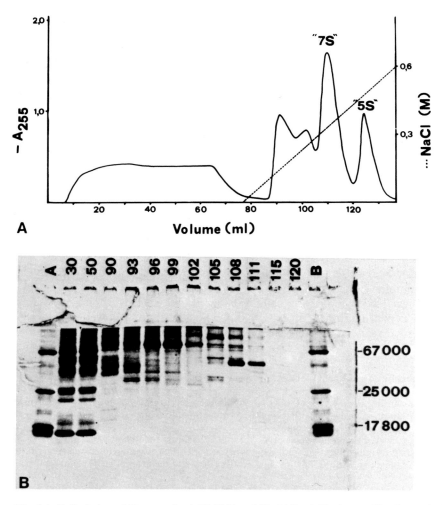

Fig. 1A, B. Isolation of *Xenopus laevis* 7S RNP and 5S rRNA. **A** Elution profile of a crude 7S RNP fraction from sucrose gradient centrifugation applied to an anion exchange column (Whatman DE52) and eluted with a 0 – 0.6M NaCl gradient. The peaks containing the 7S RNP particle and 5S rRNA are indicated. **B** Examination of aliquots from fractions of the chromatography on a 15% polyacryl-amide-SDS gel stained for protein. Molecular weight markers were applied to lanes *A* and *B*

preparation of ribosomal subunits in Tables 3 and 4. As an alternative to sucrose, glycerol may be used for the isolation of more labile RNA/protein complexes and low molecular weight RNP's. A more elegant approach for the isolation of RNP's via their specific protein component utilizes immuno-precipitation or im-muno-affinity chromatography [4]. However, a prerequisite for this is, of course, the availablility of specific antibodies. A more conventional way to separate protein, RNP, and RNA is provided by ion-exchange chromatography (Table 5 and Fig. 1). This technique is particularly useful, in combination with density gradient centrifugation, not only for RNA preparation but also for the isolation of specific proteins associated with the RNA.

Following the removal of proteins by phenolization, the RNA's may be further fractionated by gel exclusion chromatography (e.g., Sephadex). This allows the simple separation of small RNA's ($M_r < 100,000$) from larger RNA and, within the small RNA population, of RNA's which differ in molecular weight by a factor about 2 (e.g., 5S rRNA and tRNA).

In the majority of cases, fractionation still leaves a mixture of RNA's of similar molecular weight and physical properties. Sequence analysis of individual RNA species from these mixtures, especially for large RNA's, can be achieved by either cDNA cloning or isolation of the individual genes encoding the RNA's [2]. In the case of the small RNA's, isolation of single-species RNA is possible with the high resolving power of polyacrylamide gel electrophoresis as discussed below.

2.2.1 Denaturing Gel Electrophoresis

Separation of RNA's on the basis of chain length and nucleotide composition may be achieved by means of conventional denaturing polyacrylamide gel electrophoresis in the presence of 7 M urea. Length heterogeneity due to either multiple transcription termination or processing sites, or sometimes due to unspecific terminal degradation, is a major hurdle in RNA sequence determination using end-labeled RNA (see Sect. 3). However, choice of appropriate electrophoresis conditions (Table 6) allows the separation of RNA molecules even on the basis of only one nucleotide difference in chain length. For RNA molecules which form a stable, G/C-rich secondary structure, the presence of 7 M urea alone is often insufficient for complete denaturation. In these cases, electrophoresis at high temperature ($60° - 70°C$) improves the separation and may be facilitated by the use of the appropriate equipment (e.g, thermostated glass plates).

2.2.2 Native Gel Electrophoresis

In many cases RNA fractions of uniform length still contain a mixture of RNA species due to sequence polymorphisms amongst the members of a multigene

Table 6. Denaturing polyacrylamide gels

Polyacrylamide concentration (%)	Effective range of separation (nucleotides)	Migration behavior of marker dyes (nucleotides)	
25	1 – 30	7 [a]	21 [b]
20	6 – 50	8	28
12	12 – 80	14	57
8	20 – 120	19	80
6	30 – 200	26	100

[a] Bromophenol blue
[b] Xylene cyanol

Table 7. Solutions for polyacrylamide gel electrophoresis

A. Denaturing gels

Stock solutions (1 l)			
40% acrylamide		380 g	Acrylamide
		20 g	N,N'-Methylene-bis-acrylamide
10 X TEB	0.90 M	109.0 g	Tris
10 X TEB	0.90 M	55.6 g	Boric acid
10 X TEB	0.04 M	14.9 g	EDTA

Gel solutions are prepared by dilution of stock acrylamide to the desired end concentration after addition of 1/10th volume of 10 X TEB and 0.42 g urea ml^{-1} (7 M). Polymerization is initiated by the addition of 0.5 µl ml^{-1} tetramethylethylenediamine (TEMED) and 10 µl ml^{-1} 10% ammonium peroxodisulfate. Electrophoresis is carried out with 1 X TEB as electrophoresis buffer at 55 W constant power in a standard set-up (Fig. 2). The sample is loaded in a small volume of loading buffer (1 X TEB, 7 M urea, 0.05% bromophenol blue, 0.05% xylene cyanol).

B. "Native" gels

Stock solutions (1 l)		
30% acrylamide	290 g	Acrylamide
	10 g	bis-Acrylamide
10 X TEB	As above	

Gel solutions are prepared to the desired end concentration of acrylamide and 1 X TEB. Polymerization is started as above.
For the ethidium bromide/urea system the gel is made with 0.24 g ml^{-1} urea (4 M) and 2 µg ml^{-1} ethidium bromide. Ethidium bromide is present in the electrophoresis buffer at 2 µg ml^{-1} and electrophoresis is carried out at 10 mA constant current in the ethidium bromide, 0.05% bromophenol blue, 0.05% xylene cyanol.

---

family. These sequence variations lead to an alteration of the secondary/tertiary structure of the RNA which is exploited in a gel separation system employing ethidium bromide and 4 M urea [5] as detailed in Table 7. This method enables us to separate two RNA molecules of identical length but differing in sequence at only a single position.

2.2.3 Mixed Two-Dimensional Gel Electrophoresis

Any two of the conditions outlined above may be combined to a two-dimensional system in which the sample is run in a standard set-up under conditions "A" first, then the complete lane cut out and polymerized into the bottom of a gel prepared under conditions "B".

Following the electrophoresis separation of RNA molecules, they must be visualized and extracted. Visualization may be achieved by autoradiography, standard staining techniques (e.g., toluidine blue) or by UV-shadowing (Table 9). When using autoradiography with highly radioactive RNAs it is recommended not to use intensifying screens since scattering of the fluorescence these screens

Table 8. Set-up of a gel chamber

Assembling the gel plates

1. Wash the plates thoroughly with detergent and hot water rinse them in distilled water and polish with ethanol. The surfaces of the plates may be siliconized to assist gel pouring.
2. Place the larger plate (3) on a flat surface and lay side spacers (4) with the silicone block uppermost along each side of the plate.
3. Place the bottom spacer (7) along the bottom of the plate to fit snugly against the side spacers. Lay the small glass plate (5) on top of the spacer assembly and carefully check alignment and fitting of the three spacers, particularly the silicone blocks must fit tightly against the edge of the small plate.
4. Clamp the plates together with as many clips (8) as possible.

Pouring the gel

1. Support the gel at an angle of $10° - 30°$ to the horizontal and pour the gel solution down the inner surface of the large plate and fill to 1 cm above the top of the small plate.
2. Select an appropriate comb (6) and insert between the glass plates, avoiding air bubbles, and as far as required for the probe volume to be loaded. Allow the gel to polymerize.
3. Carefully remove the comb and immediately wash the pockets with distilled water. Remove the clips and bottom spacer.
4. Fill the space between the plates left by the bottom spacer with electrophoresis buffer and clamp the assembly in the electrophoresis chamber.

Table 9. UV-Shadowing

1. Wrap the gel in plastic film (e.g., Frapan, Saran Wrap) and place on any thin-layer chromatography plate which contains a fluorescent indicator in the separation layer (e.g., Merck TLC cellulose F 254).
2. Visualize the RNA bands using a hand-held UV light source $(250 - 280 \text{ nm})$ and mark their positions onto the plastic film with a laboratory marking pen.
3. Excise gel slices containing the RNA with a sharp knife or razor blade.
4. Incubate the gel slices overnight at room temperature in a suitable volume of extraction buffer (0.5 M NH_4Ac, 10 mM $MgACc$, 0.1 mM EDTA). The volume should normally be kept as low as possible, 0.4 ml are sufficient for a gel slice of dimensions 0.5×1 cm.
5. Remove the gel slice and clean up the RNA either by precipitation with ethanol (add 2.5 vol absolute ethanol and leave for several hours at $-20°C$ or 10 min powdered dry ice) by desalting on a small $(15 \times 2 \text{ cm})$ Sephadex G 50 column in water. Fractions containing the RNA (established by absorption measurement or counting radioactivity) can be directly lyophilized.

---

produce leads to loss of resolution. Two to four µg of RNA in a single band may still be detected by the UV-shadowing technique.

3 Sequencing

3.1 Labeling with ^{32}P-Phosphate

Having obtained a pure preparation of RNA, one can proceed to end-labeling and sequencing. Post-labeling methods are directed at either of the two ends of the RNA molecule. Polynucleotide kinase extracted from T4-infected *E. coli*

Table 10. Decapping and dephosphorylation of RNA

Decapping procedure

1. Lyophilize 5 μg of RNA in a 1.5 ml microcentrifuge tube.
2. Resuspend the RNA in 10 μl 10 mM NaIO$_4$ (freshly prepared)
3. Incubate for 1 h at room temperature in the dark
4. Add 20 μl 0.3 M Na-acetate pH 5.2
          75 μl ethanol
    allow precipiation for 10 min on dry ice.
5. Collect precipitate by centrifugation at 8000 rpm for 15 min.
6. Resuspend pellet in 50 μl 0.1 M Na-acetate pH 5.2 and reprecipitate by addition of 100 μl ethanol.
7. Centrifuge to collect pellet and repeat reprecipitation (step 6) *three* times.
8. Dry the last ethanol pellet and resuspend in 10 μl of 1 : 1 (vol/vol) 0.33 M aniline acetate pH 4.5: 0.15 M Na-acetate pH 4.5.
9. Incubate for 4 h at room temperature in the dark.
10. Precipitate after the addition of 40 μl Na-acetate pH 5.2 and 100 μl ethanol on dry ice for 10 min.
11. Collect the pellet and wash three times as described in step 6 above. Dry the last pellet and proceed to the dephosphorylation.

Dephosphorylation

1. Lyophilize 5 μg RNA in a microcentrifuge tube.
2. Resuspend the RNA in 10 μl buffer (50 mM Tris-HCl, ph 9.0, 1 mM MgCl$_2$, 0.1 mM ZnCl$_2$, 1 mM spermidine).
3. Add 1 μl calf intestinal alkaline phosphatase (1 : 40 diluted ammonium sulfate suspension at 4000 units ml^{-1}).
4. Incubate for 45 min at 37 °C.
5. Add 10 μl loading buffer.
6. Load the dephosphorylated RNA into one pocket of a 10% polyacrylamide, 8 M urea gel (10 cm × 10 cm × 0.5 mm).
7. Electrophorese at 30 V cm^{-1} for 1 h, or as appropriate.
8. Visualize and extract the decapped/dephosphoryated RNA from the gel as described in Section 2.2 and precipitate by addition of 1/10 vol. 3 M Na-acetate, 2 vol ethanol. Allow precipiatation for 10 min on dry ice.
9. Centrifuge to collect the RNA dry briefly under vacuum and resuspend in a few microliters of distilled water.

---

mediates the transfer of the gamma-phosphate from ATP to the 5′-hydroxyl group of an oligonucleotide [6]. Most naturally occurring RNA's have a 5′-phosphate group, many other RNA's are, in addition, capped at their 5′-termini. Thus, additional manipulation of the RNA is necessary before it can be labeled at the 5′-end. Decapping may be achieved by digestion with tobacco acid pyrophosphatase or by periodate oxidation of the 5′-end 7-methyl-guanosine and subsequent removal by aniline [7]. Decapped RNA or uncapped RNA is then dephosphorylated by alkaline phosphatase ($M_r = 140,000$) isolated from calf intenstine. Both decapping dephosphorylation result in unspecific fragmentation of the RNA. It is essential to remove the alkaline phosphatase before proceeding to the labeling reaction; however, most procedures for destruction of this enzyme, e.g., boiling in nitrilotriacetic acid, have a further detrimental effect on the RNA. Since the efficiency of 5′-labeling increases with decreasing RNA size,

Table 11. 3'- and 5'-end labeling of RNA

A. 5'-end labeling

1. Dry 25 pmol of dephosphorylated RNA in a 1.5 ml microcentrifuge tube.
2. Add 1 μl of 10× concentrated buffer (end concentration: 50 mM Tris-HCl pH 9.0, 10 mM MgCl$_2$, 5 mM DTT).
3. Add up to 8 μl of [γ-^{32}P] ATP (2000 Ci/mmol, 10 mCi/ml in stabilized aqueous solution), or more if previously lyophilized.
4. Adjust volume to 9 μl with distilled water.
5. Add 1 μl of T4-polynucleotide kinase (~ 10 Richardson units).
6. Vortex and centrifuge briefly.
7. Incubate for 30 min at 37 °C.
8. Stop the reaction by addition of 10 μl loading medium and apply to an appropriate polyacrylamide/urea gel. Electrophorese to separate excess ATP.
9. Remove the gel from the glass plates and wrap in household plastic film, stick several phosphorescent markers onto the gel with stick tape. Expose an X-ray film to the gel for between 30 s and 5 min as necessary.
10. Cut "windows" in the developed film where the RNA bands are visible.
11. Place the film on the gel, using the phosphorescent markers for precise alignment, and carefully excise slices of gel containing the labeled RNA.
12. Place the gel slices in 1.5 ml microcentrifuge tubes and add 400 μl of extraction buffer (see Table 9), elute overnight at room temperature.
13. Remove the extraction buffer to a new tube, or carefully remove the gel slice with forceps, and precipitate the RNA by addition of 1/10 vol 3 M Na-acetate, pH 6 and 2 vol of absolute ethanol. 10 – 20 μg of carrier RNA (e.g., tRNA) may be added to improve precipitation.
14. Allow precipitation to proceed for 10 min on dry ice or several hours (better, overnight) at − 20 °C.
15. Collect the labeled RNA precipitate by centrifugation at 8000 rpm in a bench-top centrifuge for 15 min.
16. Wash the pellet with 500 μl 80% ethanol, recentrifuge for 5 min and dry the RNA pellet briefly under vacuum.

B. 3'-end labeling of RNA

1. Lyophilize 25 pmol RNA in a 1.5 ml microcentrifuge tube.
2. Dissolve the RNA in 7 μl of 2× concentrated buffer (end concentration: 50 mM Hepes pH 8.3, 10 mM MgCl$_2$, 3.3 mM DTT, 10% DMSO, 10% glycerol) add 1 μl 150 μM ATP.
3. Add up to 5 μl [5'^{32}P]pCp (2000 – 3000 Ci/mmol, 10 mCi/ml in stabilized aqueous solution), adjust volume to 13 μl with distilled water.
4. Add 1 μl of T4 RNA Ligase (2 units).
5. Incubate at 4° – 11 °C overnight.
6. Proceed from step 8 in Table 11A.

removal of degradation products before labeling is expedient. Fortunately this can be easily achieved by gel electrophoresis, and indeed this also effectively separates alkaline phosphatase from the RNA, removing the requirement for other more dramatic disabling procedures. Table 10 details the decapping and dephosphorylation procedures and the electrophoresis for repurification of the treated RNA.

The labeling reaction itself is described in Table 11 A. The [γ-^{32}P] ATP should be purchased at a concentration of 10 μCi/μl in stabilized aqueous solution, since this preparation can be used as is without prior lyophilization to remove the ethanol often present and to reduce the otherwise too large volume. The amount of [γ-^{32}P] ATP used will depend upon the efficiency of the reaction, size of the RNA, and planned experiments; larger amounts, or older preparations, must be dried before use in order to maintain the end volume of 10 μl.

Labeling at the 3'-end entails addition of a nucleotide which contains a ^{32}P-labeled phosphate group. This requires that the acceptor molecule has a free hydroxyl group at the 3'-end, which is usually the case. The nucleotide addition is catalyzed by the RNA ligase ($M_r = 41,000$) isolated from phage T4-infected *E. coli*. The enzyme requires ATP and the efficiency of the reaction is influenced by sequence and structure at the 3'-end of the acceptor molecule [8]. The most efficient donor molecule for the reaction is cytidine 3', 5' bis-phosphate which may be purchased with a [5'-^{32}P] radiolabeled at high specific activities and in stabilized aqueous solution at 10 μCi/μl, which is convenient for the labeling reaction. Table 11 B gives a procedures for 3'-end labeling. The amount of pCp used will vary, but should normally be present in at least the same concentration as the RNA 3'-termini to be labeled; larger amounts must be dried before use in order to maintain the end volume of 14 μl.

The labeled RNA is separated from excess, nonincorporated isotope by gel electrophoresis. The nature of this electrophoresis will depend on the quality of the RNA used in the labeling reactions. Highly purified, single-species RNA can be loaded onto a 10 cm, 10% gel in 7M urea with subsequent electrophoresis for 1 h at 30 V cm^{-1}, which will be ample to separate the radioactive nucleotides from the labeled RNA. If the labeled RNA is expected to be heterogeneous, then an electrophoresis system of much greater resolving power will be required, the reader is referred to the previous section (Sect. 2.2) for examples of appropriate systems. The electrophoresis may be chosen so that the excess pCp or ATP is still in the gel at its completion; this simplifies the disposal of this highly radioactive waste. Where this is not possible, we recommend that a strip of DEAE paper be inserted between the glass plates at the bottom of the gel. This paper effectively binds the nucleotides as they leave the gel and reduces the activity subsequently found in the lower buffer chamber.

After electrophoresis, the gel is wrapped in plastic film and an autoradiograph taken. The radioactively labeled RNA appears as bands on the film and these may be excised with a sharp knife to produce a template for cutting the gel itself. This procedure requires that the film can be exactly positioned on the gel. This is often ensured by injecting radioactive ink into the gel at several points before autoradiography. However, we find the use of phosphorescent "spots" more convenient. These are cut from a plastic phosphorescent foil and stuck in position with sticky tape; they may be used repeatedly and allow great precision in positioning the exposed film. Having cut slices of gel containing the labeled RNA, it is eluted as described. With relatively short RNA's, such as 5S rRNA (120 nucleotides), at least 90% of the counts should be recovered, and this should be checked by "Cerenkov" counting before and after elution. Cerenkov counting is carried out in the ^{3}H channel without addition of scintillation cocktail. With larger RNA's, elution may be less efficient, which case the gel slices should first be crushed with a glass rod before adding the elution buffer. After elution, which may be carried out at raised temperatures (30° − 40°C), the gel debris are removed by filtration through glass wool which can be conveniently carried out in a disposable 1-ml pipet tip. The RNA is then precipitated as described. The rate of precipitation of RNA is dependent upon the temperature at which it is carried out; since labeling and sequencing manipulations require frequent precipitation of small amounts of RNA, a rapid precipitation technique is desirable. The use of powdered dry ice ensures quantitative precipitation within about 10 min with µg amounts of RNA. After pelleting the RNA precipitate, residual ethanol must be removed by drying under vacuum, for this purpose a vacuum centrifuge (Savant, Speed-Vac) is most convenient and almost a prerequisite for sequencing procedures.

Labeled RNA is resuspended in 50–100 µl of water and an aliquot of 1 µl counted (Cerenkow counts – about 1/3 of the true cpm); the $[^{32}P]$-RNA is stored at $-20\,°C$ until use.

3.2 Enzymatic Sequencing

This method is based upon the specificity of many RNases for particular bases, and was the first method described for rapid sequencing of RNA [9]. An end-labeled RNA is partially digested with base-specific endonucleases and the products are resolved on thin, denaturing polyacrylamide gels constructed in purpose-built chambers such as shown in Fig. 2. These products represent a spectrum of fragments ranging from the end-label to each position in the nucleotide chain, separated into pools, one for each base type. These fragments are then separated according to length which is determined by the position of the RNase cut within the sequence, thus the sequence may be read from the autoradiograms of such gels. Figure 3 shows an example of such an enzymatic sequencing.

While RNase T1 (specific for guanine) and RNase U_2 (specific for adenine) allow accurate determination of purines, the identification of the pyrimidines is more difficult. RNase PhyM cleaves after adenine and uracil with nearly equal efficiency, thus allowing assignment of uracils by comparison to the RNase U_2 cleavage pattern. Positive identification of cytosines is particularly tricky. The RNase isolated from *Bacillus cereus* (referred to hereafter as RNase B.c.) is specific for cytosine and uracil, but can be somewhat erratic in behavior. Further support for cytosines can be obtained using the RNase M1, which cuts at all residues *except* cytosines; however, products of RNase M1 digestion have altered mobility in comparison to that of the other enzymes, since M1 cleaves *before* the bases recognized and yields 5'-phosphate oligonucleotides. This calls for caution when interpreting the cleavage pattern of this enzyme, particularly in the lower region of sequence gels where the effects are greatest.

The procedure for enzymatic sequencing of end-labeled RNA is presented in Table 12. The amount of counts used will depend upon the length of the RNA., 500 cpm/nucleotide for RNases T1, U_2, PhyM, B.c., 1500 cpm/nucleotide for RNase M1, and the base-unspecific ladder provide enough activity for two gels (of 0.4 mm thickness) which need be exposed (with intensifying screens) only 24 h to X-ray films. Obviously, successful sequencing analyses can be carried out with considerably less than this simply by increasing the exposure time of the X-ray films.

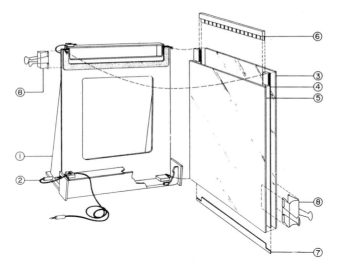

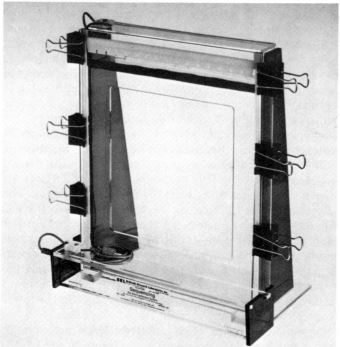

Fig. 2. The BRL sequencing gel chamber and its components. *1* stand and buffer chambers; *2* cable; *3* large glass plate; *4* side-spacer; *5* small glass plate; *6* comb; *7* bottom spacer; *8* "bulldog" clip (see Table 8 for details of gel chamber set-up)

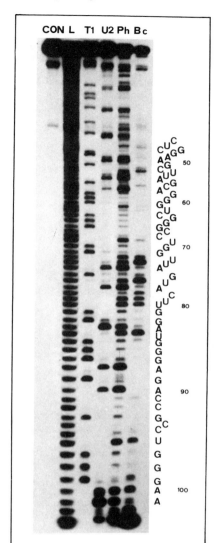
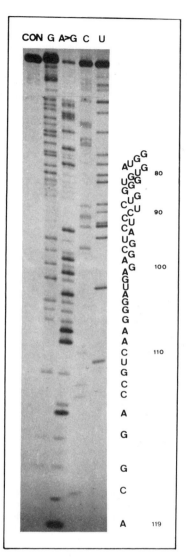

Fig. 3. Enzymatic and chemical sequencing of RNA. *Left:* the autoradiogram of a sequencing gel resolving the sequence of rabbit 5S rRNA. *CON* control incubation without enzymes; *T1* digestion with T1 RNase specific for guanine; *U* digestion with U2 RNase specific for adenine; *Ph* digestion with Phy M RNase specific for adenine and uracil; *Bc* digestion with RNase from Bacillus cereus specific for cytosine and uracil; *L* a base-unspecific ladder generated by alkaline hydrolysis of end-labeled RNA. *Right:* a chemical sequencing of 5S rRNA isolated from *E. coli*, the specificity of the reactions is given above each lane

Table 12. Enzymatic sequencing of end-labeled RNA

1. Pipet into labeled microcentrifuge tubes 5 aliquots of 50,000 cpm for RNases T1, U$_2$, PhyM, B.c. and the control 1 and 2 aliquots of 150,000 cpm for RNase M1 and the base-unspecific ladder.

2. To each aliquot add 3 μl of 1 mg/ml^{-1} solution of tRNA. If tRNA is already present in the labeled RNA preparation, from the precipitation after elution, this must be allowed for.

3. Place the tubes under vacuum and evaporate to dryness.

4. Resuspend each in 10 μl of the appropriate buffer:

   For RNases T1, PhyM and M1 and the control
   20 mM sodium citrate pH 5.0, 7 M urea, 1 mM EDTA,

   For RNase U2
   20 mM sodium citrate pH 3.5, 7 M urea, 1 mM EDTA.

   For RNase B.c.
   20 mM sodium citrate pH 5.0, 1 mM EDTA.

   For the alkaline hydrolysis
   50 mM NaOH, 1 mM EDTA.

5. Vortex thoroughly to dissolve the RNA, centrifuge briefly to collect.

6. Place all tubes except the alkaline hydrolysis in a water bath at 55 °C and incubate for 5 min. to denature the RNA.

7. Remove the tubes to room temperature and pipet 1 μl of 1:10 diluted stock enzyme solutions into the appropriate tubes, placing the drop on the wall above the RNA.

8. Vortex to mix, centrifuge to collect and quickly replace the tubes in the water bath at 55 °C.

9. Incubate the digestions for 15 min. meanwhile incubate the alkaline hydrolysis tube in a boiling water bath for 30 – 45 s, place immediately on ice and add 8 μl of loading medium.

10. Stop the enzyme digestions by placing the tubes on ice and diluting the enzyme by addition of 8 μl of loading medium.

11. Each reaction mix should be halved before loading onto sequencing gels or freezing at − 20 °C.

12. Prepare two sequencing gels of 8% and 25% polyacrylamide in 7 M urea (see Table 8) for BRL chambers. Pre-run the gels for 1 h at 45 W/gel.

13. Using a fine drawn-out capillary load one half of each sequencing digest onto each gel. The order: Control, Alkaline hydrolysis, RNase T1, RNase U$_2$, RNase PhyM, RNase B.c., RNase M1 is the most convenient for reading.

14. Electrophorese the gels at 45 – 50 W/gel. Allow bromophenol blue to migrate two-thirds down the 25% gel and xylene-cyanol to reach the middle of the 8% gel before stopping the electrophoresis.

15. After electrophoresis remove one glass plate and lay a sheet of clean, used X-ray film carefully onto the gel, press it flat. The gel will stick to the film and may then be lifted from the glass plate. Wrap the gel and film in household plastic foil, place it in an X-ray cassette with intensifying screen. Expose an X-ray film appropriately at − 80 °C.

The enzymes are purchased from Pharmacia Fine Chemicals (Sweden) and treated as recommended by the manufacturers. The 1:10 dilutions used for the sequencing reactions are stable for several weeks when stored at 4 °C. Experience often shows that variations in enzyme concentration or incubation time are required for optimal degradation of a particular RNA, presumably reflecting secondary structure effects and variation in the frequency of the bases from RNA to RNA. We, for example, use only 0.25 units of T1 when sequencing 5S rRNAs. When sequencing an RNA for the first time, it is often advisable to carry out two or three digestions with each RNase at different enzyme concentrations in order to find the best. Variation from 0.2 units to 1.5 units should be sufficient.

3.3 Chemical Sequencing

This method for sequencing RNA is based upon the modification of bases followed by strand scission at modified sites to provide a spectrum of RNA fragments which, as with enzymatic sequencing, can be resolved on polyacrylamide sequencing gels. The methodology is similar to that developed by Maxam and Gilbert for the chemical sequencing of DNA [10] and was perfected by Peattie [11]. The modification reactions are random and partial and result in opening or loss of particular bases in the RNA. Dimethylsulfate methylates the N-7 of guanosines, ring opening is then forced by reduction with sodium borohydride. Carboxyethylation of N-7 in adenine by diethylpyrocarbonate results in ring opening. This latter reaction proceeds also, but at a much slower rate, with guanine so that the reaction is actually $A \gg G$. Addition of hydrazine across the C-5 : C-6 double bond or pyrimidines results in their removal from the RNA. The specificity of this reaction can be determined by the conditions. In aqueous solution, the reaction proceeds almost exclusively with uracil whilst with anhydrous hydrazine in the presence of 3 M NaCl, modification of cytosines predominates. After modification, the damaged RNA's are extensively washed to remove reagents before proceeding to the aniline scission reaction. Here, aniline-catalyzed beta-elimination occurs only at the sites of modification and results in cleavage of the phosphodiester bonds at these sites, yielding a set of fragments which reveal the sequence when resolved on sequencing gels.

This method is considerably more time-consuming than the enzymatic method, but rewards the effort with easily read gels on which all four bases can be unambiguously identified. It is therefore to be recommended in situations where the assignment of pyrimidines is particularly difficult with the enzymatic sequencing method. One drawback is that the chemical sequencing method can only be applied to 3'-end labeled RNA, apparently due to heterogeneity on the 5'-side of the cleavages. Also, the use of reagents for nucleic acid modification adds a further hazard for the experimenter and requires special caution. All the reagents are highly poisonous. These reagents also require particular care in storage. Hydrazine undergoes auto-oxidation to yield ammonia, hydrogen, and nitrogen; in the presence of oxidizing agents it is explosive. Dimethylsulfate and hydrazine are stored tightly sealed and in the dark. Diethylpyrocarbonate decomposes to ethanol and carbon dioxide, it should be stored purged with dry nitrogen and in the cold. Aniline changes from a colorless liquid to brown on oxidation, it must be redistilled and stored frozen and in the dark: the aniline-acetate buffer should be prepared fresh. Diethylpyrocarbonate and dimethylsulfate wastes are disposed into 10 M NaOH and wastes containing hydrazine into 2 M $FeCl_3$.

Table 13 details the procedure for chemical sequencing of RNA. Considering the complexity of each reaction in comparison to the enzymatic sequencing method, it is not recommended that the modifications be conducted in parallel but rather consecutively, while the ethanol precipitations and aniline reaction can be carried out comfortably with a relatively large number of samples. Those reactions which are conducted at 90 °C should be carried out with the reaction tubes clamped between two metal racks to prevent opening due to accumulation of pressure, with subsequent loss of the labile and hazardous reagents. Since the

Table 13. Chemical sequencing of end-labeled RNA

1. In five labeled microcentrifuge tubes pipet 50,000 cpm 3'-end labeled RNA and 15 µg carrier tRNA, lyophilize. The 'control' sample is treated as in the A-reaction but without the addition of diethylpyrocarbonate.
2. *G-Reaction:* resuspend the RNA in 300 µl 50 mM sodium cacodylate pH 5.5, 1 mM EDTA.
3. Add 1 µl 50% dimethylsulfate in water.
4. Vortex, centrifuge briefly to collect
5. Incubate for 30 – 60 s at 90 °C, place on ice
6. Add   75 µl 1 M Tris-acetate pH 7.5, 1.5 M sodium acetate
            6 µl 2-mercaptoethanol
          900 µl ethanol
    mix and allow precipitation to proceed on dry ice for 10 min.
7. Centrifuge for 15 min at 8000 rpm, discard the supernatant into 10 M NaOH proceed to number 21.
8. *A Reaction and Control:* resuspend the RNA in 200 µl 50 mM Na-acetate pH 4.5, 1 mM EDTA.
9. Add 1 µl diethylpyrocarbonate (water for control), vortex and centrifuge briefly to collect
10. Incubate for 5 – 10 min at 90 °C, place on ice.
11. Add   50 µl 1.5 M Na-acetate
          750 µl ethanol
    mix and allow precipitation to proceed on dry ice for 10 min.
12. Centrifuge for 15 min at 8000 rpm, discard the supernatant into 10 M NaOH proceed to number 21.
13. *U Reaction:* resuspend the RNA in 10 µl 50% hydrazine in water (freshly prepared), vortex and centrifuge briefly to collect.
14. Incubate for 10 – 15 min on ice
15. Add 200 µl 0.3 M Na-acetate, 0.1 mM EDTA
          750 µl ethanol
    mix and allow precipitation to proceed for 10 min on dry ice.
16. Centrifuge for 15 min at 8000 rpm, discard the supernatant into 2M $FeCl_2$ proceed to number 21.
17. *C Reaction:* resuspend the RNA in 10 µl freshly prepared 3 M NaCl in anhydrous hydrazine, vortex and centrifuge briefly to collect.
18. Incubate for 20 – 30 min on ice.
19. Add 1 ml 80% ethanol (prechilled to – 20 °).
    mix and allow precipitation to proceed on dry ice for 10 min.
20. Centrifuge for 15 min at 8000 rpm, discard the supernatant into 2M $FeCl_3$ proceed to number 21.
21. *Wash Procedure:* resuspend the ethanol pellet of modified RNA in 200 µl 0.3 M Na-acetate, add 500 µl ethanol and allow precipitation to proceed on dry ice for 10 min.
22. Centrifuge at 8000 rpm for 15 min to collect pellet, discard the supernatant into the appropriate waste solution.
23. Resuspend the pellet in 200 µl 0.3 M Na-acetate add 500 µl ethanol and again precipitate on dry ice for 10 min.
24. Centrifuge at 8000 rpm for 15 min to collect pellet.
25. Add 500 µl 80% ethanol to the pellet but do not mix, recentrifuge and discard the supernatant.
26. Place the pellets under vacuum and evaporate to dryness.
    A, C and U reactions are now ready for the aniline scission, the G reaction must first go through a reduction procedure.
27. *G Reaction Only:* resuspend the dried RNA pellet in 10 µl 1 M Tris-HCl pH 8.2 add 10 µl 0.2 M $NaBH_4$ (freshly prepared)
28. Incubate for 30 min on ice.
29. Add 200 µl 0.6 M Na-acetate pH 4.5
            1 µl tRNA at 5 mg ml^{-1}
          600 µl ethanol
    mix and allow precipitation for 10 min on dry ice.

Table 13 (continued)

---

30. Centrifuge at 8000 rpm for 15 min to collect pellet.
31. Wash the RNA pellet by adding 500 μl 80% ethanol to the tube and immediately recentrifuging.
32. Place the pellet under vacuum and evaporate to dryness, the G-reaction is now also ready for the aniline scission reaction.
33. *Aniline Scission:* prepare a fresh solution of 1M aniline acetate pH 4.5 buffer by diluting 0.1 ml redistilled aniline (11 M) with 1 ml acetic acid (approx. 4.5 M, previously determined empirically) to an end pH of 4.5.
34. Resuspend each RNA pellet in 20 μl aniline-acetate pH 4.5 by thorough vortexing, centrifuge briefly to collect.
35. Incubate at 60 °C in the dark for 20 min.
36. Freeze the samples on dry ice, lyophilize to dryness
37. Resuspend the RNA in 20 μl water, freeze and lyophilize
38. Resuspend the RNA in 20 μl water, freeze and lyophilize
39. Redissolve the RNA in 10 μl loading buffer and proceed to gel electrophoresis as in Table 12 from point 11.

---

reaction products are highly uniform and secondary structure has virtually no detrimental effect on product formation, it is not necessary to construct a base-unspecific ladder. Figure 3 shows an example of a chemical sequencing analysis.

3.4 Wandering-Spot Sequence Analysis

This method [12], also known as two-dimensional mobility shift analysis, is well suited to analysis of small RNA's or RNA fragments, but, in the form described here at least, cannot be considered as a thorough sequencing technique. The method allows differentiation of the bases G and U from the bases C and A. It is, therefore, very useful for the confirmation of pyrimidine assignments made from the enzymatic sequencing method. A further use of this method lies in the determination of the site of divergence of two otherwise identical sequences.

The degradation of the end-labeled RNA is achieved by alkaline hydrolysis and the RNA fragments are separated on sequencing gels at pH 3.5 according to size and the charge on the last base of the fragment. At pH 3.5 adenine and cytosine have positive charges due, respectively, to protonation of N_1 (pK = 3.6) and N_3 (pK = 4.3). Guanine and uracil are uncharged at this pH. The terminal nucleotide carries only one negative charge on the phosphate at pH 3.5 so that overall, terminal guanosine and uridine are negatively charged while terminal cytidine and adenosine are uncharged. Thus the difference in mobility between fragments of lengths (n) and (n+1) depends upon the nature of the additional nucleotide. After electrophoresis at low pH, a strip of gel containing the "ladder" is excised and polymerized into a second gel. The second electrophoresis is carried out as usual at pH 8.3, where separation is solely according to size. Autoradiography yields a series of spots which wander from top to bottom with base-characteristic angles between successive spots. Figure 4 shows such a wandering spot analysis. The procedure for analysis by this method is presented in Table 14.

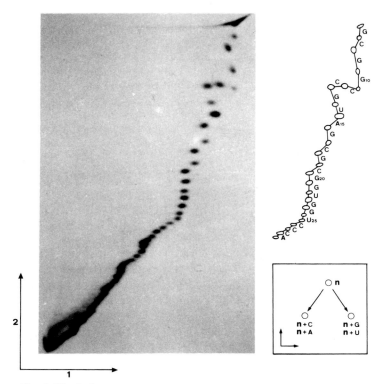

Fig. 4. Wandering-spot sequence analysis of RNA. The autoradiogram of a wandering-spot analysis of 5S rRNA isolated from *E. coli* is shown together with the interpretation based upon the mobility shift effects shown in the *inset*

Table 14. Wandering-spot sequence analysis

---

1. Lypophilize, 100,000 cpm end-labeled RNA in a microcentrifuge tube.
2. Resuspend the RNA in 10 µl 50 mM NaOH, 1 mM EDTA.
3. Incubate for 45 s in a boiling water bath.
4. Add 10 µl of loading buffer (25 mM sodium citrate pH 2.0, 7M urea, 25% sucrose, 0.2% bromophenol blue, 0.2% xylene cyanol).
5. Prepare a 10% polyacrylamide gel in 25 mM sodium-citrate pH 3.5, 7 M urea of dimensions 40 cm × 20 cm × 0.5 mm.
6. Load the alkaline hydrolysate onto the gel and electrophorese in 25 mM sodium-citrate pH 3.5 at 200 V. Stop electrophoresis when bromophenol blue has reached the middle of the gel (approx. 20 h).
7. Excise a strip of gel 16 cm × 1 cm from the "ladder" extending 4 cm below the bromophenol blue. (If sufficient counts are loaded this may be facilitated by autoradiography of the gel).
8. Place the gel strip 2 cm from the bottom of a 40 cm glass plate, place spacers into position as usual and clamp the second plate onto the first. Seal with agar as usual (see Sect. 2.2).
9. Prepare a 25% (for resolution of approx. 20 nucleotides from end) or 10% (for sequences further from the end) polyacrylamide gel in 7 M urea as described in Sect. 2.2. To 2 ml of gel solution add TEMED and ammonium persulfate and pour a little into the assembled gel mold and allow to polymerize and thus fix the gel strip in position. Add TEMED and ammonium persulfate to the remaining gel solution and pour into the mold. Allow to polymerize.
10. Electrophorese in TEB at 1000 V until bromophenol (25% gel) or xylene cyanol (10% gel) has reached the middle of the gel, autoradiograph.

3.5 End-Nucleotide Determination

Establishing the first and last nucleotides of an RNA from sequencing gels is often difficult. Enzymatic digestion yields the 5′-end nucleotide as a 3′, 5′-bis-phosphate which, due to its higher charge to mass ratio, migrates particularly fast and may be lost from sequencing gels. Chemical sequencing of 3′-end label-ed RNA identifies the 3′-end nucleotide through the pCp which remains after scission; again this may be lost from gels. Enzymatic digestion of 3′-end labeled RNA cannot reveal the 3′-end nucleotide, since the fragment which would identify it is Cp, which carries no radiolabel.

In order to firmly establish the nature of the end nucleotides it is, therefore, necessary to use a further technique which again bases on the specificity of certain RNases. RNase T2 digests RNA to 5′ phosphate nucleotides, when allowed to proceed to completion with 3′-end labeled RNA a mixture of nucleotides is produced, of which one type only can carry a radiolabel, namely that arising from the 3′-end. P1 RNase digests RNA to 5′-phosphate nucleotides and is therefore used with 5′-end labeled RNA.

The mixture of nucleotides can be separated by thin-layer chromatography, subsequent autoradiography of the chromatography plate identifies the labeled nucleotide as shown in Fig. 5. Details of the procedures are given in Table 15. While rapid results are achieved with 1000 cpm, the identification of end nucleo-tides can be achieved with as little as 100 cpm with appropriately long exposure times.

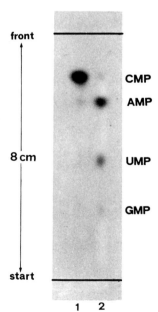

Fig. 5. End-nucleotide determination. The autoradiogram of a PEI-cellulose plate is shown, probe *1* was a P1 digest of [5′-^{32}P] 5S rRNA from *Equisetum arvense*, probe *2* was a T2 digest of the same RNA labeled at the 3′-end. The migration of the nucleotide of this RNA can be established as C whilst the 3′-end labeled material is obviously heterogeneous

Table 15. End-nucleotide determination

1. Lyophilize 1000 cpm 3'-end and 5'-end labeled RNA with 1 – 10 µg carrier RNA in microcentri-
    fuge tubes.
2. Resuspend 3'-end labeled RNA in 20 mM NH$_4$-acetate pH 4.5, add T2 RNase to 0.1 unit/µg
    RNA incubate for 3 h at 37 °C.
3. Resuspend 5'-end labeled RNA in 50 mM NH$_4$-acetate pH 5.3, add P1 RNase to 1 µg/µg RNA
    incubate for 5 h at 37 °C.
4. Cut pre-washed (2 M formic acid/pyridine pH 2.2, distilled water) poly-ethyline-imine-cellulose
    chromatography plates (polygram CEL 300 PEI, Macherey-Nagel) into 10 cm × 10 cm squares.
    Draw a soft pencil line 1 cm from one edge and mark positions for loading samples.
5. Load the digested RNA smaples and 0.8 A$_{260}$ units of AMP, CMP, GMP, and UMP as standards
6. Standard the plate in a beaker containing 0.5% formic acid and run until the solvent front is
    0.5 cm from the top, remove the plate and allow to dry.
7. Return the plates to the beaker which now contains 0.15 M Li-formate pH 3.0. Run again until
    the front reaches 0.5 cm from the top edge, dry the plate.
8. Examine the plate under UV and mark in pencil the positions of the four standards. Wrap the
    plate in plastic film, stick two or three phosphorescent spots onto the plate and expose to an X-
    ray film overnight.
9. Develop the film and, using the phosphorescent spots for orientation, mark the positions of the
    standards on the film.

3.6 Interpretation of Sequencing Gels

The techniques presented in this chapter were originally heralded under titles
such as "rapid-read-off-sequencing". Whilst these methods do represent an
enormous improvement on previous sequencing techniques, they are perhaps not
so facile as one might have thought, and the band patterns which are the end-
point of the more or less laborious practical work require interpretation rather
than mere translation onto paper. The points which must be kept in mind, and
problems which may occur will be dealt with below.

3.6.1 RNA Heterogeneity

The requirement for single-species RNA has been emphasized and the effect of
RNA heterogeneity on sequencing gels is devastating. Where assignments cannot
be made at the majority of positions because all enzymes cut with apparently
equal efficiency, the material is probably heterogeneous. Examination of the un-
cut material at the top of the lanes may reveal more than one band, in this case
the next step is clear – repurification of the material via gel electrophoresis. A
common problem is length heterogeneity at the labeled end; in this case each
enzymatic cleavage leads to two (or more) labeled fragments which differ only by
one nucleotide. Such heterogeneity manifests itself as *double-banding* in gels,
again electrophoretic purification is called for. It is, unfortunately, possible that
the multiple RNA's have the same length, in this case separation according
to secondary structure must be attempted, several possibilities are given in Section 2.2.

3.6.2 Enzymatic Digestion

An indication of the specificity of the enyzmes used in sequencing has been given in Section 3.2. It is important to note that the enzymes do cut unspecifically, so that it is the major band at each position which must be taken, this is usually clear for the purines, but is particularly exacting for the pyrimidines, where identification is based on comparison of cleavage by two enzymes. Uracils should yield bands in both PhyM and B.c. lanes, cystosines only in the B.c. lane. Here, weak cleavage by PhyM at cytosines and the irregularity of the B.c. digestion may cause confusion. When cytosines occur in clusters only the first is cut, the others showing no, or only very weak, cleavage.

At regions of particularly strong secondary structure, the enzymes may cleavage weakly or not at all, while the alkaline hydrolysis proceeds as usual. This leads to *gaps in the enzyme lanes* on sequencing gels which are not found in the alkaline hydrolysis. Attempts to denature the RNA more effectively before enzymatic digestion may be successful; however, the most effective approach may be to turn to chemical sequencing.

3.6.3 Electrophoresis

Stable secondary structures may lead to a further effect of considerable significance – *band compression*. This effect manifests itself as a particularly strong band in all lanes *including* the alkaline hydrolysis. This strong band consists, in fact, of several fragments which should normally be resolved below it but which, due to formation of hairpin-loop structures, mimic fragments of shorter length. These hairpin loops may correspond to structures also present in the native RNA molecule, although this must not necessarily be the case. The number of residues "missed" through this phenomenon can not be assessed. Running sequence gels at higher temperatures often helps to force denaturation of such regions, an electrophoresis chamber with thermostated glass plates is commercially available (LKB, Sweden). In some cases sequencing with RNA labeled at the opposite end, and thus attacking the region from the other side, may be sufficient to complete the sequence.

Whilst these latter effects are annoying to those who merely wish to obtain a sequence, the investigator who is interested in RNA secondary structure will welcome this "free" structural information.

One last effect which must be mentioned is "smiling". This describes the characteristic appearance of sequence gels when the distribution of heat during the electrophoresis is uneven with the center of the gel much hotter than the edges. Since an increase in temperature reduces the viscosity of the gel, fragments in the center of the gel migrate faster than those at the edges. Smiling may be merely unaesthetic or can make the interpretation of some gels impossible. Cures are to use lower voltages, with the risk of incomplete denaturation or turn to commercial chambers which have thermostated plates to distribute the heat evenly or which use thick glass plates (5 mm) to prevent heat loss at the edges.

3.6.4 Chemical Sequencing

The products of chemical sequencing are highly uniform and are produced irrespectively of the secondary structure of the RNA, so that interpretation of chemical sequencing gels is usually easy. Such problems as do arise can usually be traced back to either insufficient washing between modification and scission or to deterioration of the modification reagents.

References

1. Rickwood D, Hames BD (1985) Gel electrophoresis of nucleic acids; a practical approach. IRL Press, Oxford, England, Washington, DC
2. Maniatis T, Fritsch EF, Sambrook J (1982) Molecular cloning. A laboratory handbook. Cold Spring Harbor Lab, New York
3. Murray HG, Thompson WF (1980) Nucleic Acids Res 8:4321 – 4325
4. Payvar F, Schimke RT (1979) Eur J Biochem 101:271
5. Wakefield L, Gurdon JB (1983) EMBO J 2:1613 – 1619
6. Szekely M, Sanger F (1969) J Mol Biol 43:607 – 617
7. Rose JK, Lodish HF (1976) Nature 262:32 – 37
8. England TE, Uhlenbeck OC (1978) Nature 275:560 – 561
9. Donis-Keller H, Maxam AM, Gilbert W (1977) Nucleic Acids Res 4:2527 – 238
10. Maxam AM, Gilbert W (1977) Proc Natl Acad Sci USA 74:560 – 564
11. Peattie DA (1979) Proc Natl Acad Sci USA 76:1760 – 1764
12. Lockard RE, Alzner-Deweerd BA, Heckman JE, MacGee J, Tabor MW, RajBhandary UL (1978) Nucleic Acids Res 5:37 – 56

8.2 Synthetic Genes as a Powerful Tool for Protein Structure-Function Analysis

RONALD FRANK and HELMUT BLÖCKER[1]

Contents

1 Introduction

Proteins effect their versatile biological functions by folding into defined tertiary structures. A folded protein chain assembles precisely the relevant atomic groups that participate in substrate binding and catalysis, and provides the necessary stability and flexibility of the active conformation. The principles for building such complex structures from selected amino acid sequences are still poorly understood. Advanced techniques such as X-ray crystallography and other spectroscopic methods (NMR, IR, CD) have brought a deeper insight into the three-dimensional architecture and physicochemical properties of proteins. However, conclusions and theories derived from these studies require experimental verification.

Despite important progress in peptide chemistry, the chemical synthesis of model proteins remains limited to relatively small molecules (<50 AA). These are useful tools for studies on isolated folding domains, but their physicochemical behavior differs considerably from that of larger proteins [1]. De novo or semi-synthesis of larger protein chains is still too complex to be of general value.

Recent advances in genetic engineering techniques and chemical synthesis of oligodeoxyribonucleotides have opened an alternative route to circumvent the multiple difficulties raised by the complex nature of proteins. The structure of DNA is much more regular and the chemical synthesis of DNA fragments has be-

1 DNA Synthesis Group, GBF (Gesellschaft für Biotechnologische Forschung mbH), Mascheroder Weg 1, D-3300 Braunschweig, FRG

Advanced Methods in Protein Microsequence Analysis
Ed. by B. Wittmann-Liebold et al.
© Springer-Verlag Berlin Heidelberg 1986

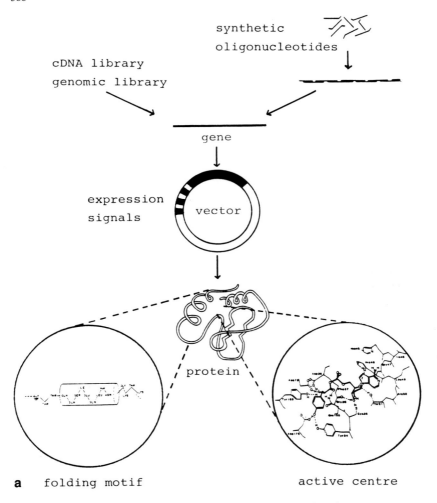

Fig. 1a, b. Basic strategies in protein engineering via genetic engineering

come routine in many laboratories. An extensive set of enzymes is available for the manipulation of DNA sequences, and cloning allows propagation and selection of a single molecule. The basic techniques employed are outlined in Fig. 1. A prerequisite is a gene which codes for the amino acid sequence of the protein to be investigated. In those cases where the protein is a natural one, its gene can be isolated from a cDNA or genomic library. Highly sensitive screening methods using synthetic oligonucleotides as hybridization probes allow the identification of any gene if only a tiny part of the protein or DNA sequence is known. The gene is then inserted into a vector (small-size genetic element capable of self-replication) with its start codon correctly positioned behind suitable regulatory sequences (ex-

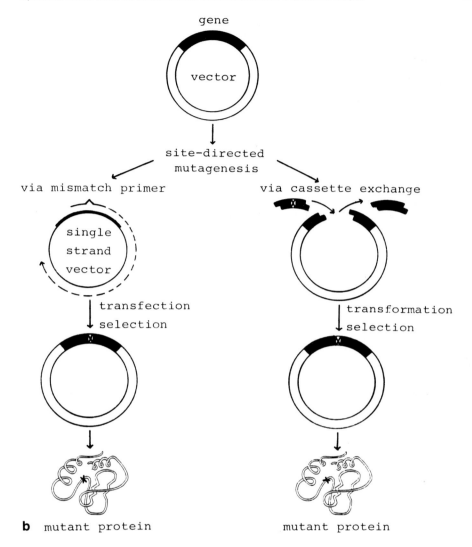

gene

vector

site-directed
mutagenesis

via mismatch primer   via cassette exchange

single
strand
vector

transfection   transformation
selection   selection

b mutant protein   mutant protein

pression signals). All manipulations of the initially isolated DNA fragment carrying the gene are greatly facilitated with the help of synthetic oligonucleotides as
linkers or adaptors. After cell transformation of the vector the protein synthesis
machinery will then produce the target protein.

   As in natural evolution, new protein variants can be obtained by mutagenesis
of the parent gene (Fig. 1b). Efficient mismatch primer and cassette strategies
have been developed to introduce any possible mutation at a predefined site. Single amino acid exchanges, as well as insertions and deletions of any size, allow
for a systematic investigation of, for example, individual amino acid side-chain
participation in a catalytic center, the influence of amino acid substitutions on

folding pathways or the structural and functional role of whole protein domains. Undoubtedly, the rapid and economic availability of synthetic DNA as mismatch primers and double-strand cassettes is the basis for the versatility of the site-directed mutagenesis concept.

From the beginning of chemical DNA synthesis, the construction of complete genes has been a challenge to chemists. To date a large number of functional genes have been assembled from synthetic oligonucleotides up to a length of over 600 base pairs. With regard to protein engineering studies, synthetic genes may offer more rapid access to the desired gene; all necessary features for expression and further manipulation are already included, whenever the primary structure of the gene or gene product is known. Similarly, virtually any other protein sequence can be encoded in the synthetic gene and produced biologically. This aspect is particularly important for de novo creation of new protein structures. Gene synthesis in combination with genetic engineering techniques is a powerful tool for the study of complex protein structure-function model systems.

Tailoring of a new protein for optimal adaption to an assigned function still seems a long way away. However, the methodology is at hand and a few steps have already been made. In the following we will discuss strategies known at present for rapid gene synthesis.

2 Construction of DNA Double Strands

In this section we will describe the strategies currently available for the construction of DNA double strands which need manipulations other than the simple combination of two complementary oligonucleotides. Such DNA segments may be used as complete functional genes, as long adaptors to fuse an isolated gene properly to the vector's regulatory signals if suitable restriction sites are not present immediately around the start codon, or as long "cassettes" for site-directed mutagenesis.

Classification of strategies discussed below is arbitrary and is made in such a way as to emphasize the philosophy behind them. Many intermediate variations are possible and have been used.

2.1 The Short Fragment Approach

The classical strategy developed by Agarwal et al. [2] makes use of a class of enzymes that catalyze the sealing of single-strand breaks (nicks) in double-stranded DNA: the DNA ligases. The predefined DNA sequence which is to be synthesized is divided into short fragments (15 to 20 nucleotides in length) that overlap each other by at least five nucleotides (Fig. 2). Base pairing (hybridization) between complementary overlaps will assemble the fragments correctly, end to end, like two rows of bricks in a wall. The nicks that appear in turns on either strand are sealed upon incubation with the enzyme to yield the continuous double strand. It should be pointed out that the nicked double strand as shown in Fig. 2 is physi-

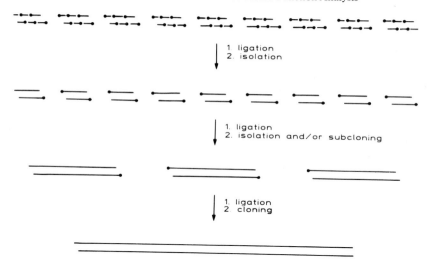

Fig. 2. Construction of a long DNA double strand by the short-fragment approach. *Dots* indicate 5'-phosphorylated ends

cally not existent under reaction conditions. Hybridization over such short overlaps is a dynamic process and the portion of double strands is dependent on the thermal stability of individual pairings. Temporarily formed duplices are then stabilized by complexation with the enzyme and ligated. The fragments must have 5'-phosphorylated and 3'-nonphosphorylated ends. Only this configuration at the nicks is substrate for the enzyme. Highly purified DNA ligase, free of DNase and phosphatase contaminants, is available from many commercial suppliers. Predominantly DNA ligase from T4 infected *E. coli* (T4 DNA ligase) is used.

Some important requirements have to be fulfilled for a successful construction:

1. A careful design of the fragments must ensure uniqueness of overlaps within a set of fragments ligated. This will reduce or practically exclude base pairing and ligation of other than the desired combinations. For long DNA, build up from more than ten fragments, the planning is conveniently done by computational methods.
2. Purity of the fragments especially with regard to base modified contaminants is an important point to be considered (see 4 and Sect. 3). Such impurities will result in mutations and can make the search for the correct gene sequence tedious.
3. Complete 5'-phosphorylation of the fragments is crucial. Synthetic oligonucleotides have a free 5'-hydroxyl end that has to be enzymatically phosphorylated with polynucleotide kinase (T4 PNK) and ATP. This reaction allows introduction of a radioactive label by the use of $[\gamma\text{-}^{32}P]$ ATP which greatly facilitates monitoring of the phosphorylation and subsequent ligation. Conveniently all the synthetic fragments to be ligated can be combined and phosphorylat-

ed all together. As T4 PNK and T4 DNA ligase both have the same cofactor and buffer requirements, no isolation of the radiolabeled fragments is necessary. However, completion of the phosphorylation has to be ensured before proceeding with the ligation (see also 5). Simple and rapid assays are described in the literature [3].

4. Ligation should be performed at the highest possible temperature (T4 DNA ligase will even endure 40 °C for a short time). The correct overlaps are designed to have the highest thermal stability (see 1) and high ligation temperatures will disfavor most of the pairings between unwanted fragments as well as failure sequences and $n-1$ or $n+1$ contaminants. The temperature effect on product distribution is demonstrated in Fig. 3B.

5. Correct stoichiometry of the fragments is essential. T4 DNA ligase is known to catalyze ligation also at mismatched overlaps, double strand breaks (blunt ends) and small gaps. Thus, if a fragment is underrepresented, numerous side reactions may occur at the remaining ends. Even if correct amounts of oligonucleotides were combined, an incomplete 5'-phosphorylation will cause an insufficient number of active components to be present.

6. The final DNA double strands need suitable ends for recombination with restriction sites of the cloning vector. Fragments forming these ends are commonly included without a 5'-phosphate to protect them from self-recombination. However, these ends are critical sites with respect to problems mentioned under 5. We always observed a much cleaner ligation by allowing self-recombination of ends and recutting of multimers with the corresponding restriction endonucleases. Blunt ends present special problems, as ligation of blunt ended double strands does not always result in restriction sites that can be recut.

The work expenditure on a long gene, as outlined in Fig. 2, greatly depends on the number of subsegments that have to be prepared. The aim is to include a maximum number of fragments into the first ligation step (but see also Sect. 2.4). Considering all the precautions mentioned above, we found that very long DNA double strands (up to 300 base pairs) can be obtained in high yields from up to 40 fragments in a quick one-step ligation reaction. Purity of the product is often high enough for direct cloning without any further purification. The mutation rate presently observed with this cloned DNA is much lower than the previous rate of 1 in 500. Figure 3 shows a few examples of gene construction following this approach.

2.2 The Long Fragment Approach

Since the availability of longer synthetic oligonucleotides (30 – 50-mers) DNA double strands can be assembled from fewer fragments (Fig. 4). Overlaps become much longer and mismatches caused by base-modified failure sequences or wrong pairings will not have a pronounced effect on thermal duplex stability. As a consequence, the correct sequences are no longer favored by high ligation temperatures. This strategy therefore relies on fragments of high purity to reduce

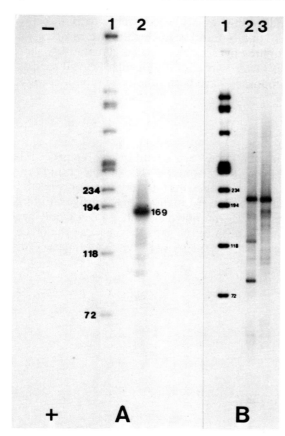

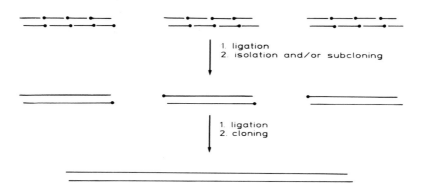

Fig. 3A,B. Examples for one-step ligations of large numbers of short fragments. Samples were taken directly from ligation mixtures and analyzed on nondenaturing 10% polyacrylamide-gels (20 cm × 40 cm ×0.04 cm). Gels were exposed to X-ray films with amplifying screens. A lane 1: ∅χ174 RF DNA-fragments, $5'$-^{32}P-labeled; lane 2: ligation of 22 fragments to yield a 169 bp segment (0.1 nmol of each fragment, 66 mM Tris.HCl pH8, 6 mM MgCl$_2$, 10 mM DTE, 0.2 mM [γ-^{32}P] ATP of 1 Ci mmol^{-1}, reaction volume 20 μl, incubation was 30 min at 37 °C with 6U T4 PNK, 2 min at 100 °C, 6 h at 30 °C with 2U T4 DNA ligase, 6 h at 15 °C with additional 2U T4 DNA ligase, 15 min at 65 °C, adjustment of buffer conditions for cleavage of multimers with restriction endonucleases by dilution to 40 μl, 6 h at 37 °C with 5U of each enzyme). B lane 1: see A, lane 1; lane 2: ligation of 28 fragments to yield a 211 bp segment (reaction conditions see A, lane 2, volume was 40 μl); lane 3: as lane 2 but first incubation with ligase was 6 h at 37 °C

Fig. 4. Construction of a long DNA double strand by the long-fragment approach. Dots indicate $5'$-phosphorylated ends

mutation rates. Preferably, the ligation sites are not placed opposite to the center of the complementary fragments. Double-strand segments are preformed separately from pairs of fragments with long overlaps and ligated via relatively short overlaps (5 – 7 nucleotides) such as DNA restriction fragments. Several successful gene synthesis following this approach have been reported in the literature.

2.3 Fill-In Approaches

A further reduction of fragment numbers has been attempted by the procedure shown in Fig. 5A, basically proposed by Rossi et al. [4]. The double strand is no longer completely chemically synthesized. Fragments (40 – 60-mers) only partially overlap to give a specific and stable enough template/primer configuration. A DNA polymerase (Klenow fragment of *E. coli* polymerase I, T4 DNA polymerase, reverse transcriptase) is then used to fill in the gaps left between the neighboring fragments. The remaining nicks are then closed by DNA ligase. This

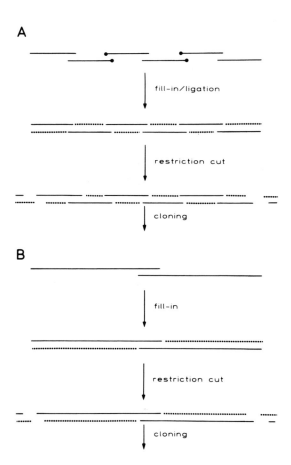

Fig. 5A, B. Construction of a long DNA double strand by fill-in approaches. *Dots* indicate 5'-phosphorylated ends, *dashed lines* indicate enzymatically synthesized DNA

rather elegant strategy, however, currently results in too many mutations (deletions predominantly) to be of practical use.

Very recently the synthesis of rather long oligonucleotides (over 100-mers) has become feasible. A variation of the fill-in approach (Fig. 5B) was successfully used to prepare DNA double strands. Two oligonucleotides are designed to overlap only with their 3'-end regions. DNA polymerase then completes the full length duplex. At this stage, any base modification or failure sequence inevitably will be fixed as a mutation. After cutting at the ends with suitable restriction endonucleases, accompanied with a loss of about 10 base pairs, the segment can be cloned. Although the DNA obtained by this approach currently does not show a distinct band but rather a smear on an analytical gel, it has been shown that the target sequence can be sorted out in a reasonable time. Long genes may be assembled by recombination of several subcloned pieces.

2.4 General Remarks

It has been pointed out several times that contaminated oligonucleotide fragments can result in a considerable accumulation of mutated sequences. Although rapid DNA sequencing techniques are available and many clones can be analyzed, a high mutation rate will complicate the search for the correct sequence and may increase the work load in this part of the project. Each chosen methodology (including the chemistry) gives rise to a certain mutation rate. Therefore, the length of double strands constructed, prior to the first cloning and analysis, should be carefully adapted to the frequency of mutations expected. Even if the methodology would allow the synthesis of much longer pieces of DNA, the length should be kept below the length that gives rise to the statistical appearance of one error, e.g., about 200 base pairs for a mutation rate of 1 in 300. To combine a complete gene from subcloned pieces that are easily identified (sequencing of one or two clones) may be more efficient than to construct the whole gene in one step.

Genes obtained by the strategies described in Sections 2.1 and 2.2 basically represent pieces of DNA built up from small cassettes. New gene variants for protein engineering are easily made, in high yields, by exchange, insertion or deletion of two or more fragments. No other technique, e.g., mismatch-primer mutagenesis, is needed to obtain specifically altered gene sequences. Genes constructed by filling-in of long fragments are more suited for site-directed mutagenesis as described in Fig. 1B. A mass production of such long fragments to build new variants would be very uneconomic.

An almost unlimited number of sequence permutations are available for the creation of new proteins and the change of even only a few target sites within a natural protein can rapidly exceed a hundred desired mutations. The enzymatic construction of DNA double strands, including the cloning, isolation, and sequence analysis, can be achieved in about 2 to 3 weeks. Many gene variants can be handled simultaneously.

3 Chemical Synthesis of Oligodeoxyribonucleotides

Two very efficient methods for the chemical assembly of oligodeoxyribonucleo-
tide chains have evolved from three decades of intensive research by many laboratories all over the world: the phosphotriester route and more recently the phosphoramidite route (Fig. 6). The chemistry has been described already in many review articles [5, 6] and a detailed compilation of strategies and work protocols
are given in References [7]. Here, we will raise only a few points regarding
reasonable ways of producing synthetic DNA fragments with special emphasis on
their use in gene construction.

Rapid and economic production of oligonucleotides is exclusively achieved by
the solid phase technique, initially developed by Merrifield for the synthesis of
peptides [8]. The type I building block (Fig. 6) is linked via its 3'-protecting
group to the surface of a polymeric carrier. Beads of silica gel or porous glass are
most widely used. Benefits of this technique are: separation of the reaction product (the elongated chain) is achieved by simple washing operations; high excesses
of reagents can be used to drive reactions to completion; working on an economic small scale is possible and desirable as only nmol amounts of the target sequences are needed. Basically the two to four steps required for the coupling of
one nucleotide unit (Fig. 6) are repeated until the chain is finished. The present
chemistry allows reaction times in the range of a few seconds and one elongation
cycle can be completed in about 6 min.

It seems unlikely that further improvements in the chemistry will significantly
reduce cycle times. Thus technical performance of the chemical reactions greatly
determines the efficiency (number of cycles per unit of time) of oligonucleotide
assembly. The support must be embedded into a suitable reaction vessel that allows coupling under anhydrous conditions and efficient washing. Small sintered

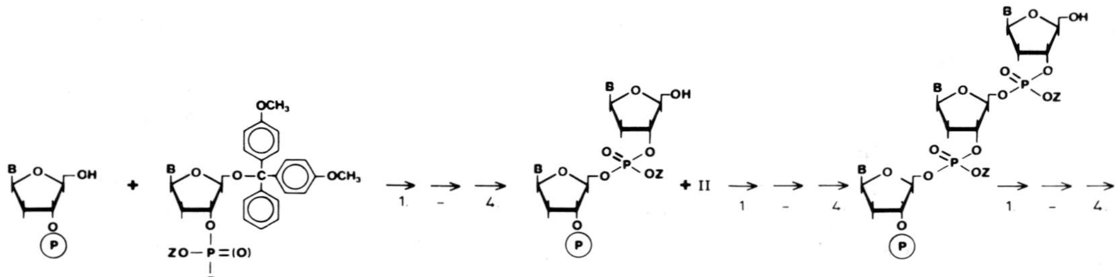

P = Carrier $-\overset{O}{\overset{\|}{C}}-CH_2-CH_2-\overset{O}{\overset{\|}{C}}-$

B = protected nucleobase A,C,G,T

R = phosphate activating group

Z = phosphate protecting group

1.   coupling
(2.) oxidation
(3.) capping
4.   detritylation

Fig. 6. Chemical synthesis of oligodeoxyribonucleotides. Phosphotriester method: $R = -O^-$ activated with a benzene-sulfonic acid derivative (chloride, azolide) and catalyzed with a base such as pyridine or N-methylimidazole; Z = o-chlorophenyl. Phosphoramidite method: R = dialkyloamino activated with tetrazole; Z = methyl or β-cyanoethyl

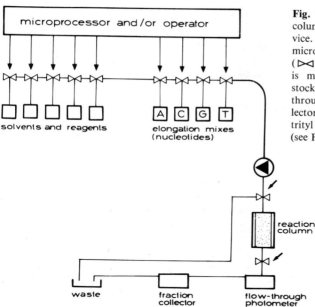

Fig. 7. Schematic drawing of a one column flow-through synthesis device. *Arrows* point to manually or microprocessor controlled valves (⋈). Flow of solvents and reagents is maintained by overpressure on stock bottles or a pump ◀. A flow-through photometer or fraction collector may be connected to monitor trityl colors at the detritylation steps (see Fig. 6, step 4)

glass funnels or centrifuge tubes closed with rubber septa or syringes are the most simple equipment. However, application of reagents and solvents requires a lot of manual work and time. A column-type reactor is easily connected to a solvent delivery system that feeds reagents and solvents from selected stock bottles (Fig. 9). Injection of activated building blocks and switching of solvent selector valves are the only manual operations needed. Continuous flow-through of solvents is very efficient and washing times can be reduced considerably. With these manual devices a limited number of parallel syntheses can be performed depending on the skill of the operator.

The column device already comprises the basic features of a fully automated synthesizer (Fig. 7). Only the appropriate electronically controlled valves have to be built in and connected to a microprocessor. The fastest of the currently available synthesizers finish one elongation cycle every 6 min and include various options such as parallel syntheses of up to four sequences, synthesis of mixed sequences and on-line cleavage from the support. Some manufacturers guarantee reliable repetitive coupling yields of >98%. Recently it has been shown by gel analysis that the synthesis of oligonucleotides over one hundred nucleotides in length results in a band of expected mobility (Fig. 8). However, prices of these machines are still very high and also operating costs may soon exhaust a limited budget, as only approved high quality reagents and solvents have to be used. In certain cases the manufacturer's guarantee is only valid if their own (sometimes very costly) chemicals are exclusively used.

A new route for the mass production of oligonucleotides was introduced with the segmental support approach [9]. Mechanically and chemically stable segments of support material such as cellulose paper or conventional materials sealed in membranes or kept in distinct column compartments allow the combi-

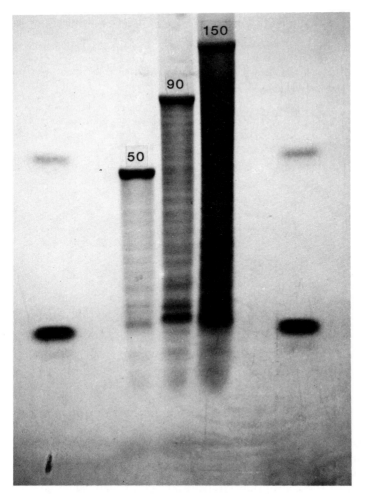

Fig. 8. Three preparations of long oligonucleotides by the β-cyanoethyl phosphoramidite chemistry on a Model 8600 DNA synthesizer, analyzed by electrophoresis on a denaturing polyacrylamide gel and UV shadowing. [The photograph was kindly provided by Biosearch Inc. (New Brunswick Scientific Co., Inc.)]

nation of different sequences that require addition of the same building block into one common reaction (Fig. 9). In this way large numbers of individual oligonucleotides can be synthesized simultaneously with only four separate coupling reactions at each elongation cycle. The total work load depends only on the length of the sequences and not on their number. Such multi-syntheses are conveniently done with an inexpensive four-column type manually operated solvent delivery system. We use small cellulose discs as solid support and routinely include up to 150 sequences in one multi-synthesis. This segmental support technique is particularly advantageous for economic production of oligonucleotide fragments used in the classical short fragment approach to gene construction.

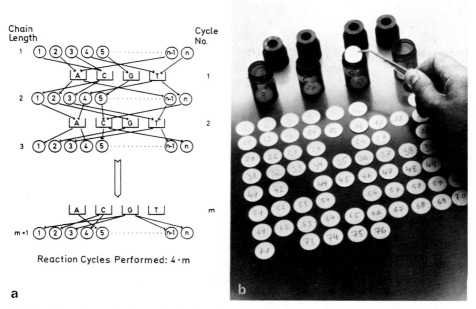

Fig. 9. a Flow scheme for the synthesis of *n* oligonucleotides of chain length *m* by the segmental support technique. *Circles* indicate disc-shaped support segments on which oligonucleotide sequences characterized by numbers are assembled. **b** Sorting the discs into *A, C, G,* and *T* groups

Today, the most widely followed concept of protecting group combination is still the one that developed from the pioneering work of Agarwal et al. [2]. However, it does not provide complete protection, and several side reactions are observed, e.g., modification of guanine and thymine at enolizable carboxyl functions, reactions at the second proton of mono-acylated exocyclic amino functions, depurination due to repeated acidic treatment (Fig. 10). Accumulation of side products in particular limits the practical length in oligonucleotide assembly. Substantial progress (although not the final answer) has been achieved in the development of new protecting groups. We await such new building blocks, for DNA synthesis, soon to be commercially available. The use of the latter will not only make chemical synthesis more reliable and increase yields but also greatly facilitate purification especially of longer oligonucleotides.

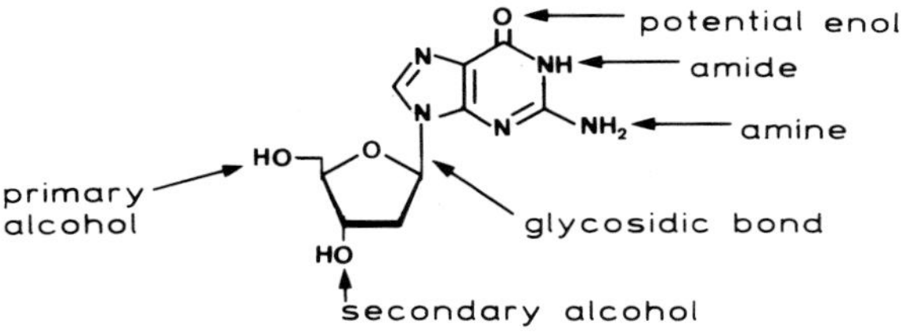

Fig. 10. Reactive sites of a deoxynucleoside, dG

Deprotection, isolation, and characterization of the assembled oligonucleotide fragments contribute considerably to the over-all expenditure of any DNA
synthesis project. Many manipulations, such as extractions, evaporations, chromatographic operations, desaltings and measurements, have to be done with
each fragment separately. There is still a demand for simplified work-up protocols that allow simultaneous handling of many preparations. Introduction of β-
cyanoethyl phosphoramidites [10] and additional protecting groups on G and T
in the phosphotriester route [11] allows the use of ammonia as the only alkaline
deprotection reagent. Ammonia is particularly convenient as it can be removed
by evaporation. Chain scission by nucleophilic attack of OH^- on phosphorous
is, under appropriate conditions, apparently no longer a practical problem at
least for shorter oligonucleotides up to 20 nucleotides in length [12].

Isolation of the target oligonucleotides from deprotection mixtures still ac-
counts for more than half of the labor and drastically increases with the number
of samples to be handled. Straightforward complete deprotection followed by
preparative gel electrophoresis or HPLC does not generally yield pure products.
Many impurities such as base modified sequences, only differing slightly in their
physicochemical properties from the target sequence, are observed. For safety
reasons a two-step purification based on two different separation principles may
be considered (for example reverse phase and ion exchange chromatography).
If a capping step was included in the synthesis, the lipophilicity of the 5'-dimeth-
oxy-trityl ether, remaining after alkaline deprotection, can be exploited to separate the target oligonucleotide from protecting groups and truncated chains by
reversed phase chromatography. HPLC is very effective but time consuming.
Alternatively, many samples (5 to 10) are easily purified simultaneously with
almost the same efficiency by flush chromatography on small disposable columns. By using an adsorbent which has both hydrophobic and ion exchange
capacity, we have recently extended this simple chromatographic procedure and
also included the acid-catalyzed cleavage of the trityl group. This procedure can
easily be automated and future autosynthesizers may have built in such a work-
up option [13].

An extensive sequence analysis of large numbers of oligonucleotides is still
beyond the limit of practicability. Most very long oligonucleotides are not obtained in sufficiently pure state to allow sequence analysis. Therefore, researchers tend to rely on the chemistry and sequence their DNA after cloning. This situation may change with further improvements in solid phase Maxam-Gilbert sequencing [14] and FAB mass spectrometry [15]. However, some analytical data
should be collected to avoid complete failure. Polyacrylamide gel electrophoresis
after radiolabeling with T4 PNK and $[\gamma\text{-}^{32}P]$ ATP is a suitable method for easily
obtaining valuable information about the purity and to some extent the identity
of the synthesized chains [16]. Coelectrophoresis of homo-oligomeric chain
length markers allow a correct sizing of oligonucleotide sequences up to 50-mers
according to length and base composition (Fig. 11). Only those samples that differ considerably from expected mobilities need be further subjected to sequence
analysis.

To summarize the arguments discussed above, the following criteria may be
considered when choosing between a short-fragment and a long-fragment ap-

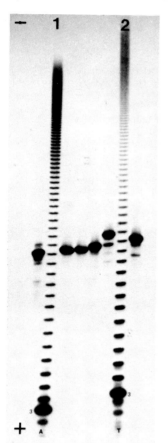

Fig. 11. Sizing of synthetic oligonucleotides by electrophoresis on a 20% denaturing (7 M urea) polyacrylamide-gel. *Lane 1:* A-ladder $(pdA)_{3-n}$; *lane 2:* T-ladder $(pdT)_{3-n}$

proach to gene construction. Short fragments: their synthesis is less prone to failures; large numbers are needed, but these are conveniently prepared by the segmental support technique; they can be easily obtained in high purity and reliably characterized; further purification is achieved through ligation selectivity; reusability is high for construction of gene variants. Long fragments: an extremely good chemistry with reliable high repetitive yields and very few side reactions is required; fewer fragments are required, which reduces work-up labor; purification to near homogeneity is difficult. Available know-how, equipment, budget, and annual demand are further important criteria. The methodology is still rapidly advancing and automation of the segmental support technique and work-up procedures as well as better protection strategies will soon bring further advantages.

Acknowledgments. The work of all members of our group, M. Becker, K. Giesa, W. Heikens, B. Kornak, H. Krause, G. Kurth, A. Meyerhans and K. Schwellnus, is greatly acknowledged. We thank V. Wray for linguistic advice and B. Seeger-Kunth for typing the manuscript.

References

1. Mutter M (1985) Angew Chem. 97:639 – 654
2. Agarwal KL, Büchi H, Caruthers MH, Gupta N, Khorana HG, Kleppe K, Kumar A, Ohtsuka E, RajBhandary UL, van de Sande JH, Sgaramella V, Weber H, Yamada T (1970) Nature 227:27 – 34
3. Blöcker H, Frank R, Köster H (1978) Liebigs Angew Chem 1978:991 – 1006
4. Rossi JJ, Kierzek R, Huang T, Walker PA, Itakura K (1982) J Biol Chem 257:9226 – 9229
5. Ohtsuka E, Ikehara M, Söll D (1982) Nucleic Acids Res 10:6553 – 6570
6. Reese CB (ed) (1984) Tetrahedron Symp-In-Print No 13. Tetrahedron 40:1 – 163
7. Gait MJ (ed) (1984) Oligonucleotide synthesis. IRL, Eynsham
8. Merrified RB (1963) J Am Chem Soc 85:2149 – 2154
9. Frank R, Heikens W, Heisterberg-Moutsis G, Blöcker H (1983) Nucleic Acids Res 11:4365 – 4377
10. Sinha ND, Biernat J, McManus J, Köster H (1984) Nuclei Acids Res 12:4539 – 4557
11. Matsuzaki J-I, Hotoda H, Sekine M, Hata T (1984) Tetrahedron Lett 25:4019 – 4022
12. Blöcker H et al. (in preparation)
13. a) Meyerhans A, Heisterberg-Moutsis G, Kurth G, Blöcker H, Frank R (1985) Nucleosides Nucleotides 4:245
    b) Eur Patent Appl 85110454.7
14. Rosenthal A, Schwertner S, Hahn V, Hunger H-D (1985) Nucleic Acids Res 13:1173 – 1184
15. Grotjahn L, Frank R, Blöcker H (1982) Nucleic Acids Res 10:4671 – 4678
16. Frank R, Köster H (1979) Nucleic Acids Res 6:2069 – 2087

8.3 Generation and Interpretation of Fast Atom Bombardment Mass Spectra of Modified Peptides

KLAUS ECKART[1]

Contents

1 Introduction

Until recently the possibility of mass spectrometric investigations was limited by the volatility of the investigated compounds. Due to the high polarity and low volatility of peptides, derivatization was necessary to obtain volatile derivatives, but new difficulties like incomplete reactions and/or side reactions and fragmentation were involved, and last but not least, the time needed for the derivatization was considerable.

In the 1970's two methods were employed to obtain mass spectra of non-volatile compounds.

1. Field Desorption Mass Spectrometry (FDMS) [1] and
2. Particle Induced Desorption Mass Spectrometry (PIDMS) [2].

Successful applications of FDMS are rare due to the problems of generating truly activated emitters and the well known short ion current lifetimes. More success was achieved by PIDMS, but the low yield of ions limited this otherwise very promising approach for its successful application when using commercially available sectorfield and quadrupole mass spectrometers. A major breakthrough resulted when PIDMS was modified to Fast Atom Bombardment Mass Spectrometry. Application of a liquid matrix [3], usually glycerol, produces stable ion currents in high yield having lifetimes in the $5-10$ min range; this ion stability

---

1 Institut für Organische Chemie, Technische Universität Berlin, Straße des 17. Juni 135, D-1000 Berlin 12
Present address: Max-von-Pettenkofer-Institut, Bundesgesundheitsamt, Thielallee 88 – 92, D-1000 Berlin 33

Advanced Methods in Protein Microsequence Analysis
Ed. by B. Wittmann-Liebold et al.
© Springer-Verlag Berlin Heidelberg 1986

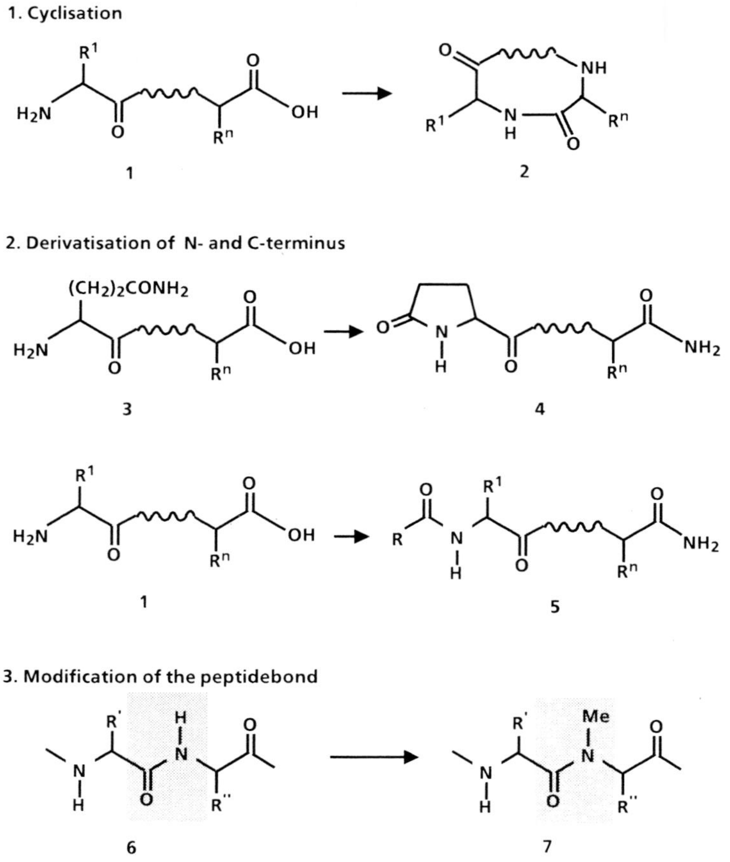

1. Cyclisation

1   2

2. Derivatisation of N- and C-terminus

3   4

1   5

3. Modification of the peptidebond

6   7

Fig. 1. Peptide modifications

allows the combination of this powerful ionization technique with the Tandem Mass Spectrometry (MSMS) method [4].

The strategy outlined in this chapter concerns N- and C-termini modified peptides and peptide bond modifications (Fig. 1). First, using the results of amino acid analysis, the information on the molecular weight of the FAB mass spectrum will be interpreted in terms of the modifications mentioned above. Second, it will be demonstrated how the application of the MSMS technique to the molecular ion and fragment ions of the peptide leads to the complete sequence information of the peptide being analyzed.

The amount of peptide needed is 10^{-7} mol or less and the time required for the mass spectrometric analysis of a single peptide is less than 4 h.

2 Sample Preparation

Preparation of the FAB sample is the most important step for recording the FAB mass spectrum because sensitivity and quality of the spectrum are strongly dependent on the matrix conditions.

The mass spectrometer used is any commercially available sectorfield or quadrupole instrument with a FAB ion source equipped with a discharge atom gun. As source of atoms for bombardment, Xenon are the best choice. The Xenon atoms are produced by charge exchange with neutral gas in the atom gun [5]. The sample is centred on the FAB target of the solid probe. No differences in the quality and sensitivity in the spectra are noted whether the material is gold, silver, copper, or stainless steel. It is only necessary to deposit a drop of $0.1 - 0.3 \, \mu l$ at the target on the point of beam incidence. This point of application can be recognized by alteration in the target surface.

The ion source parameters are as follows:

Pressure: 5×10^{-6} Torr (generated by the Xenon)
Temperature: 20 °C (not heated).

Discharge FAB source:

Accelerating voltage: $7 - 8$ kV
Emission current: $0.05 - 0.1$ mA
Both values are connected with each other and are dependent on the Xenon pressure.

Before switching on the FAB discharge source, it is important to remove traces of air from the gas inlet system by flushing it once with Xenon (this is not necessary if the gas inlet system is closed between the periods of measurements).

Preparation of the FAB sample begins with dissolving the peptide. Depending on the solubility, typical solvents are methanol, acidified methanol, dimethyl-sulfoxide, and for extremely insoluble peptides a commercially available solution of 33% HBr in glacial acetic acid (Fluka) may be used. Formic acid, glacial acetic acid, and trifluoracetic acid may also be used to solubilize proteins or peptides.

For synthetic peptides the following technique for dissolving is useful:

1. A small amount $(5 - 10 \, \mu g)$ is removed with the tip of a spatula from the sample container.
2. Approximately $2 - 4 \, \mu l$ solvent is withdrawn from the solvent bottle with a micro capillary.
3. The solution is blown out onto the sample; the sample now in solution is withdrawn into the capillary.
4. This capillary is dipped into a drop of glycerol (bidistilled, commercially available); a ratio of $1.2 \, \mu l$ per $1 \, \mu l$ of sample volume is the best.
5. Mixing is achieved by blowing out the fluids into a small bottle (i.e., Eppendorf test tubes 3815) and keeping the top of the capillary in the mixed drop.
6. This solution is introduced onto the FAB target to the mass spectrometer and subjected to ionization.

7. The recording conditions depend on the mass spectrometer in use and the details are normally included with the operation manuals supplied by the manufacturers.

3 Interpretation of the FAB Mass Spectra

Mass spectrometric sequencing of peptides should begin with an amino acid analysis for the following reasons:

1. Amino acid analysis provides evidence that the sample is a peptide.
2. The content of detectable amino acids can be determined. This information can in general not be obtained by FABMS; although fragments corresponding to individual amino acid constituents can be observed [6].
3. The advantages of preliminary amino acid analyses are (a) that small amounts of sample are needed; (b) at low cost; and (c) relatively short analysis time, which cannot be achieved by FAB mass spectrometry.

The protocol for the interpretation of the FAB mass spectrum in conjunction with the results of amino acid analysis is outlined in Fig. 2. The first question is whether one is dealing with a blocked peptide or not. The answer can be obtained by Edman degradation in combination with the amino acid analysis of the peptide. Alternatively, the M_{diff} value described below supplies this information. A M_{diff} value of 18 daltons suggests the presence of a nonblocked peptide with a C-terminal carboxyl group.

From the amino acid analysis the calculated molecular weight $M_{calc.}$ is obtained by adding the molecular weights of the amino acid residues from the corresponding amino acid constituents (cf. Table 1). This value supplies the smallest possible molecular weight without any group at the termini of the peptide (Fig. 3). For this reason the "real" molecular weight of the peptide must be equal or greater than this value. The molecular weight $M_{exp.}$ from the FAB mass spectrum contains all the additional information which cannot be obtained from the amino acid analysis. The difference between both molecular weights, i.e., the $M_{diff.}$ value, contains the information about the constitution of the peptide termini. Some important $M_{diff.}$ values are listed in Table 2.

Differentiation between asparagine and aspartic acid and, analogously, between glutamine and glutamic acid is normally impossible by amino acid analysis. This can be achieved by interpretation of the $M_{diff.}$ value. For this reason $M_{calc.}$ is calculated by using only asparagine and glutamine and not the corresponding free acids. Occurrence of one acid instead of the amide can be determined by an upward shift of $M_{diff.}$ by 1 dalton. Such shifts are not always due to terminal modifications as described below and can therefore often be used to determine the number of side chains with the free acid function.

It is useful to devide the terminal modifications into three types. These are illustrated in Table 2. The first is a cyclic peptide resulting from head to tail linkage (Figs. 1 and 2), which can be recognized by $M_{diff.} = 0$ dalton. Replacement of a carboxamide side chain by a carboxylic acid results in $M_{diff.} = 1$ dalton and therefore the $M_{diff.}$ value is an equivalent to the number of the free carboxylic

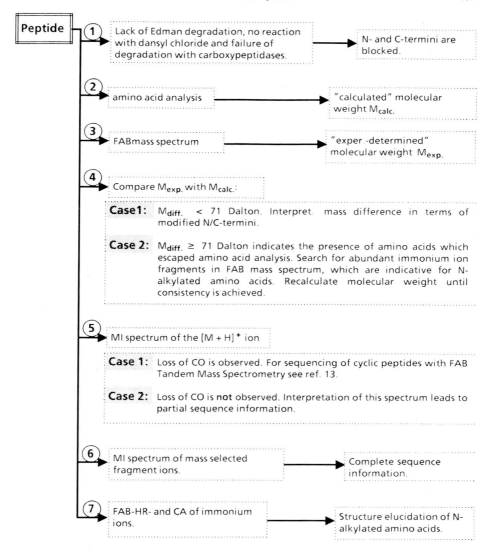

Peptide		
①	Lack of Edman degradation, no reaction with dansyl chloride and failure of degradation with carboxypeptidases.	N- and C-termini are blocked.
②	amino acid analysis	"calculated" molecular weight $M_{calc.}$
③	FABmass spectrum	"exper.-determined" molecular weight $M_{exp.}$

④ Compare $M_{exp.}$ with $M_{calc.}$:

Case 1: $M_{diff.}$ < 71 Dalton. Interpret mass difference in terms of modified N/C-termini.

Case 2: $M_{diff.}$ ≥ 71 Dalton indicates the presence of amino acids which escaped amino acid analysis. Search for abundant immonium ion fragments in FAB mass spectrum, which are indicative for N-alkylated amino acids. Recalculate molecular weight until consistency is achieved.

⑤ MI spectrum of the $[M + H]^+$ ion

Case 1: Loss of CO is observed. For sequencing of cyclic peptides with FAB Tandem Mass Spectrometry see ref. 13.

Case 2: Loss of CO is **not** observed. Interpretation of this spectrum leads to partial sequence information.

⑥	MI spectrum of mass selected fragment ions.	Complete sequence information.
⑦	FAB-HR- and CA of immonium ions.	Structure elucidation of N-alkylated amino acids.

Fig. 2. Protocol for sequencing of modified (blocked) peptides with combined amino acid analysis and FAB Tandem Mass Spectrometry

acid side chains. Replacement of a carboxamide by a methylester results in $M_{diff.} = 15$ daltons. This modification is normally a synthetic one which is a side product of NH_3/methanol cleavage reaction during solid phase synthesis. For this reason this modification normally exists only at the C-terminus.

A second modification is the special instance of a linear peptide e.g., a peptide with an N-terminal pyroglutamic acid residue. This modification occurs only with the Glx amino acid. From a formalistic point of view peptide 4 (Fig. 1) is generated by ring closure of the N-terminal glutamine with loss of NH_3 (− 17 dal-

Table 1. Molecular weights and structures of the amino acids detectable with conventional amino acid analysis

schematic structure of an amino acid residue

No.	amino acid	structure of R	molecular weight	
			of the amino acid residue	of the amino acid
1	Ala	$-CH_3$	71	89
2	Arg	$-(CH_2)_3-NH-C=NH$ $\quad\quad\quad\quad\quad NH_2$	156	174
3	Asn	$-CH_2CONH_2$	114	132
4	Asp	$-CH_2COOH$	115	133
5	Cys	$-CH_2SH$	103	121
6	Gln	$-(CH_2)_2-CONH_2$	128	146
7	Glu	$-(CH_2)_2-COOH$	129	147
8	Gly	$-H$	57	75
9	His		137	155
10	Ile		113	131
11	Leu	$-CH_2-CH-(CH_3)_2$	113	131
12	Lys	$-(CH_2)_4-NH_2$	128	146
13	Met	$-(CH_2)_2-S\,CH_3$	131	149
14	Orn	$-(CH_2)_3-NH_2$	114	132
15	Phe	$-CH_2-C_6H_5$	147	165
16	Ser	$-CH_2\,OH$	87	105
17	Thr	$-CH(CH_3)OH$	101	119
18	Trp		186	204
19	Tyr	$-CH_2-C_6H_4-pOH$	163	181
20	Val	$-CH-(CH_3)_2$	99	117

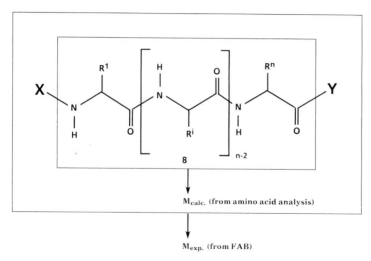

M$_{calc.}$ (from amino acid analysis)

M$_{exp.}$ (from FAB)

Fig. 3. Detection of the terminal groups X and Y of the peptide

Table 2. Some possible and useful M$_{diff.}$ values

M$_{diff.}$ [a] [dalton]	N- and/or C-modifications [a]		
	cyclic peptide[b]	linear pGlu peptide[b]	linear peptide
0	Asx = Asn, Glx = Gln	Asx = Asn, Glx = Gln	——
1[c]	CONH$_2$ ⇒ COOH	CONH$_2$ ⇒ COOH	——
15[c]	CONH2 ⇒ COOCH$_3$	CONH2 ⇒ COOCH$_3$	——
17	——	——	Asx = Asn, Glx = Gln C-terminus = CONH$_2$
18	——	——	Asx = Asn, Glx = Gln CONH$_2$ ⇒ COOH
45	——	——	N-terminus = N-formyl C-terminus = CONH$_2$
46	——	——	N-terminus = N-formyl CONH$_2$ ⇒ COOH
59	——	——	N-terminus = N-acetyl C-terminus = CONH$_2$
60	——	——	N-terminus = N-acetyl CONH$_2$ ⇒ COOH

[a] Appearance of more than one modification in a peptide result in an additional increment to the M$_{diff.}$ values.

[b] For the mass spectrometric differentiation of both cases see [7]. The sequencing of cyclic peptides with FAB Tandem Mass Spectrometry has been described in [13].

[c] The replacement of CONH$_2$ might also occur in the side chains of the amino acids asparagine and glutamine. The position can be determined by the fragmentation pattern of the peptide under FAB conditions.

tons) and replacement of the C-terminal OH group from peptide 3 by the NH$_2$ group (− 1 dalton). For the reason that peptide 3 must have M$_{diff.}$ = 18 daltons, the two modifications produce M$_{diff.}$ = 0 dalton. Replacements in the side chain result in the same M$_{diff.}$ values as described above for the cyclic peptides. It should be mentioned that one carboxamide modification in contrast to the cyclic peptides can occur at the C-terminus. The nature of the C-terminus is available from the fragmentations at the C-terminus of the peptide. Distinction between linear and cyclic peptides using Tandem Mass Spectrometry had been described [7, 8]. Often cyclic and linear peptides can be distinguished by their chromatographic behavior due to the lower polarity of cyclic peptides.

The third type concerns modifications of a linear peptide. Due to the fact that the linear peptide 1 has $M_{diff.} = 18$ daltons a linear peptide with a C-terminal carboxamide must have $M_{diff.} = 17$ daltons. N-terminal acylation generates the corresponding upward shift for the N-terminal fragment ions, i.e., N-formyl 28 daltons or N-acetyl 42 daltons. The $M_{diff.}$ value consists of the C- and N-terminal group of the peptide as listed in Table 2.

Molecular weight information of peptides can be obtained easily by FABMS. Usually the positive ion FAB mass spectrum is recorded and this technique supplies the $[M + H]^+$ species of the investigated peptide. Due to the possibility of terminal groups the molecular ion must appear in the mass region above $M_{calc.}$. $M_{exp.}$ is obtained by subtraction of 1 dalton from the value of the $[M + H]^+$ ion. If, however, there is any doubt about the $[M + H]^+$ ion, the molecular weight can be checked by recording the negative ion FAB mass spectrum, where the $[M - H]^-$ ion is usually observed. If this option is not available the addition of NaCl to the glycerol matrix in the positive ion mode generates exclusively the cationized $[M + Na]^+$ ion of the molecule and not of fragment ions derived from fragmentations in the mass spectrometer

In a similar manner the presence of N-alkylated amino acids can be recognized. Due to the molecular weight of the smallest possible N-alkylated amino acid, e.g. sarcosine, $M_{diff.}$ values > 70 daltons have to be interpreted in terms of the presence of such an amino acid, which escaped the amino acid analysis. The higher basicity of the N-alkylated nitrogen causes the higher probability of appearance of the corresponding fragment ion in the mass spectrum of the peptide (Fig. 4). The mass of this fragment ion can be obtained by subtraction of 27 dalton from the molecular weight of the corresponding amino acid residue (+ 1 dalton for protonation and − 28 dalton for the CO loss). The fragment ion of the N-alkylated amino acid often gives rise to the base peak in the FAB mass spectrum and so these constituents can be recognized easily. After having obtained this information, the $M_{diff.}$ value has to be recalculated with this additional amino acid and interpreted in terms of modified termini as mentioned above.

This approach can in principle also be used for the detection of unusual amino acids not modified by N-alkylation. Recognition of the corresponding fragment ion is somewhat more difficult to achieve due to the lower appearance probability of this fragment ion.

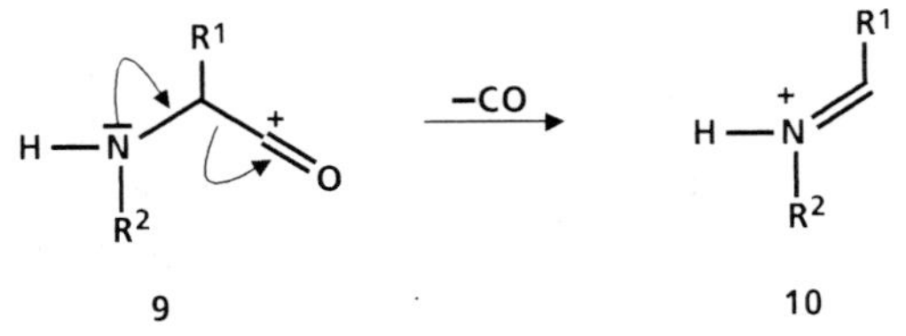

R1 = amino acid side chain
R2 = H, alkyl

Fig. 4. Fragmentation of amino acid residues under FAB conditions

4 Sequencing of Peptides

In the FAB mass spectra of oligopeptides fragment ions corresponding to the peptide bond cleavage can often be observed. From these fragment ion sequence information can be obtained. Complete and unambiguous sequencing is often hindered by the low significance of these ions above background and sometimes by the lack of some fragment ions which are necessary to arrive at a complete sequence.

To overcome this serious difficulty, application of the MSMS methodology (Fig. 5) is very helpful. The most serious problem in practice is the availability of an appropriate mass spectrometer. The detailed description of such experiments will therefore be excluded. Pertinent information may be obtained from the literature in which successful applications to adipokinetic hormones [9, 10] and synthetic [pGlu^6]SP$_{6-11}$ analogs [7] are described in detail [8]. Application of the MSMS methodology to the molecular ion and one or two mass selected fragment ions provides, in most cases, the complete sequence information of peptides up to ten amino acids. The interpretation begins with attempts to match the observed losses of neutral species to the known amino acid residues caused by peptide bond cleavage. Sometimes, however, other fragmentations, i.e., in the side chain of the amino acids, may occur.

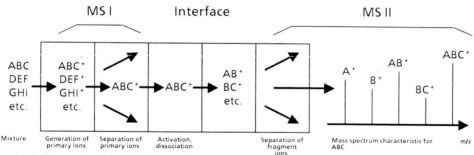

Fig. 5. Schematic of the MSMS experiment

Partial sequencing can often be achieved by interpretation of the FAB mass spectrum analogously to the MSMS spectra. For a reliable interpretation it is important to know the main pathways of the peptide fragmentation. Due to the higher intensity of the spectra of positive charged ions only the interpretation of these ions will be described. The fragmentation behavior is less dependent from the question whether the spectrum is a normal FAB spectrum or a MSMS spectrum of a mass selected ion. To provide an overview, the fragmentation is displayed in Fig. 6.

Figure 6 shows the possible backbone fragmentations of linear peptides. The most important one is the cleavage of the amide bond generating the N-terminal fragment ion B$_2$ and the C-terminal ion Y$_2$. Other fragment ions can sometimes be observed, but they are lower in intensity. However, consecutive cleavage reactions in the side chains are also possible:

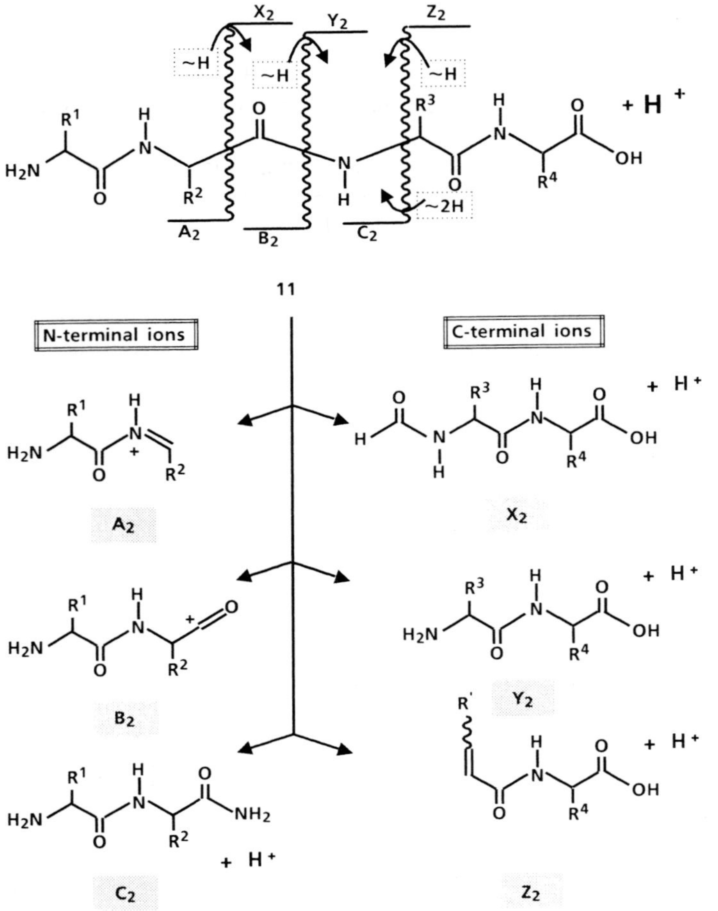

Fig. 6. Fragmentation and nomenclature of peptide fragment ions [14]

1. loss of NH_3 from Arg Asn, Gln, Lys and Orn;
2. loss of H_2O from Asp, Glu, Ser, Thr and Tyr;
3. loss of CN_2H_4 from Arg;
4. loss of $CONH_3$ from Asn and Gln;
5. loss of $COOH_2$ from Asp and Glu;
6. loss of CN_3H_5 from Arg.

   The abundance of fragment ions due to side chain reactions is strongly dependent on the nature of the peptide.

   In the FAB mass spectrum the interpretation should only refer to ions with a signal-to-noise ratio better than 3 related to the chemical noise and ions with a mass lower than 200 dalton should be excluded. The first step is calculation of all possible B_2 fragment ions, starting with dipeptides and ending with peptides of n-1 amino acids. The weights of these ions will be calculated by adding the weights of the amino acid residues and adding one for the proton. If the peptide is N-terminal blocked, the weight of the N-terminal group must be included. After calculation matching of the fragments with the FAB mass spectrum leads to possible partial sequences. Inconsistencies may occur due to ambiguities, i.e., Asn = ProNH$_2$. Such ambiguities can be resolved as described in the next step.

This step starts with the calculation of the corresponding C-terminal fragment ions. These ions will be calculated in the same manner as the N-terminal ions above but the terminal group is different. For NH$_2$, 17 daltons must be used, for OH, 18 daltons and for OCH$_3$, 32 daltons. If the C-terminal group is unknown inspection of the mass range 200 dalton below the molecular ion can provide the information about the C-terminal amino acid and their C-terminal group. Matching of the possible C-terminal fragment ions with the FAB mass spectrum provides additional partial sequence information. Often two alternative fragment series remove ambiguities from the interpretation. If this is not achieved Tandem Mass Spectrometry can solve such problems as described in the literature [7, 8]. Often it is very helpful to search for other possible backbone and side chain fragment ions to justify the proposed sequence.

It is obvious that computer programs with a permutation algorithm can be very helpful for the calculation of the fragment ions. Such programs can be designed using personal computers with basic as programming language.

The structures of the N-alkylated amino acids can be elucidated by application of the MSMS technique to the fragment ions as shown in Fig. 4. Often these spectra contain enough information for a structure elucidation of N-alkylated amino acids. This can be simplified by obtaining the elemental composition of the investigated fragment ion by High Resolution FAB Mass Spectrometry [11]. The proposed structures can be proved by comparing the MSMS spectrum of the unknown amino acid with the spectrum of a reference ion generated by loss of CO$_2$H$_2$ from a reference amino acid under FAB conditions. The common rule is that equal spectra recorded under collisional activation conditions suggest equal structures of the ions in question [12]. The applicability has been demonstrated with two otherwise difficult to analyze examples:

1. differentiation between N-methylleucine and N-methylisoleucine [7],
2. differentiation between N-methylvaline and N-isobutylglycine [8].

Application of the approach described above leads to MSMS spectra, which allow the unequivocal structure elucidation of each amino acid.

Another common problem concerns the distinction of isomeric C/N-alkylated amino acids, i.e., alanine and sarcosine, during sequencing of the peptide. These amino acids cannot be distinguished by Edman sequencing and the same problem occurs by FABMS sequencing due to the same mass increment of both amino acids. The differences in stability of peptide bonds at N/C-alkylated amino acids can be used for differentiation of the positions of both amino acids. Investigation of the effect of N-alkylation on the peptide bond stability under FAB conditions shows a strong labilization effect [7, 8]. This leads to a much higher intensity of fragment ions generated by cleavage of the peptide bonds adjacent to sarcosine as well as alanine; so a clear determination of the position of each amino acid is possible.

5 Summary

The technique of sample preparation for FAB has been described. Demonstration of the principles of the interpretation of the obtained FAB mass spectra combined with the result of the amino acid analysis proved to be extremely useful in recognizing modifications of the peptide. Problems associated with incomplete structure information which often is provided by the FAB mass spectrum can be overcome by using the MSMS methodology.

The MSMS technique in conjunction with high resolution FAB mass spectrometry is very useful for the structure elucidation of N-alkylated amino acids, which escape conventional amino acid analysis.

Acknowledgments. I am indebted to Prof. Dr. Helmut Schwarz, who gave me the opportunity to work on this project as part of my Ph. D. Thesis. The financial support of the Deutsche Forschungsgemeinschaft (Sonderforschungsbereich 9, Projekt A9) and the Fonds der Chemischen Industrie is gratefully acknowledged.

References

1. Beckey HD (1977) Principles of field ionization and field desorption mass spectrometry. Pergamon, Oxford
2. MacFarelane RD (1982) Acc Chem Res 15:268
3. Gower JL (1985) Biomed Mass Spectrom 12:191
4. McLafferty FW (1983) Tandem mass spectrometry. Wiley, New York
5. Franks J (1983) Int J Mass Spectrom Ion Phys 46:343
6. Barber M, Bordoli RS, Sedgwick RD, Tyler AN (1982) Biomed Mass Spectrom 9:208
7. Eckart K, Schwarz H, Gilon C, Chorev M (1986) Eur J Biochem 157:209
8. The complete data and conditions are available from the Ph. D. thesis: Eckart K (1985) Dissertation, Tech Univ Berlin
9. Eckart K, Schwarz H, Ziegler R (1985) Biomed Mass Spectrom 12:623
10. Ziegler R, Eckart K, Schwarz H, Keller R (1985) Biochem Biophys Res Commun 133:337
11. Gilliam GM, Landis PW, Occolowitz JL (1984) Anal Chem 56:2285
12. Levsen K, Schwarz H (1983) Mass Spectrom Rev 2:77
13. Eckart K, Schwarz H, Tomer KB, Gross ML (1985) J Am Chem Soc 107:6765
14. Roepstorff P, Fohlman J (1984) Biomed Mass Spectrom 11:604

Subject Index